AF352896

Catalytic Biomass to Renewable Biofuels and Biomaterials

Catalytic Biomass to Renewable Biofuels and Biomaterials

Special Issue Editors

Yi-Tong Wang
Zhen Fang

MDPI • Basel • Beijing • Wuhan • Barcelona • Belgrade • Manchester • Tokyo • Cluj • Tianjin

Special Issue Editors
Yi-Tong Wang
North China University of Science and Technology
China

Zhen Fang
Nanjing Agricultural University
China

Editorial Office
MDPI
St. Alban-Anlage 66 4052 Basel, Switzerland

This is a reprint of articles from the Special Issue published online in the open access journal *Catalysts* (ISSN 2073-4344) (available at: https://www.mdpi.com/journal/catalysts/special_issues/Biomass_Renewable_Biofuels_Biomaterials).

For citation purposes, cite each article independently as indicated on the article page online and as indicated below:

LastName, A.A.; LastName, B.B.; LastName, C.C. Article Title. *Journal Name* **Year**, *Article Number*, Page Range.

ISBN 978-3-03936-312-4 (Hbk)
ISBN 978-3-03936-313-1 (PDF)

© 2020 by the authors. Articles in this book are Open Access and distributed under the Creative Commons Attribution (CC BY) license, which allows users to download, copy and build upon published articles, as long as the author and publisher are properly credited, which ensures maximum dissemination and a wider impact of our publications.
The book as a whole is distributed by MDPI under the terms and conditions of the Creative Commons license CC BY-NC-ND.

Contents

About the Special Issue Editors

Yi-Tong Wang is an associate professor in the School of Metallurgy and Energy, North China University of Science and Technology. Dr. Wang obtained her PhD at University of Chinese Academy of Sciences (CAS, China). Her research focuses on high value-added utilization of solid waste, catalytic material synthesis, magnetic material preparation, and biomass conversion into biofuels. Dr. Wang has published more than 15 research papers in biomass conversion and has served as a peer reviewer for major scientific journals.

Zhen Fang is a professor and the leader of the biomass group at Nanjing Agricultural University. He is the inventor of the "fast hydrolysis" process. He is listed in the "Most Cited Chinese Researchers" in energy for 2014–2019 (Elsevier-Scopus). Professor Fang specializes in thermal/biochemical conversion of biomass, synthesis and applications of nanocatalysts, pretreatment of biomass for biorefineries, and supercritical fluid processes. He obtained his PhDs from China Agricultural University and McGill University. Professor Fang is Editor-in-Chief of the Springer Book Series Biofuels and Biorefineries and is Associate Editor of Biotechnology for Biofuels and The Journal of Supercritical Fluids. He has more than 20 years of international research experience in renewable energy and green technologies at top universities and institutes around the world, including 1 year in Spain (University of Zaragoza), 3 years in Japan (Biomass Technology Research Center, AIST; Tohoku University), and more than 8 years in Canada (McGill). He worked in industry for 7 years as an engineer in energy, bioresource utilization, and engine design before moving to academia.

Preface to "Catalytic Biomass to Renewable Biofuels and Biomaterials"

Biomass energy is the oldest and most sustainable resource known to humankind, but it has been developed inefficiently so that it has been unable to replace fossil resources as an energy until now. Biomass comes directly or indirectly from the photosynthesis of green plants and is the only renewable carbon resource. Biomass can be converted into solid, liquid, and gas fuels and biomaterials by combustion, pyrolysis, chemical reaction, and fermentation. The by-products from the production process can also be converted into high value-added products to be applied in various fields. Due to the diversity of biomass types, the methods for biomass conversion are also abundant. At present, the cost of clean production of biofuels and materials is relatively high. However, the development of biomass catalytic conversion is propelled through the design of catalytic materials, optimization of chemical processes, product development, and utilization of by-product resources as high value-added chemicals. This book focuses on the introduction of biomass chemical conversion, co-combustion technology, and biological conversion technology in order to provide guidance for the development and application of biomass energy. Each individual article was contributed by experts or professionals in related fields and was peer-reviewed and edited for content and consistency in terminology. This book is titled *Catalytic Biomass to Renewable Biofuels and Biomaterials* and contains 12 articles.

Yi-Tong Wang, Zhen Fang
Special Issue Editors

Editorial

Catalytic Biomass to Renewable Biofuels and Biomaterials

Yi-Tong Wang [1] and Zhen Fang [2,*]

[1]   School of Metallurgy and Energy, North China University of Science and Technology, 21 Bohai Avenue, Tangshan 063210, China; wangyt@ncst.edu.cn

[2]   Biomass Group, College of Engineering, Nanjing Agricultural University, 40 Dianjiangtai Road, Nanjing 210031, China

*   Correspondence: zhenfang@njau.edu.cn

Received: 21 April 2020; Accepted: 26 April 2020; Published: 28 April 2020

As the only renewable carbon source, biomass can be converted into biofuels, chemicals, and biomaterials, such as ethanol, butanol, glucose, furfural, biochar, and bio-oils, and is considered as a substitute for fossil oil. The Special Issue "Catalytic Biomass to Renewable Biofuels and Biomaterials" focuses on the introduction of biomass chemical conversion, co-combustion technology, and biological conversion technology to provide guidance for the development and application of biomass energy. This issue includes eleven papers: one review and ten research articles. Glycerol, the main by-product in biodiesel production was acetylated to prepare value-added products from the research results of Bartoli et al. [1]. The conversion of glycerol to acetyl derivatives was facilitated by a heterogeneous catalyst generated from the thermal hydrolysis of biosolids obtained from a municipal wastewater treatment facility. This work can help improve the process economics of renewable diesel production. Based on the most difficult step of the fractionation for lignocellulose-based biorefineries, Hochegger et al. investigated the potential of a base-catalyzed organosoly process as a fractionation technique for European larch sawdust. One particular experimental set (443 K, 30% v/v MeOH, 30% w/w NaOH) resulted in the most promising results, with a cellulose recovery of 89%, delignification efficiency of 91%, and pure lignin yield of 82%. The isolated lignin fractions showed promising glass transition temperatures ($\geq$424 K) with high thermal stabilities and preferential O/C and H/C ratios. This, together with high contents of phenolic hydroxyl ($\geq$1.83 mmol/g) and carboxyl groups ($\geq$0.52 mmol/g), indicated a high valorization potential [2]. Sanahuja-Parejo et al. reported the catalytic co-pyrolysis of grape seeds and waste tires for the production of high-quality bio-oils in a pilot-scale auger reactor using different low-cost Ca-based catalysts. The results demonstrated that this upgrading strategy was suitable for the production of better-quality bio-oils with major potential for use as drop-in fuels. In addition, owing to the CO_2-capture effect inherent to these catalysts, a more environmentally friendly gas product was produced, comprising H_2 and CH_4 as the main components [3]. Diesel and jet fuel range cycloalkanes were obtained in ~84.8% overall carbon yield with cyclopentanone and furfural, which can be produced from hemicellulose. Carbon yield of 86.1% diesel and jet fuel range cycloalkanes was gained over Pd/H-ZSM-5 catalyst under solvent-free conditions. The cycloalkane mixture obtained with a high density (0.82 g mL^{-1}) and a low freezing point (241.7 K) can be mixed into diesel and jet fuel to increase their volumetric heat values or payloads [4]. Lavarda et al. reported the research for gel-type and macroporous cross-linked copolymers employed in the catalytic hydrolysis of wheat straw pretreated in 1-ethyl-3-methylimidazolium acetate to obtain sugars [5]. Long et al. prepared a highly active, sustainable, and facile catalytic system composed of K_2CO_3, Ph_2SiH_2, and bio-based solvent 2-methyltetrahydrofuran to reduce HMF (hydroxymethylfurfural). At a low temperature of 25 °C, HMF could be completely converted to 2,5-bis(hydroxymethyl)furan (BHMF) in a good yield of 94% after 2 h, based on siloxane in situ formed via hydrosilylation [6]. Xu et al. synthesized a series of $Ca(OH)_2/Al_2O_3$ catalysts for selectively producing N-containing chemicals from polyethylene terephthalate (PET) via catalytic fast pyrolysis with ammonia process, which can ease environmental problems from PET [7]. The study on the co-combustion of sludge and wheat straw was presented by Xue et al. with results of the blended fuel helping to remedy the defect of each individual component and also to promote the combustion. Addition of sludge could raise the melting point of wheat straw ash and reduce the slagging tendency. Co-combustion restrained the release of K and transferred it into aluminosilicate and phosphate as well. The release and leaching toxicity of the two heavy metals (Pb and Zn) in the co-combustion were weakened effectively by wheat straw [8]. Benevenuti et al. reported the production of biofuels from fermentation of gases by *Clostridium carboxidivorans*. In comparison to ATCC®2713 medium, TYA (Tryptone-Yeast extract-Arginine) increased cell growth by 77%, reducing the cost by 47% and TPYGarg (Tryptone-Peptone-Yeast extract-Glucose-Arginine) increased ethanol production more than four-times, and the cost was reduced by 31% [9]. The role of humic acid's chemical structure derived from different biomass feedstocks on Fe(III) bioreduction activity was reported by Wei et al. The role that the different compost-derived humic acids played in the dissimilatory Fe(III) bioreduction and chemical structures accelerating Fe reduction was discussed. This study can aid in searching sustainable humic acid-rich composts for wide-ranging applications to catalyzing redox-mediated reactions of pollutants in soils [10]. In a review, Liberato et al. introduced some organisms such as *Clostridium carboxidivorans*, *C. ragsdalei*, and *C. ljungdahlii* used to catalyze the production of biofuels, including ethanol and butanol, and several chemical preparations, such as acetone, 1,3-propanediol, and butyric acid. In this review, literature data showing the technical viability of these processes are presented, evidencing the opportunity to investigate them in a biorefinery context [11].

We are honored to be the Guest Editors of this Special Issue and would like to thank all the authors of this Special Issue, the reviewers for providing us with their valuable comments, the Editor-in-Chief, and all the staff of the *Catalysts* Editorial Office. We wish to acknowledge the financial support from the Natural Science Foundation of China (No. 21878161) and Nanjing Agricultural University (68Q-0603).

References

1. Bartoli, M.; Zhu, C.; Chae, M.; Bressler, D. Glycerol Acetylation Mediated by Thermally Hydrolysed Biosolids-Based Material. *Catalysts* **2020**, *10*, 5. [CrossRef]
2. Hochegger, M.; Trimmel, G.; Cottyn-Boitte, B.; Cézard, L.; Majira, A.; Schober, S.; Mittelbach, M. Influence of Base-Catalyzed Organosolv Fractionation of Larch Wood Sawdust on Fraction Yields and Lignin Properties. *Catalysts* **2019**, *9*, 996. [CrossRef]
3. Sanahuja-Parejo, O.; Veses, A.; López, J.; Murillo, R.; Callén, M.; García, T. Ca-based Catalysts for the Production of High-Quality Bio-Oils from the Catalytic Co-Pyrolysis of Grape Seeds and Waste Tyres. *Catalysts* **2019**, *9*, 992. [CrossRef]
4. Wang, W.; Sun, S.; Han, F.; Li, G.; Shao, X.; Li, N. Synthesis of Diesel and Jet Fuel Range Cycloalkanes with Cyclopentanone and Furfural. *Catalysts* **2019**, *9*, 886. [CrossRef]
5. Lavarda, G.; Morales-delaRosa, S.; Centomo, P.; Campos-Martin, J.; Zecca, M.; Fierro, J. Gel-Type and Macroporous Cross-Linked Copolymers Functionalized with Acid Groups for the Hydrolysis of Wheat Straw Pretreated with an Ionic Liquid. *Catalysts* **2019**, *9*, 675. [CrossRef]
6. Long, J.; Zhao, W.; Xu, Y.; Li, H.; Yang, S. Carbonate-Catalyzed Room-Temperature Selective Reduction of Biomass-Derived 5-Hydroxymethylfurfural into 2,5-Bis(hydroxymethyl)furan. *Catalysts* **2018**, *8*, 633. [CrossRef]
7. Xu, L.; Na, X.; Zhang, L.; Dong, Q.; Dong, G.; Wang, Y.; Fang, Z. Selective Production of Terephthalonitrile and Benzonitrile via Pyrolysis of Polyethylene Terephthalate (PET) with Ammonia over Ca(OH)$_2$/Al$_2$O$_3$. *Catalysts* **2019**, *9*, 436. [CrossRef]
8. Xue, Z.; Zhong, Z.; Zhang, B. Experimental Studies on Co-Combustion of Sludge and Wheat Straw. *Catalysts* **2019**, *9*, 182. [CrossRef]

9.   Benevenuti, C.; Botelho, A.; Ribeiro, R.; Branco, M.; Pereira, A.; Vieira, A.; Ferreira, T.; Amaral, P. Experimental Design to Improve Cell Growth and Ethanol Production in Syngas Fermentation by Clostridium carboxidivorans. *Catalysts* **2020**, *10*, 59. [CrossRef]

10.   Wei, Y.; Wei, Z.; Zhang, F.; Li, X.; Tan, W.; Xi, B. Role of Humic Acid Chemical Structure Derived from Different Biomass Feedstocks on Fe(III) Bioreduction Activity: Implication for Sustainable Use of Bioresources. *Catalysts* **2019**, *9*, 450. [CrossRef]

11.   Liberato, V.; Benevenuti, C.; Coelho, F.; Botelho, A.; Amaral, P.; Pereira, N.; Ferreira, T. *Clostridium* sp. as Bio-Catalyst for Fuels and Chemicals Production in a Biorefinery Context. *Catalysts* **2019**, *9*, 962. [CrossRef]

© 2020 by the authors. Licensee MDPI, Basel, Switzerland. This article is an open access article distributed under the terms and conditions of the Creative Commons Attribution (CC BY) license (http://creativecommons.org/licenses/by/4.0/).

Article

Experimental Design to Improve Cell Growth and Ethanol Production in Syngas Fermentation by *Clostridium carboxidivorans*

Carolina Benevenuti [1], Alanna Botelho [1], Roberta Ribeiro [1], Marcelle Branco [2], Adejanildo Pereira [1], Anna Carolyna Vieira [1], Tatiana Ferreira [2] and Priscilla Amaral [1,*]

[1]   Department of Biochemical Engineering, School of Chemistry, Universidade Federal do Rio de Janeiro, 21941-909 Rio de Janeiro, Brazil; carolbenevenuti@hotmail.com (C.B.); lanna.mbotelho@gmail.com (A.B.); robertadosreisribeiro@gmail.com (R.R.); adejanildosp@gmail.com (A.P.); annacgv@gmail.com (A.C.V.)

[2]   Department of Organic Process, School of Chemistry, Universidade Federal do Rio de Janeiro, 21941-909 Rio de Janeiro, Brazil; cellebranco@gmail.com (M.B.); tatiana@eq.ufrj.br (T.F.)

*   Correspondence: pamaral@eq.ufrj.br; Tel.: +55-21-3938-7623

Received: 10 December 2019; Accepted: 27 December 2019; Published: 1 January 2020

Abstract: Fermentation of gases from biomass gasification, named syngas, is an important alternative process to obtain biofuels. Sequential experimental designs were used to increase cell growth and ethanol production during syngas fermentation by *Clostridium carboxidivorans*. Based on ATCC (American Type Culture Collection) 2713 medium composition, it was possible to propose a best medium composition for cell growth, herein called TYA (Tryptone-Yeast extract-Arginine) medium and another one for ethanol production herein called TPYGarg (Tryptone-Peptone-Yeast extract-Glucose-Arginine) medium. In comparison to ATCC® 2713 medium, TYA increased cell growth by 77%, reducing 47% in cost and TPYGarg increased ethanol production more than four-times, and the cost was reduced by 31%. In 72 h of syngas fermentation in TPYGarg medium, 1.75-g/L of cells, 2.28 g/L of ethanol, and 0.74 g/L of butanol were achieved, increasing productivity for syngas fermentation.

Keywords: syngas fermentation; *Clostridium carboxidivorans*; biomass production; ethanol; experimental design; anaerobic process

---

1. Introduction

The world growing energy demand has incited important discussions about the current energy system based on fossil fuels. Alternative energy sources from renewable resources are needed mainly due to ever-higher amounts of greenhouse gas emissions and oil price fluctuation [1].

In this context, biofuels, especially ethanol and butanol, represent potential substitutes of fossil fuels, such as gasoline. Currently, those alcohols are produced through fermentation processes, mostly using sugarcane and corn as the raw material. The major disadvantage of these processes is the competition between food and fuel production [2]. Lignocellulosic feedstocks from agricultural waste or directly from energy crops emerge as alternative raw materials for biofuel production as they are inexpensive, renewable, and do not affect food supply [3,4].

The biochemical conversion of lignocellulosic feedstocks into biofuels by anaerobic bacteria, such as *Clostridium* species, produces mainly acetone, butanol, and ethanol (ABE), through ABE fermentation [5]. However, a complex pre-treatment is required to obtain simple sugars from the polymeric structure of lignocellulose [6]. An alternative destination for lignocellulosic feedstocks involves biomass gasification, which eliminates the pre-treatment step. In this case, full use of lignocellulosic biomass, including lignin, is possible through thermochemical conversion. Besides, urban solid residues can also be used as feedstock for gasification, with the advantage of reducing environmental impacts. The biomass gasification results in synthesis gas mainly composed of CO (carbon monoxide), CO_2 (carbon dioxide), and H_2 (hydrogen). Synthesis gas, also named syngas, obtained from gasification, can be converted into ethanol and higher alcohols as butanol and hexanol, through a hybrid thermo/biochemical process. This conversion can be performed by *Clostridum sp.*, being *Clostridium ljungdahlii*, *C. carboxidivorans*, *C. autoethanogenum*, and *C. rasgdalei* the most studied species. These species use the acetyl-CoA metabolic pathway, also known as Wood–Ljungdahl pathway, for acetyl-CoA synthesis, energy conservation, and alcohol and acid production [7].

A current problem of anaerobic bioprocesses is low cell density, which reduces productivity [8]. Usually, media composition used in CO/CO_2 fermentation is similar to sugar-based media for *Clostridium* sp. growth and butanol production. Numerous and expensive components are mixed to provide essential metals, vitamins, minerals, and nitrogen needed for cell growth and metabolism [9]. As consequence, the preparation of those media requires intensive labor and is expensive. Replacing these compounds by cheaper and complex sources is an important challenge. ATCC® medium 2713 (Wilkins Chalgren Anaerobic Medium) is indicated by American Type Culture Collection (ATCC) for *Clostridium carboxidivorans* activation and growth and ATCC® medium 1754 (without fructose) is another common medium used for syngas fermentation by *Clostridium carboxidivorans*, reported by several authors [10–14]. Low specific growth rates, ranging from 0.005 to 0.08 h^{-1} were reported for those media with gaseous or soluble substrates [5,15–17]. Studies using glucose-rich medium or only carbon monoxide as the carbon source have been carried out to evaluate the effect of the carbon source on cell growth and solvents production [5,15,17–21] and the maximum values for cell density and specific growth rate were 0.55 g dry weight of cells/L and 0.231 h^{-1}, respectively. In addition, the effects of trace metals on product formation were assessed [1,22–26] and the maximum ethanol production reported was 3.5 g/L after 72 h of fermentation [22].

Therefore, the goal of this work was to obtain a low-cost culture medium, reduced also in number of components, by a sequence of experimental designs in order to improve cell growth and solvents production by *Clostridium carboxidivorans* using glucose and syngas as carbon sources.

2. Results

2.1. Nutrient Screening by Placket–Burman (PB) Design

Syngas fermentation by *C. carboxidivorans* must be supplemented with nutrient media containing nitrogen, vitamins, and ions sources. The American Type Culture Collection (ATCC) indicates ATCC® medium 2713 Wilkins Chalgren Anaerobic Medium for *C. carboxidivorans* growth, which is composed of tryptone, 10 g/L; gelatin peptone, 10 g/L; yeast extract, 5 g/L; glucose, 1 g/L; sodium chloride, 5 g/L; L-arginine, 1 g/L; sodium pyruvate, 1 g/L; menadione, 0.5 mg/L; and hemin, 5 mg/L. These components were screened by a PB experimental design using ATCC medium concentrations as the higher level (+1) and absence (zero) as lower level (−1).

For ethanol production, glucose was maintained at 1.0 g/L because of its importance as a simple sugar to induce microbial metabolism, and the results are present in Table 1.

Ethanol production ranged between 0.03 and 1.92 g/L (Table 2). Analysis of variance (ANOVA) and the significance of the results were verified using an *F*-test at 10% of significance. The components that influence ethanol production can be observed in the Pareto diagram with the bars that extend beyond the red vertical line (Figure 1). Tryptone, gelatin peptone, and L-arginine influence *C. carboxidivorans* ethanol production positively.

Table 1. Placket–Burman design matrix with real and coded values for *C. carboxidivorans* syngas fermentation with 1 g/L glucose. Response variables: ethanol production—P_{EtOH}(g/L).

Run	Real Values (Corresponding Coded Levels)								P_{EtOH}
	TRYP [1]	PEP [2]	YE [3]	NaCl [4]	L-ARG [5]	PYR [6]	MEN [7]A	HEM [8]	
1	10(1)	0(−1)	5(1)	0(−1)	0(−1)	0(−1)	0.5(1)	5(1)	0.79
2	10(1)	10(1)	0(−1)	5(1)	0(−1)	0(−1)	0(−1)	5(1)	0.90
3	0(−1)	10(1)	5(1)	0(−1)	1(1)	0(−1)	0(−1)	0(−1)	0.69
4	10(1)	0(−1)	5(1)	5(1)	0(−1)	1(1)	0(−1)	0(−1)	0.78
5	10(1)	10(1)	0(−1)	5(1)	1(1)	0(−1)	0.5(1)	0(−1)	1.21
6	10(1)	10(1)	5(1)	0(−1)	1(1)	1(1)	0(−1)	5(1)	1.87
7	0(−1)	10(1)	5(1)	5(1)	0(−1)	1(1)	0.5(1)	0(−1)	0.51
8	0(−1)	0(−1)	5(1)	5(1)	1(1)	0(−1)	0.5(1)	5(1)	0.25
9	0(−1)	0(−1)	0(−1)	5(1)	1(1)	1(1)	0(−1)	5(1)	0.03
10	10(1)	0(−1)	0(−1)	0(−1)	1(1)	1(1)	0.5(1)	0(−1)	0.99
11	0(−1)	10(1)	0(−1)	0(−1)	0(−1)	1(1)	0.5(1)	5(1)	0.51
12	0(−1)	0(−1)	0(−1)	0(−1)	0(−1)	0(−1)	0(−1)	0(−1)	0.04
13 (C)	5(0)	5(0)	2.5(0)	2.5(0)	0.5(0)	0.5(0)	0.25(0)	2.5(0)	1.65
14 (C)	5(0)	5(0)	2.5(0)	2.5(0)	0.5(0)	0.5(0)	0.25(0)	2.5(0)	1.92
15 (C)	5(0)	5(0)	2.5(0)	2.5(0)	0.5(0)	0.5(0)	0.25(0)	2.5(0)	1.79

[1] TRYP: tryptone concentration (g/L); [2] PEP: peptone concentration (g/L); [3] YE: yeast extract concentration (g/L); [4] NaCl: sodium chloride concentration (g/L); [5] L-ARG: L-arginine concentration (g/L); [6] PYR: sodium pyruvate concentration (g/L0; [7] MEN: menadione concentration (mg/L); [8] HEM: hemin concentration (mg/L); (C) central point.

Table 2. Placket–Burman design matrix with real and coded values for *C. carboxidivorans* syngas fermentation with 1 g/L glucose. Response variables: biomass concentration—$[X_{24}]$ (g d.w.cells/L)—and maximum ethanol production—P_{EtOH} (g/L).

Run	Real Values (Corresponding Coded Levels)									$[X_{24}]$
	TRYP [1]	PEP [2]	YE [3]	NaCl [4]	GLU [5]	ARG [6]	PYR [7]	MEN [8]	HEM [9]	
1	0(−1)	0(−1)	0(−1)	0(−1)	0(−1)	0(−1)	0(−1)	0(−1)	5(+1)	0.044
2	10(+1)	0(−1)	0(−1)	0(−1)	1(+1)	0(−1)	1(+1)	0.5(+1)	0(−1)	0.400
3	0(−1)	10(+1)	0(−1)	0(−1)	1(+1)	1(+1)	0(−1)	0.5 (+1)	0(−1)	0.423
4	10(+1)	10(+1)	0(−1)	0(−1)	0(−1)	1(+1)	1(+1)	0(−1)	5(+1)	0.735
5	0(−1)	0(−1)	5(+1)	0(−1)	1(+1)	1(+1)	1(+1)	0(−1)	0(−1)	0.407
6	10(+1)	0(−1)	5(+1)	0(−1)	0(−1)	1(+1)	0(−1)	0.5(+1)	5(+1)	0.714
7	0(−1)	10(+1)	5(+1)	0(−1)	0(−1)	0(−1)	1(+1)	0.5(+1)	5(+1)	0.606
8	10(+1)	10(+1)	5(+1)	0(−1)	1(+1)	0(−1)	0(−1)	0(−1)	0(−1)	0.868
9	0(−1)	0(−1)	0(−1)	5(+1)	0(−1)	1(+1)	1(+1)	0.5(+1)	0(−1)	0.202
10	10(+1)	0(−1)	0(−1)	5(+1)	1(+1)	1(+1)	0(−1)	0(−1)	5(+1)	0.481
11	0(−1)	10(+1)	0(−1)	5(+1)	1(+1)	0(−1)	1(+1)	0(−1)	5(+1)	0.286
12	10(+1)	10(+1)	0(−1)	5(+1)	0(−1)	0(−1)	0(−1)	0.5(+1)	0(−1)	0.805
13	0(−1)	0(−1)	5(+1)	5(+1)	1(+1)	0(−1)	0(−1)	0.5(+1)	5(+1)	0.301
14	10(+1)	0(−1)	5(+1)	5(+1)	0(−1)	0(−1)	1(+1)	0(−1)	0(−1)	0.591
15	0(−1)	10(+1)	5(+1)	5(+1)	0(−1)	1(+1)	0(−1)	0(−1)	0(−1)	0.796
16	10(+1)	10(+1)	5(+1)	5(+1)	1(+1)	1(+1)	1(+1)	0.5(+1)	5(+1)	0.984
17 (C)	5(0)	5(0)	2.5(0)	2.5(0)	0.5(0)	0.5(0)	0.5(0)	0.25(0)	2.5(0)	0.701
18 (C)	5(0)	5(0)	2.5(0)	2.5(0)	0.5(0)	0.5(0)	0.5(0)	0.25(0)	2.5(0)	0.692
19 (C)	5(0)	5(0)	2.5(0)	2.5(0)	0.5(0)	0.5(0)	0.5(0)	0.25(0	2.5(0)	0.582

[1] TRYP: tryptone concentration (g/L); [2] PEP: peptone concentration (g/L); [3] YE: yeast extract concentration (g/L); [4] NaCl: sodium chloride concentration (g/L); [5] GLU: glucose (g/L); [6] ARG: L-arginine concentration (g/L); [7] PYR: sodium pyruvate concentration (g/L); [8] MEN: menadione concentration (mg/L); [9] HEM: hemin concentration (mg/L); (C) central point.

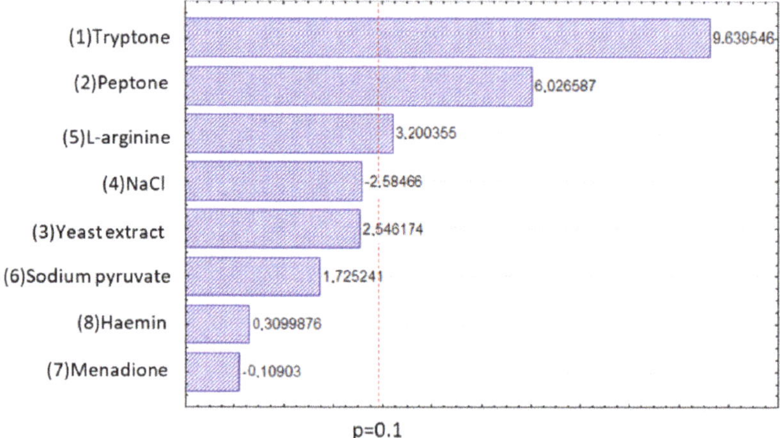

Figure 1. Pareto diagram for the estimated effect of each variable of the Placket–Burman design for *C. carboxidivorans* syngas fermentation with 1 g/L glucose: maximum ethanol production. The point at which the effects estimates were statistically significant (at $p = 0.1$) is indicated by the broken vertical line.

Biomass concentration after 24 h ($[X_{24}]$) (g d.w. cells/L) of fermentation was included as dependent variable in another PB design, in this case, including glucose as independent variable. The results are shown in Table 2.

Cell concentration ranged from 0.04 to 0.98 g d.w. cells/L, while deviation of center point was inferior to 10%, indicating a good reproducibility. Analysis of variance (ANOVA) and the significance of the results were verified using Fisher's statistical test (*F*-test) at 10% of significance. The Pareto diagram (Figure 2) shows tryptone, peptone, yeast extract, and L-arginine concentrations as the variables that influences cell growth, indicating a positive effect. Considering yeast extract is not only a vitamin source, but also a compound that contains carbon, nitrogen, and other elements, it is possible that it can replace other compounds for cell growth, as the results show.

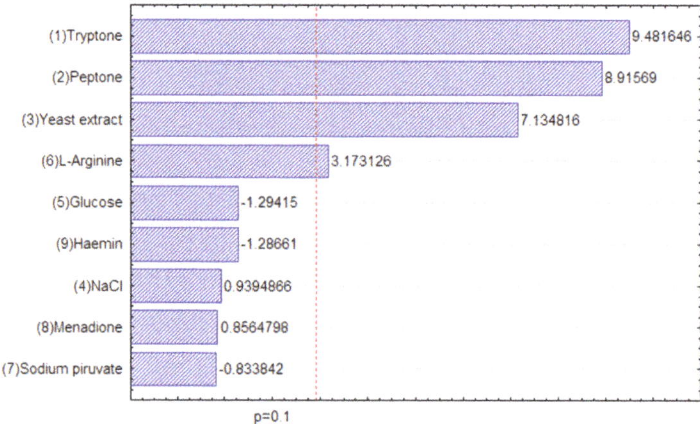

Figure 2. Pareto diagram for the estimated effect of each variable of the Placket–Burman design for *C. carboxidivorans* syngas fermentation on cell concentration ($[X_{24}]$). The point at which the effects estimates were statistically significant (at $p = 0.1$) is indicated by the broken vertical line.

Therefore, tryptone, peptone, yeast extract, and L-arginine were relevant to biomass production and they were all considered to optimize cell growth medium. For ethanol, tryptone, gelatin peptone, and L-arginine were investigated in the next steps, but yeast extract was also included since cell growth is also important during ethanol production.

2.2. Increasing Biomass Production by Central Composite Rotational Design (CCRD)

A central composite rotational design (CCRD) was considered in order to increase biomass production during syngas fermentation, excluding the components that were not relevant to cell growth according to PB design ($p > 0.1$). For this design, the concentrations of the selected variables were adjusted, as indicated by the effect of each one of them. Cell concentrations after 24 h of syngas fermentation by *C. carboxidivorans* at different conditions of CCRD are shown in Table 3.

Table 3. Central composite rotational design (CCRD) matrix with real and coded values and cell concentration after 24 h of syngas fermentation by *C. carboxidivorans*—[X_{24}] (g d.w. cells/L).

Run	Real Values (Coded Levels)				[X_{24}]
	TRY [1]	PEP [2]	YE [3]	ARG [4]	
1	8(−1)	8(−1)	3(−1)	0.8(−1)	0.99
2	8(−1)	8(−1)	3(−1)	1.2(+1)	0.85
3	8(−1)	8(−1)	7(+1)	0.8(−1)	1.05
4	8(−1)	8(−1)	7(+1)	1.2(+1)	1.05
5	8(−1)	12(+1)	3(−1)	0.8(−1)	1.24
6	8(−1)	12(+1)	3(−1)	1.2(+1)	1.03
7	8(−1)	12(+1)	7(+1)	0.8(−1)	0.88
8	8(−1)	12(+1)	7(+1)	1.2(+1)	1.29
9	12(+1)	8(−1)	3(−1)	0.8(−1)	0.99
10	12(+1)	8(−1)	3(−1)	1.2(+1)	1.22
11	12(+1)	8(−1)	7(+1)	0.8(−1)	1.17
12	12(+1)	8(−1)	7(+1)	1.2(+1)	1.36
13	12(+1)	12(+1)	3(−1)	0.8(−1)	1.02
14	12(+1)	12(+1)	3(−1)	1.2(+1)	1.02
15	12(+1)	12(+1)	7(+1)	0.8(−1)	1.29
16	12(+1)	12(+1)	7(+1)	1.2(+1)	1.26
17	6(−2)	10(0)	5(0)	1(0)	0.83
18	14(+2)	10(0)	5(0)	1(0)	1.23
19	10(0)	6(−2)	5(0)	1(0)	1.12
20	10(0)	14(+2)	5(0)	1(0)	1.20
21	10(0)	10(0)	1(−2)	1(0)	1.02
22	10(0)	10(0)	9(+2)	1(0)	1.35
23	10(0)	10(0)	5(0)	0.6(−2)	1.06
24	10(0)	10(0)	5(0)	1.4(+2)	1.07
25 (C)	10(0)	10(0)	5(0)	1(0)	1.06
26 (C)	10(0)	10(0)	5(0)	1(0)	1.16
27 (C)	10(0)	10(0)	5(0)	1(0)	1.06

[1] TRY: tryptone concentration (g/L), [2] PEP: gelatin peptone (g/L); [3] YE: yeast extract concentration (g/L); [4] ARG: L-arginine concentration (g/L). (C) Central point.

Table 3 shows that the cell concentration ranged from 0.83 to 1.35 g d.w. cells/L, both values higher than those found in the Placket–Burman design (Table 2), showing an improvement in biomass production by the statistical design. Analysis of variance (ANOVA) and the significance of the results were verified using Fisher's statistical test (F-test) at 10% of significance. The ANOVA (Table 4) showed that tryptone and yeast extract were relevant ($p < 0.1$) to cell growth. The complete predicted model was not significant (data not shown) and therefore it was improved by eliminating non-significant terms ($p > 0.1$). The ANOVA for the reduced model is presented in Table 4.

Table 4. Analysis of variance (ANOVA) for central composite rotational design (CCRD) for biomass production of syngas fermentation by *Clostridium carboxidivorans* after 24 h.

Factor	SS [1]	Df [2]	MS [3]	F-Value [4]	p-Value [5]
(1)Tryptone(L)	0.126702	1	0.126702	37.56348	0.025604
(3)Yeast extract(L)	0.117852	1	0.117852	34.93986	0.027448
(4)L-arginine(L)	0.008221	1	0.008221	2.43742	0.258861
1L by 3L	0.029087	1	0.029087	8.62357	0.099034
3L by 4L	0.029224	1	0.029224	8.66407	0.098638
Lack of Fit	0.225563	19	0.011872	3.51964	0.244182
Pure Error	0.006746	2	0.003373		
Total SS	0.543396	26			

[1] SS: sum of squares; [2] Df: degree of freedom; [3] MS: mean square; [4] F-value: Test comparing model variance with residual (error) variance; [5] p-value: significant <0.1; (L) linear term; (R^2 = 0.58).

Table 4 shows that there was no lack of adjustment ($p > 0.1$) for the predicted model, which indicates that it is adequate to describe the process. Despite the non-significance of L-arginine, it was considered in the model as the linear interaction between this component and yeast extract was significant (Table 4). The first-order polynomial model proposed for cell growth including all the linear, quadratic, and linear interaction coefficients that were significant to this variable is presented in Equation (1).

$$[X_{24}] = 1.54158 - 0.01697 \times T - 0.17840 \times YE - 0.44168 \times A + 0.01066 \times T \times YE + 0.10684 \times YE \times A \tag{1}$$

where T, YE, and A represent real values of tryptone, yeast extract and L-arginine concentrations (g/L), respectively.

From the presented model, the response surface plots were obtained (Figure 3).

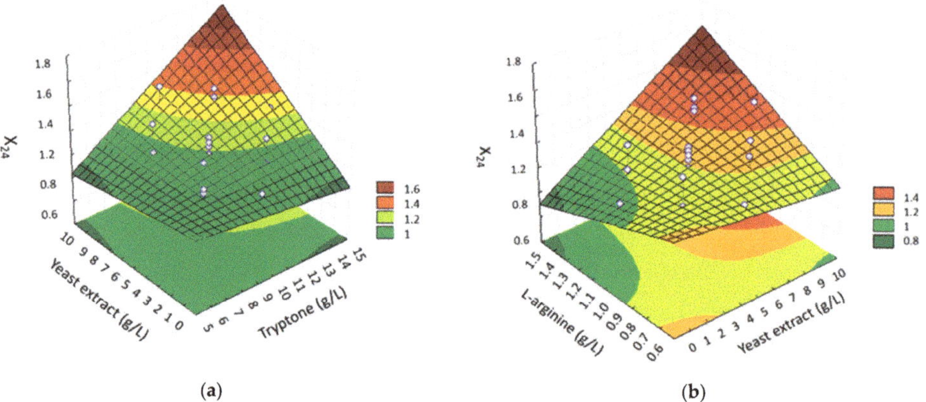

(a)   **(b)**

Figure 3. Response surface plots for *C. carboxidivorans* syngas fermentation with cell concentration ($[X_{24}]$) as function of: (**a**) tryptone and yeast extract and (**b**) yeast extract and L-arginine.

Due to the non-significance of the quadratic terms of the model, it was not possible to obtain the conditions that would result in maximum value for cell growth. However, new experiments were performed in order to validate the obtained model by setting tryptone, yeast extract at higher levels (14 and 9 g/L, respectively), peptone at a lower level (6 g/L) or absence and L-arginine at a higher (1.4 g/L), lower (0.6 g/L) levels, or absence, as shown in Table 5. Indeed, absence or presence of peptone (comparing medium A with C and D with E) did not influence cell growth as predicted by the model. Furthermore, the absence of L-arginine reduced cell growth (medium B), also as predicted by the model.

The best condition was obtained with L-arginine at higher level (medium E), similar to the predicted by the model under these same conditions (1.77 g d.w. cells/L). As a result, we propose a new medium for *C. carboxidivorans* cell growth during syngas fermentation - TYA medium: Tryptone 14 g/L; yeast extract 9 g/L; and L-arginine 1.4 g/L.

Table 5. Biomass production in syngas fermentation by *Clostridium carboxidivorans* after 24 h.

Component	Medium				
	A	**B**	**C**	**D**	**E**
Tryptone	14	14	14	14	14
Yeast extract	9	9	9	9	9
L-arginine	0.6	0	0.6	1.4	1.4
Peptone from gelatin	6	6	0	6	0
Model prediction [X_{24}]	1.35	1.04	1.35	1.77	1.77
[X_{24}] (g d.w. cells/L) [1]	1.18 ± 0.02	1.09 ± 0.05	1.19 ± 0.04	1.69 ± 0.02	1.73 ± 0.06

[1] Experimental value.

2.3. Improving Ethanol Production by Fractional Factorial Design (FFD)

Considering the results of the PB design for ethanol production, a fractional factorial design (FFD) was proposed to further investigate the influence of the significant variables on ethanol production during syngas fermentation (tryptone, gelatin peptone, and L-arginine). Yeast extract was also included in this design because of its effect on cell growth. Glucose was maintained constant (1 g/L) as in the PB design. Results of ethanol production after 144 h of syngas fermentation by *C. carboxidivorans* at different conditions of FFD are present in Table 6.

Table 6. Fractional factorial design matrix with real and coded values and ethanol production after 144 h of syngas fermentation by *Clostridium carboxidivorans*—P_{EtOH} (g/L).

Run	Real Values (Corresponding Coded Levels)				P_{EtOH}
	TRY [1]	**PEP [2]**	**YE [3]**	**L-ARG [4]**	
1	2(−1)	2(−1)	1(−1)	0.2(−1)	0.287
2	12(1)	12(1)	1(−1)	1.2(1)	0.350
3	2(−1)	2(−1)	1(−1)	1.2(1)	0.340
4	12(1)	12(1)	1(−1)	0.2(−1)	0.257
5	2(−1)	2(−1)	7(1)	1.2(1)	0.216
6	12(1)	12(1)	7(1)	0.2(−1)	0.671
7	2(−1)	2(−1)	7(1)	0.2(−1)	0.677
8	12(1)	12(1)	7(1)	1.2(1)	1.614
9 (C)	7(0)	7(0)	4(0)	0.7(0)	0.985
10 (C)	7(0)	7(0)	4(0)	0.7(0)	0.967

[1] TRYP: tryptone concentration (g/L); [2] PEP: peptone concentration (g/L); [3] YE: yeast extract concentration (g/L); [4] L-ARG: L-arginine concentration (g/L); (C) central point.

Ethanol production ranged between 0.216 and 1.614 g/L (Table 6), increasing the lower value of the PB design (Table 1), but maintaining the higher value, indicating that the concentrations studied in the PB design were already close to maximum response. All variables of the FFD were relevant to ethanol production during syngas fermentation by *C. carboxidivorans*, including yeast extract (Figure 4). Different concentrations of yeast extract were used in FFD in comparison with PB design, which might explain this result. All variables affected the response positively, which means that higher concentrations stimulate ethanol production. Therefore, a culture medium composed of all those components at the higher levels (+1) of concentration is proposed for syngas fermentation, herein called TPYGarg, containing: 12 g/L of tryptone, 12 g/L of peptone from gelatin, 7 g/L of yeast extract, 1.2 g/L of L-arginine, and 1 g/L of glucose.

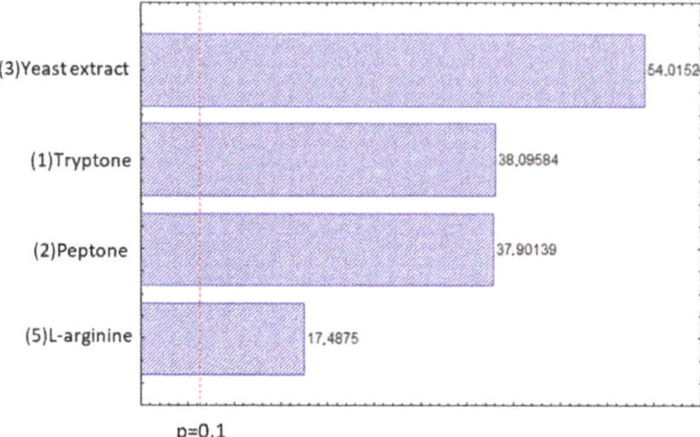

Figure 4. Pareto diagram for the estimated effect of each variable of the Fractional Factorial Design on ethanol production—P_{EtOH}—in the syngas fermentation by *Clostridium carboxidivorans*. The point at which the effects estimates were statistically significant (at $p = 0.10$) is indicated by the broken vertical line.

2.4. Tryptone-Peptone-Yeast Extract-Glucose-Arginine Medium (TPYGarg) for Ethanol Production

2.4.1. Cell Growth

During ethanol production, cell growth is important since more cells produce more ethanol. Therefore, *C. carboxidivorans* growth in TPYGarg and ATCC® 2713 media was monitored for comparison. A short lag phase of approximately 2 h for TPYGarg and 4 h for ATCC® 2713, is depicted in Figure 5a. Even though 40% more biomass is achieved for TPYGarg after 8 h of growth, in relation to ATCC® 2713 medium, both growth profiles meet after 24 h (Figure 5a). Specific growth rate obtained for TPYGarg ($0.82 \ h^{-1}$) was higher than that obtained for ATCC® 2713 ($0.64 \ h^{-1}$). These values are much higher than those recently reported for ABE fermentation ($0.080 \ h^{-1}$) and HBE fermentation ($0.086 \ h^{-1}$) by Fernández-Naveira et al. [18].

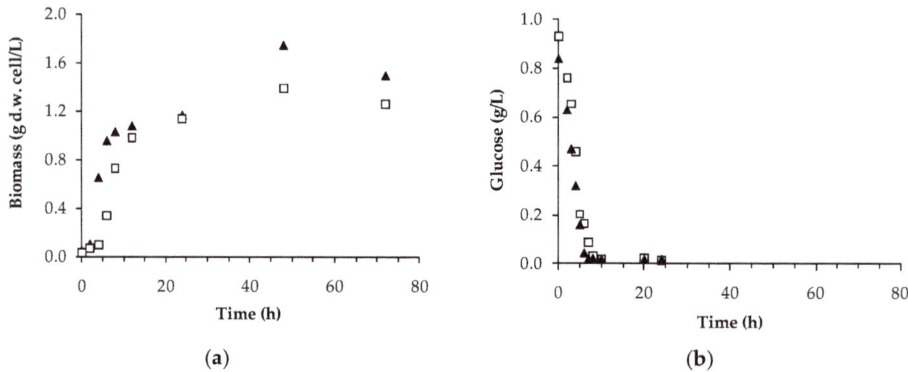

Figure 5. *C. carboxidivorans* growth (**a**) and glucose consumption (**b**) on Tryptone-Peptone-Yeast extract-Glucose-Arginine medium (TPYGarg) (▲) and ATCC® 2713 (□).

Glucose consumption was also monitored during syngas fermentation. Figure 5b shows that after 7 h of fermentation glucose was completely depleted from TPYGarg medium while total glucose consumption occurred after 8 h of fermentation on ATCC® 2713 medium. It is possible to observe that cells are still growing (Figure 5) even after glucose is completely consumed, mainly on TPYGarg medium.

2.4.2. Metabolites Production

Two phases are observed during *C. carboxidivorans* growth on TPYGarg and ATCC® 2713 with no pH control (Figure 6). At the beginning of fermentation (4 to 5 h), a decrease in pH values before glucose depletion is detected, and this is followed by an increase in pH levels on both media. These phases are described in literature as acidogenesis and solventogenesis [27].

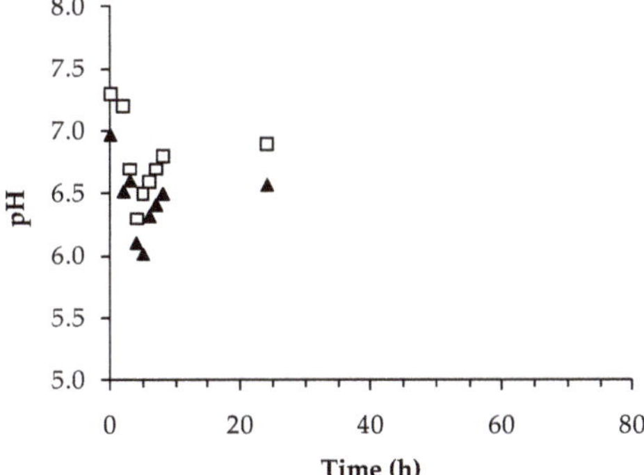

Figure 6. pH during syngas fermentation by *C. carboxidivorans* on Tryptone-Peptone-Yeast extract-Glucose-Arginine medium (TPYGarg) (▲) and ATCC® 2713 (□).

In the present study, with glucose and CO/CO_2 as carbon sources, acid production starts immediately after seed culture inoculation and along with bacterial biomass increase, as Figure 7 depicts. Most abundant acids obtained were acetic and lactic acids and their concentrations still increase even after glucose depletion, suggesting that those acids can be produced from carbon sources other than glucose, as CO and CO_2 from syngas, or from tryptone and peptone. Maximum concentrations of acetic acid were 3.57 g/L and 2.85 g/L after 72 h of fermentation on TPYGarg and ATCC® 2713 media, respectively. For lactic acid, 3-times more acid is produced after 8 h on TPYGarg in relation to ATCC® 2713, but at the end (70 h) a similar value is obtained on ATCC® 2713.

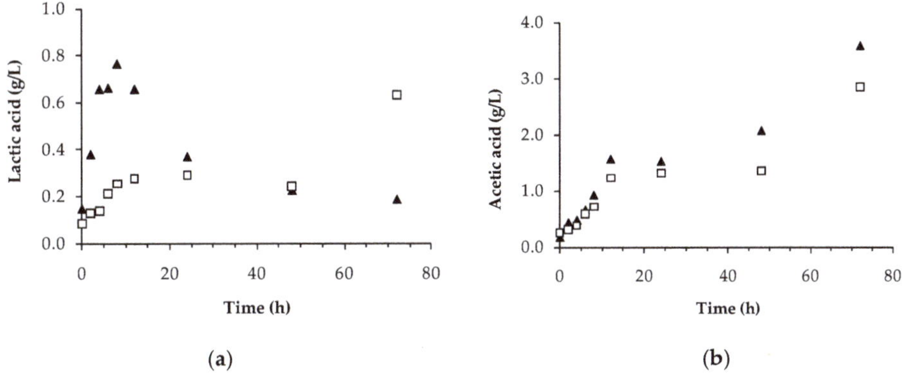

(a) **(b)**

Figure 7. *Cont.*

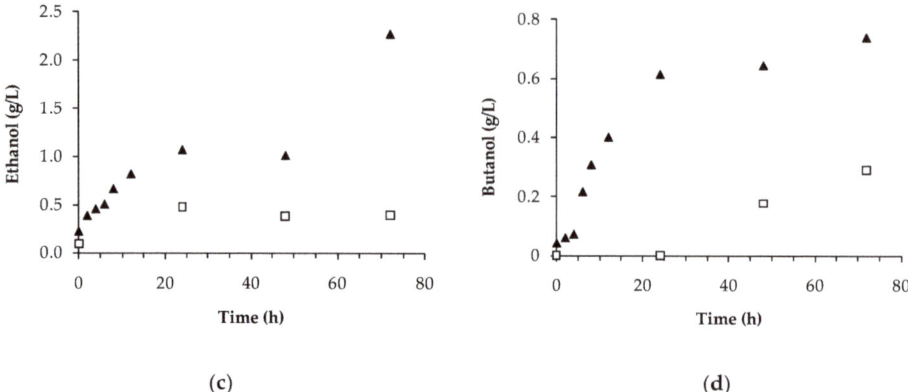

Figure 7. Lactic acid (**a**), acetic acid (**b**), ethanol (**c**), and butanol (**d**) production during syngas fermentation by *C. carboxidivorans* on Tryptone-Peptone-Yeast extract-Glucose-Arginine medium (TPYGarg) (▲) and ATCC® 2713 (□).

Maximum ethanol production on TPYGarg was 5.7-times higher than on ATCC® 2713 after 72 h of fermentation (Figure 7c). Ethanol productivity (considering 24 h) was also improved on TPYGarg (0.045 g/L·h) in comparison to ATCC® 2713 (0.020 g/L·h). Fernández-Naveira et al. [5] reported an inferior ethanol productivity (0.03 g/L·h) using the same microorganism and a medium containing 30 g/L of glucose, 1 g/L of yeast extract, minerals, trace metal, and vitamin solutions, with no syngas addition. This ethanol productivity was higher than those reported on syngas fermentation [14,19,24], which might be related to medium composition. Regarding butanol production, TPYGarg performance was also better than ATCC® 2713, with a butanol production almost 2.5-times higher. For butanol productivity, Fernández-Naveira et al. [5] reported 0.007 g/L·h, while 0.04 g/L·h was obtained in this study.

Maximum biomass and solvents production on ATCC® 2713 and TPYGarg over 72 h of syngas fermentation with *C. carboxidivorans* are presented in Table 7. Biomass concentration on ATCC® 2713 was 26% inferior in comparison to TPYGarg. Ethanol and butanol production were 4.75- and 2.55-times higher, respectively, on TPYGarg medium in comparison to ATCC® 2713.

Table 7. Cell concentration ([X_{24}] (g d.w. cells/L)), ethanol and butanol production (P_{EtOH} and P_{But}, respectively, in g/L) and productivities (Q_{EtOH} and Q_{But}, respectively, in g/L•h) after 72 h of syngas fermentation by *Clostridium carboxidivorans*.

Medium	Maximum Values in 72 h				
	[$X_{máx}$]	P_{EtOH}	P_{But}	Q_{EtOH}	Q_{But}
ATCC® 2713 [1]	1.38	0.47	0.29	0.020	0.004
TPYGarg [2]	1.75	2.28	0.74	0.045	0.039

[1] 10 g/L tryptone, 10 g/L peptone from gelatin, 5 g/L yeast extract; 1 g/L glucose, 5 g/L NaCl, 1 g/L ʟ-arginine, 1 g/L sodium pyruvate, 0.0005 g/L menadione and 0.005 g/L hemin; [2] 12 g/L tryptone, 12 g/L peptone from gelatin, 7 g/L yeast extract, 1 g/L glucose, and 1.2 g/L ʟ-arginine.

2.4.3. Cost Comparison

The competitiveness of new technologies compared to those already consolidated is very important to enable its implementation on industrial scale. As important as metabolite production and growth analysis, a cost-effectiveness study represents a fundamental step. A cost prospect was calculated for ATCC® 2713, TYA, and TPYGarg media. As shown in Table 8, TYA and TPYGarg are 47% and 31% cheaper than ATCC®2713 and promote higher cell growth and ethanol production, respectively.

Table 8. Medium composition and cost of ATCC® 2713, Tryptone, yeast extract, and L-arginine (TYA) and Tryptone-Peptone-Yeast extract-Glucose-Arginine (TPYGarg).

Compounds	Price	ATCC® 2713		TYA		TPYGarg	
	USD/kg [1]	g/L [2]	US$/L [3]	g/L [2]	USD/L [3]	g/L [2]	USD/L [3]
Tryptone	94.2	10	0.94	14	1.32	12	1.13
Peptone	70	10	0.70	-	-	12	0.84
Yeast extract	48.2	5	0.24	9	0.43	7	0.34
D (+) Glucose	9.96	1	0.01	-	-	1	0.01
NaCl	10.56	5	0.05	-	-	-	-
Sodium pyruvate	1430.00	1	1.43	-	-	-	-
Menadione	1630.00	0.0005	≈0	-	-	-	-
Hemin	5960.00	0.005	0.03	-	-	-	-
L-arginine	46	1	0.48	1.4	0.06	1.2	0.06
Total (USD/L) [4]			3.52		1.82		2.37

[1] Cost of each component was calculated based on Sigma–Aldrich prices (Accessed on October 2019); [2] amount of component (g) per liter of medium; [3] Cost of component per liter of medium; and [4] Total cost of medium per liter.

3. Discussion

In the exploratory study to evaluate medium components effect on cell growth, tryptone, peptone, yeast extract, and L-arginine concentrations were considered significant for biomass production. Zhang et al. [14] reported that trace metals are required for *C. carboxidivorans* growth and these elements might be present in yeast extract. Peptone and tryptone are both sources of amino acids. However, as they are obtained from different protein sources (peptone is pancreatic digested gelatin and tryptone is the pancreatic digested casein), they might have complementary functions for cell growth. Another important observation is that glucose was not significant for cell growth, which might be related to the presence of CO and CO_2, indicating that syngas was used as the carbon source. Although the best result was obtained for the composition of the ATCC® medium 2713 (Table 1, run 16), the statistical evaluation shows that some components are not significant to cell growth, which means that its addition to medium composition just adds costs.

Considering the positive effect of these components on cell growth, the experimental design to optimize this medium was planned with higher concentrations. In that case, peptone was not significant to cell growth, which shows that it was only significant in PB design because tryptone concentration was not enough. The medium proposed for *C. carboxidivorans* cell growth after optimization, is reduced to three components, besides syngas. ATCC® medium 2713 is composed of nine components and many other media reported in literature have more than that, for example BDM—base defined medium (22 components) and P7Mt medium (20 components) [17]. In those laborious media, cell growth for *C. carboxidivorans* ranges from 0.4 to 0.6 g/L [5,17,18]. Therefore, it is evident that the medium proposed in the present work not only has the advantage of being easy to prepare but also yields a higher biomass.

For ethanol, experimental design revealed that a culture medium composed of 12 g/L of tryptone, 12 g/L of peptone from gelatin, 7 g/L of yeast extract, 1.2 g/L of L-arginine, and 1 g/L of glucose was the best for syngas fermentation, herein called TPYGarg. The development of a culture medium with no yeast extract, but with vitamins (biotin, pantothenic, and p-aminobenzoic acid), ammonia, trace metals, minerals, cysteine, sodium sulfide, and resazurin was reported by Phillips et al. [17]. After 200 h, around 3.0 g/L of ethanol (approximately, 0.015 g/L·h) was obtained in a medium 27-times cheaper than the standard one [14]. Despite cheaper, this medium still requires 20 components and demands time to prepare it. A ATCC® 1754 modified medium was proposed by Ramachandryia [28], using cotton seed extract (CSE) and morpholinoethanesulfonic acid (MES), with no yeast extract, resulting in 2.78 g/L of ethanol after 350 h of fermentation (approximately, 0.008 g/L·h). A modification in the ATCC® 1754 medium with MES as buffer was also reported, leading to 6.1 g/L of ethanol in a two stirred-tank bioreactor (in series) [21], without considering the high cost of MES. Enhancement of alcohol synthesis was reported using CO-rich off-gas fermentation by *C. carboxidivorans* by adding molybdenum/iron/cupper, aiming at the enzymes of the microorganism metabolism [22]. A final ethanol concentration of 3.5 g/L was obtained after 72 h of fermentation (approximately, 0.049 g/L·h) [19], the higher productivity found for this process in literature so far, but without considering the culture medium cost.

Cell growth, glucose consumption, and medium pH were monitored during syngas fermentation in TPYGarg and high biomass was obtained as well as the detection of two phases: pH reduction in the beginning followed by pH increase. In classical ABE fermentation using glucose as sole carbon source or in HBE fermentation using gaseous substrate, the bioconversion by Clostridia occurs in two steps. First, the exponential bacterial growth and organic acids production can be observed, known as the acidogenesis phase. Then, those acids are converted into solvents as ethanol, butanol, and acetone in the ABE process, and hexanol, butanol, and ethanol in the HBE process, known as the solventogenic phase. The acidogenic phase usually shows greater efficiency at neutral or slightly high pH values, leading to the medium acidification due to acid production. Concerning ABE fermentation, studies have reported that the solventogenic phase is stimulated during medium acidification [5]. On the other hand, HBE fermentation studies have demonstrated that alcohol production usually occurs near pH 6.0 [29]. Below this pH level, low alcohol production and a decreasing growth capacity are usually attributed to the "acid crash" phenomenon [30].

In the present study, solvent production started while glucose was being consumed and cells were growing. Although, it is expected that solventogenic phase only starts after medium acidification as result of acid production, the production of organic acids and solvents occurred simultaneously in the present study. Formation of organic acids occurs during exponential growth, decreasing the culture pH level, making it unfavorable for the cell population and resulting in a metabolic change at the end of the exponential phase. Therefore, those organic acids are partly re-assimilated and rebuilt into neutral products as the solvents. However, not all species and strains of solventogenic clostridia have the same behavior. Lipovsky et al. [31] and Branska et al. [32] verified butanol production associated with cell growth by clostridia via the acetone–butanol–ethanol (ABE) pathway.

When using syngas as the only carbon source for *C. carboxidivorans* fermentation, a mixture of organic acids as acetic, butyric, and hexanoic acids was expected, and further partial conversion of those acids into respective alcohols, following HBE fermentation process. Using glucose as the only carbon source, ABE fermentation was expected, producing lactic, acetic, formic, and propionic acids, which then could be converted into acetone, ethanol, and butanol. However, it was observed that *C. carboxidivorans* does not follow classical ABE or HBE fermentations when both carbon sources were used (glucose and CO/CO_2), since acids and alcohols production were detected at the same time on TPYGarg and ATCC® 2713 media.

Higher growth rates and alcohol productivities obtained in the mixotrophic fermentation (glucose and CO/CO_2) when compared to heterotrophic (glucose) and autotrophic (CO/CO_2) processes reported in the literature [33], is probably the result of the two metabolic pathways synergy, glycolysis and Wood–Ljungdahl. The glycolysis converts glucose into pyruvate, producing 2 mol of pyruvate, two ATP, and four electrons. The pyruvate is converted to acetyl-CoA, an important intermediate for biofuel production, which results in the release of 2 mol of CO_2 and 4 additional electrons [33]. The autotrophic metabolism, known as Wood–Ljungdahl (WL) pathway, consists of two branches. In the methyl branch, CO_2 is reduced to a methyl group consuming six electrons and one ATP. While in the carbonyl branch, CO_2 is reduced to CO consuming two electrons. The CO and the methyl group formed in this first step of WL pathway are converted into 1 mol of acetyl-CoA [33–35]. Thereby, the glycolysis and WL are complementary pathways as CO_2 and electrons produced in the glycolysis are fully utilized by WL to produce one additional mol of acetyl-CoA increasing its yield by 50% in relation to glucose metabolism. Besides, the syngas feed contributes to enhance the metabolic flow of the WL pathway, as CO, CO_2, and H_2 are introduced [33].

4. Material and Methods

4.1. Microorganism and Culture Medium

Clostridium carboxidivorans DSM15243 was obtained from Deutsche Sammlung von Mikroorganismen und Zellkulturen GmbH (Braunschweig, Germany). The strain was anaerobically activated in ATCC® 2713 medium. ATCC® 2713 contains the following components (per liter): tryptone, 10 g; peptone from gelatin, 10 g; yeast extract, 5 g; L-arginine, 1 g; sodium pyruvate, 1 g; menadione, 0.5 mg; hemin, 5 mg; and glucose, 1 g. Next, the cells were grown in 100 mL serum glass bottles (Wheaton, NJ, USA) containing 50 mL of Brain Heart Infusion (BHI) medium from Sigma-Aldrich and syngas. Syngas was provided by White Martins Praxair Inc. (Joinville–SC, Brazil) and it is composed of 25% CO, 43.9% H_2, 10.02% CO_2, 10.05% N_2, and 11.01% methane.

4.2. Syngas Fermentation

All fermentations were performed in 50 mL serum glass bottles containing 30 mL of each culture medium studied. Media composition is detailed in Tables 1–3, Table 5, and Table 6. After media preparation, syngas was flushed in the liquid phase for 5 min. The glass bottles were sealed with gas impermeable butyl rubber septum stopper and aluminum seal and sterilized in autoclave at 0.5 atm for 20 min. After sterilization, seed culture was aseptically inoculated in all glass bottles to achieve 0.05 g d.w. cells/L. Syngas was added in the headspace at 1.22 atm and the bottles were incubated horizontally [36] at 37 °C and 150 rpm in TECNAL TE-420 shaker. Fermented culture mediums were sampled for optical density (OD) measurement and high-performance liquid chromatography (HPLC) analysis.

4.3. Experimental Designs to Increase Cell Growth and Ethanol Production

4.3.1. Screening Medium Components by Plackett-Burman Design

Plackett–Burman (PB) design was applied to screen culture medium components that influence biomass and ethanol production during syngas fermentation. Eight components and their concentrations were based on ATCC® 2713 medium (tryptone, 10 g/L; gelatin peptone, 10 g/L; yeast extract, 5 g/L; glucose, 1 g/L; sodium chloride, 5 g/L; L-arginine, 1 g/L; sodium pyruvate, 1 g/L; menadione, 0.5 mg/L; and hemin, 5 mg/L). The concentration of each nutrient at different levels (minimum, −1; central, 0 and maximum, +1) can be found in Tables 1 and 2. A 12-run PB design plus 3 central level (0) was performed for ethanol production, setting glucose was at 1.0 g/L (Table 1) and a 16-run PB design with 3 central level (0) was performed for biomass production. All experiments were conducted in random order and duplicates.

Biomass concentration after 24 h of fermentation—[X_{24}] and ethanol concentration after 144 h of fermentation—[P_{EtOH}] were used as response variables.

4.3.2. Central Composite Rotational Design (CCRD) to Increase Cell Growth

A central composite rotational design (CCRD) was proposed after the PB screening in order to maximize biomass production during syngas fermentation. Cell concentration after 24 h of fermentation—[X_{24}] was the response variable. The parameters set and used as independent variables were: tryptone, peptone, yeast extract, and L-arginine concentrations as shown in Table 3, with high (+1) and low levels (−1), as well as axial (−2, +2), and central (0) levels. A CCRD with 24 runs plus 3 central level assays were performed, totalling 27 tests (Table 3). All experiments were conducted in random order and duplicates. The results obtained were evaluated by Analysis of Variance (ANOVA) and the effects were considered significant when $p < 0.10$.

4.3.3. Fractional Factorial Design (FFD) to Improve Ethanol Production

A fractional factorial design (FFD) was used to evaluate the relevant components influencing ethanol production during syngas fermentation. Four parameters (Tryptone, Peptone from gelatin, yeast extract, and L-arginine concentrations) were screened in this study to determine the most significant input factors for ethanol production by a two-level fraction factorial design (2^{4-1}) with three central points. The response variable was ethanol production after 144 h of fermentation—P_{EtOH}. Table 6 shows the values representing the levels for each parameter studied. All experiments were conducted in random order and duplicates.

4.4. Analytical Methods

4.4.1. Cell Growth

Optical density (OD) was measured at 600 nm using a UV-VIS spectrophotometer (Bell SP 2000 UV). Cell concentration (g dry weight cells/L) was estimated using a standard curve.

4.4.2. Product and by-Products Quantification

Metabolites were analyzed by HPLC (High performance Liquid Chromatography) from Shimadzu equipped with Aminex® HPX-87H, 300 × 7.8 mm (Bio-Rad Laboratories Ltd., São Paulo–SP, Brazil) column and RI (refractive index) detector (Shimadzu®, Kyoto, Japan). The mobile phase was H_2SO_4 5 mM, 0.6 mL/min flow rate, and 20 µL injection volume. The column temperature was set at 55 °C.

4.5. Statistical Analysis

The statistical analysis was performed using STATISTICA 7.1 software (StatSoft, Inc., Tulsa, OK, USA). Analysis of variance (ANOVA) and the significance of the results were verified using Fisher's statistical test (F-test) at 10% of significance.

5. Conclusions

Experimental design is a powerful tool for reducing culture medium components. In the present study, a PB design followed by CCRD indicated that tryptone, yeast extract, and L-arginine were the most important culture medium components for *C. carboxidivorans* growth from syngas. The proposed medium for *C. carboxidivorans* growth, herein named TYA (14 g/L tryptone, 9 g/L yeast extract, and 1.4 g/L L-arginine) is cheaper (47%) than ATCC® 2713 and reduced in components. For ethanol production, peptone and glucose were also relevant. After applying an experimental design sequence, a culture medium for syngas fermentation by *C. carboxidivorans* was proposed and named TPYGarg (12 g/L tryptone, 12 g/L peptone from gelatin, 7 g/L yeast extract, 1.2 g/L L-arginine, and 1 g/L glucose). Almost five-times higher ethanol concentration and more than two-times higher butanol concentration were produced on TPYGarg in comparison to ATCC® 2713. TPYGarg represents a 31% reduction in cost of culture medium for syngas fermentation by *C. carboxidivorans* compared to ATCC® 2713. The increase in ethanol productivity with this new medium in comparison to literature reports is outstanding.

Author Contributions: C.B., A.B., T.F., and P.A. conceived and planned the experiments. C.B., A.B., M.B., and R.R. carried out the experiments. A.P. and A.C.V. were responsible for analyses and A.P. conducted the statistical analysis. C.B. wrote the manuscript with support from T.F. and P.A. T.F. and P.A. supervised the project. All authors provided critical feedback and helped shape the research, analysis, and manuscript. All authors have read and agreed to the published version of the manuscript.

Funding: This research received external funding from Faperj Project no. E-26/010.002984/2014 (Pensa Rio 2014).

Conflicts of Interest: The authors declare no conflict of interest.

References

1. Ramió-Pujol, S.; Ganigué, R.; Bañeras, L.; Colprim, J. Impact of formate on the growth and productivity of Clostridium ljungdahlii PETC and Clostridium carboxidivorans P7 grown on syngas. *Int. Microbiol.* **2015**, *17*, 195–204.
2. Munasinghe, P.C.; Khanal, S.K. Biomass-derived syngas fermentation into biofuels. *Biofuels* **2011**, *101*, 79–98.
3. Yang, T.C.; Kumaran, J.; Amartey, S.; Maki, M.; Li, X.; Lu, F.; Qin, W. Biofuels and bioproducts produced through microbial conversion of biomass. In *Bioenergy Research: Advances and Applications*; Elsevier: Amsterdam, The Netherlands, 2014; Chapter 5; pp. 71–93.
4. Sun, Y.; Cheng, J. Hydrolysis of lignocellulosic materials for ethanol production: A review. *Bioresour. Technol.* **2002**, *83*, 1–11. [CrossRef]
5. Fernández-Naveira, Á.; Veiga, M.C.; Kennes, C. Glucose bioconversion profile in the syngas-metabolizing species Clostridium carboxidivorans. *Bioresour. Technol.* **2017**, *244*, 552–559. [CrossRef] [PubMed]
6. Li, D.; Meng, C.; Wu, G.; Xie, B.; Han, Y.; Guo, Y.; Song, C.; Gao, Z.; Huang, Z. Effects of zinc on the production of alcohol by Clostridium carboxidivorans P7 using model syngas. *J. Ind. Microbiol. Biotechnol.* **2018**, *45*, 61–69. [CrossRef] [PubMed]
7. Ljungdahl, L. The Autotrophic Pathway of Acetate Synthesis in Acetogenic Bacteria. *Annu. Rev. Microbiol.* **1986**, *40*, 415–450. [CrossRef] [PubMed]
8. Coelho, F.; Nele, M.; Ribeiro, R.; Ferreira, T.; Amaral, P. Clostridium carboxidivorans' Surface Characterization Using Contact Angle Measurement (CAM). *Chem. Eng. Trans.* **2016**, *50*, 277–282.
9. Wilkins, M.R.; Atiyeh, H.K. Microbial production of ethanol from carbon monoxide. *Curr. Opin. Biotechnol.* **2011**, *22*, 326–330. [CrossRef]
10. Cheng, C.; Li, W.; Lin, M.; Yang, S.-T. Metabolic engineering of Clostridium carboxidivorans for enhanced ethanol and butanol production from syngas and glucose. *Bioresour. Technol.* **2019**, *284*, 415–423. [CrossRef]
11. Liou, J.S.C.; Balkwill, D.L.; Drake, G.R.; Tanner, R.S. Clostridium carboxidivorans sp. nov., a solvent-producing clostridium isolated from an agricultural settling lagoon, and reclassification of the acetogen Clostridium scatologenes strain SL1 as Clostridium drakei sp. nov. *Int. J. Syst. Evol. Microbiol.* **2005**, *55*, 2085–2091. [CrossRef]
12. Rajagopalan, S.; P. Datar, R.; Lewis, R.S. Formation of ethanol from carbon monoxide via a new microbial catalyst. *Biomass Bioenergy* **2002**, *23*, 487–493. [CrossRef]
13. Zhang, J.; Yu, L.; Lin, M.; Yan, Q.; Yang, S.-T. n-Butanol production from sucrose and sugarcane juice by engineered Clostridium tyrobutyricum overexpressing sucrose catabolism genes and adhE2. *Bioresour. Technol.* **2017**, *233*, 51–57. [CrossRef] [PubMed]
14. Zhang, J.; Taylor, S.; Wang, Y. Effects of end products on fermentation profiles in Clostridium carboxidivorans P7 for syngas fermentation. *Bioresour. Technol.* **2016**, *218*, 1055–1063. [CrossRef] [PubMed]
15. Ramió-Pujol, S.; Ganigué, R.; Bañeras, L.; Colprim, J. Incubation at 25 °C prevents acid crash and enhances alcohol production in Clostridium carboxidivorans P7. *Bioresour. Technol.* **2015**, *192*, 296–303.
16. Lewis, R.S.; Tanner, R.S.; Huhnke, R.L. Huhnke Indirect or Direct Fermentation of Biomass to Fuel Alcohol. U.S. Patent No. 11/441,392, 25 May 2006.
17. Phillips, J.R.; Atiyeh, H.K.; Tanner, R.S.; Torres, J.R.; Saxena, J.; Wilkins, M.R.; Huhnke, R.L. Butanol and hexanol production in Clostridium carboxidivorans syngas fermentation: Medium development and culture techniques. *Bioresour. Technol.* **2015**, *190*, 114–121. [CrossRef] [PubMed]
18. Fernández-Naveira, Á.; Abubackar, H.N.; Veiga, M.C.; Kennes, C. Efficient butanol-ethanol (B-E) production from carbon monoxide fermentation by Clostridium carboxidivorans. *Appl. Microbiol. Biotechnol.* **2016**, *100*, 3361–3370. [CrossRef] [PubMed]
19. Jang, Y.S.; Malaviya, A.; Cho, C.; Lee, J.; Lee, S.Y. Butanol production from renewable biomass by clostridia. *Bioresour. Technol.* **2012**, *123*, 653–663. [CrossRef]
20. Munasinghe, P.C.; Khanal, S.K. Evaluation of hydrogen and carbon monoxide mass transfer and a correlation between the myoglobin-protein bioassay and gas chromatography method for carbon monoxide determination. *RSC Adv.* **2014**, *4*, 37575–37581. [CrossRef]
21. Doll, K.; Rückel, A.; Kämpf, P.; Wende, M.; Weuster-Botz, D. Two stirred-tank bioreactors in series enable continuous production of alcohols from carbon monoxide with Clostridium carboxidivorans. *Bioprocess Biosyst. Eng.* **2018**, *41*, 1403–1416. [CrossRef]

22. Shen, S.; Gu, Y.; Chai, C.; Jiang, W.; Zhuang, Y.; Wang, Y. Enhanced alcohol titre and ratio in carbon monoxide-rich off-gas fermentation of Clostridium carboxidivorans through combination of trace metals optimization with variable-temperature cultivation. *Bioresour. Technol.* **2017**, *239*, 236–243. [CrossRef]

23. Saxena, J.; Tanner, R.S. Effect of trace metals on ethanol production from synthesis gas by the ethanologenic acetogen, Clostridium ragsdalei. *J. Ind. Microbiol. Biotechnol.* **2011**, *38*, 513–521. [CrossRef] [PubMed]

24. Mohammadi, M.; Najafpour, G.D.; Younesi, H.; Lahijani, P.; Uzir, M.H.; Mohamed, A.R. Bioconversion of synthesis gas to second generation biofuels: A review. *Renew. Sustain. Energy Rev.* **2011**, *15*, 4255–4273. [CrossRef]

25. Wan, N.; Sathish, A.; You, L.; Tang, Y.J.; Wen, Z. Deciphering Clostridium metabolism and its responses to bioreactor mass transfer during syngas fermentation. *Sci. Rep.* **2017**, *7*, 1–11. [CrossRef] [PubMed]

26. Ahlawat, S.; Kaushal, M.; Palabhanvi, B.; Muthuraj, M.; Goswami, G.; Das, D. Nutrient modulation based process engineering strategy for improved butanol production from Clostridium acetobutylicum. *Biotechnol. Prog.* **2018**, *35*, e2771. [CrossRef] [PubMed]

27. Fernández-Naveira, Á.; Veiga, M.C.; Kennes, C. Effect of pH control on the anaerobic H-B-E fermentation of syngas in bioreactors. *J. Chem. Technol. Biotechnol.* **2017**, *92*, 1178–1185. [CrossRef]

28. Ramachandriya, K.D.; Kundiyana, D.K.; Wilkins, M.R.; Terrill, J.B.; Atiyeh, H.K.; Huhnke, R.L. Carbon dioxide conversion to fuels and chemicals using a hybrid green process. *Appl. Energy* **2013**, *112*, 289–299. [CrossRef]

29. Abubackar, H.N.; Fernández-naveira, Á.; Veiga, M.C.; Kennes, C. Impact of cyclic pH shifts on carbon monoxide fermentation to ethanol by Clostridium autoethanogenum. *Fuel* **2016**, *178*, 56–62. [CrossRef]

30. Maddox, I.S.; Steiner, E.; Hirsch, S.; Wessner, S.; Gutierrez, N.A.; Gapes, J.R.; Schuster, K.C. Acid Crash and Acidogenic ABE Fermentation 95 The Cause of "Acid Crash" and "Acidogenic Fermentations" During the Batch Acetone-Butanol-Ethanol (ABE-) Fermentation Process Fermentation Symposium JMMB Research Article. *J. Mol. Microbiol. Biotechnol* **2000**, *2*, 95–100.

31. Lipovsky, J.; Patakova, P.; Paulova, L.; Pokorny, T.; Rychtera, M.; Melzoch, K. Butanol production by Clostridium pasteurianum NRRL B-598 in continuous culture compared to batch and fed-batch systems. *Fuel Process. Technol.* **2016**, *144*, 139–144. [CrossRef]

32. Branska, B.; Pechacova, Z.; Kolek, J.; Vasylkivska, M.; Patakova, P. Flow cytometry analysis of Clostridium beijerinckii NRRL B-598 populations exhibiting different phenotypes induced by changes in cultivation conditions. *Biotechnol. Biofuels* **2018**, *11*, 1–16. [CrossRef]

33. Fast, A.G.; Schmidt, E.D.; Jones, S.W.; Tracy, B.P. Acetogenic mixotrophy: NOVEL options for yield improvement in biofuels and biochemicals production. *Curr. Opin. Biotechnol.* **2015**, *33*, 60–72. [CrossRef] [PubMed]

34. Redl, S.; Diender, M.; Jensen, T.Ø.; Sousa, D.Z.; Nielsen, A.T. Exploiting the potential of gas fermentation. *Ind. Crops Prod.* **2017**, *106*, 21–30. [CrossRef]

35. Jones, S.W.; Fast, A.G.; Carlson, E.D.; Wiedel, C.A.; Au, J.; Antoniewicz, M.R.; Papoutsakis, E.T.; Tracy, B.P. CO_2 fixation by anaerobic non-photosynthetic mixotrophy for improved carbon conversion. *Nat. Commun.* **2016**, *7*, 1–9. [CrossRef] [PubMed]

36. Ribeiro, R.R.; Coelho, F.; Ferreira, T.F.; Amaral, P.F.F. A New Strategy for Acetogenic Bacteriacell Growth and Metabolites Production Using Syngas in Lab Scale. *IOSR J. Biotechnol. Biochem.* **2017**, *03*, 27–30. [CrossRef]

© 2020 by the authors. Licensee MDPI, Basel, Switzerland. This article is an open access article distributed under the terms and conditions of the Creative Commons Attribution (CC BY) license (http://creativecommons.org/licenses/by/4.0/).

 *catalysts* 

Article

Glycerol Acetylation Mediated by Thermally Hydrolysed Biosolids-Based Material

Mattia Bartoli, Chengyong Zhu, Michael Chae and David C. Bressler *

Department of Agricultural, Food and Nutritional Science, University of Alberta, 410 Ag/For Building, Edmonton, AB T6G 2P5, Canada; mattia.bartoli@polito.it (M.B.); chengyon@ualberta.ca (C.Z.); mchae@ualberta.ca (M.C.)

* Correspondence: dbressle@ualberta.ca; Tel.: +1-780-492-4986

Received: 15 November 2019; Accepted: 16 December 2019; Published: 18 December 2019

Abstract: Crude glycerol is the main by-product of many renewable diesel production platforms. However, the process of refining glycerol from this crude by-product stream is very expensive, and thus does not currently compete with alternative processes. The acetylation of glycerol provides an intriguing strategy to recover value-added products that are employable as fuel additives. In this work, the conversion of glycerol to acetyl derivatives was facilitated by a heterogeneous catalyst generated from the thermal hydrolysis of biosolids obtained from a municipal wastewater treatment facility. The reaction was studied using several conditions including temperature, catalyst loading, acetic acid:glycerol molar ratio, and reaction time. The data demonstrate the potential for using two distinct by-product streams to generate fuel additives that can help improve the process economics of renewable diesel production.

Keywords: biosolids; esterification; glycerol conversion; heterogeneous catalysis

---

1. Introduction

Biodiesel and renewable diesel commodities are one of the main driving forces of the green economy [1] as they decrease the emission of CO_2 compared to fossil fuels [2]. Currently, biodiesel production is mainly performed through transesterification of triglycerides using methyl or ethyl alcohols, which leads to the massive production of glycerol as a by-product [3]. Other technologies that are capable of generating renewable diesel, such as thermal conversion of lipids, also generate glycerol as a by-product [4–6]. The Organization for Economic Co-operation and Development of the Food and Agriculture Organization of the United Nations (OECD-FAO) reports that global biodiesel production will increase from 36 to 39 billion litres between 2017 and 2027 [7]. Since the amount of crude glycerol recovered as a by-product is roughly 10% of the biodiesel generated [8], annual global production of crude glycerol is anticipated to reach 3.9 billion litres by 2027. Currently, producing refined glycerol from crude by-product streams is far more expensive than traditional processes [9–11] even if glycerol itself has many uses [12], such as an additive for cosmetic [13,14] and pharmaceutical [15] industries and as cattle feed [16,17]. Nonetheless, glycerol is an attractive feedstock for several conversion procedures, including oxidation [18], hydrogenolysis [19], etherification [20], esterification [21,22], glycerol carbonate synthesis [23–25] and biological conversions [24]. In particular, glycerol acetyl derivatives have received increasing attention due to their wide applications in many fields, from polymer production to fuel additive manufacturing [26–30]. The glycerol acetylation process produces three different compounds, monoacetins, diacetins and triacetin, whose yields are affected by many process and chemical parameters [31,32]. Mixtures of diacetins and triacetin are valuable diesel or gasoline additives leading to enhanced cold resistance, improved viscosity and anti-knocking properties [33,34]. Similarly, monoacetins could be converted to solketal derivatives that show the same properties of di- and triacetin, with low consumption of acetic acid required for their production [35].

Furthermore, acetins mixtures are used by the pharmaceutical industry as filming [36] and as dispersant agents [37] and as a solvent for hydrophobic molecules and essential oils [38,39].

Glycerol acetylation is generally performed under catalytic conditions using heterogeneous acidic catalysts (i.e., ion exchange resins [40,41], heteropolyacids [42], zeolites [43], clays [44], biochar [45]) as well as both acetic acid [46] and acetic anhydride [47]. Acetic anhydride is a very effective acetylation agent under mild conditions (room temperature with short reaction time) and leads to the chemoselective formation of triacetin with a high release of heat. Despite this, acetic acid is preferentially used for glycerol acetylation due to its lower cost and the strict regulations associated with acetic anhydride use [48]. The acetylation procedure carried out using acetic acid as the acylating agent leads to the formation of various acetyl esters of glycerol and generally requires higher temperatures and a high acetic acid/glycerol ratio [33]. Acetic acid consumption and product purification has been optimized using several continuous processes based on distillation as reported in many studies [49–52].

Recently, Bartoli et al. [23] described a novel catalyst derived from the solid residue recovered after the thermal hydrolysis of biosolids. Thermal hydrolysis of biosolids has been shown to dramatically improve settling rates, and thus offers a potential alternative to current disposal strategies that employ natural settling in large biosolids lagoons [53]. To help offset the high costs that would be associated with the thermal hydrolysis of biosolids at large-scale, valorisation of the metal-rich solid residues recovered from hydrolysates was explored. Bartoli et al. [23] demonstrated that this material functioned as an effective inter- and intramolecular esterification agent for the production of glycerol carbonate and glycidol through urea glycerolysis.

Following up on this study, the same biosolids-based catalyst was employed and evaluated for glycerol acetylation using acetic acid. The esterification reaction was studied under several conditions of reaction time, temperature, acetic acid/glycerol ratio, and catalyst loading. Here, we provide a proof-of-concept that our biosolids-based catalyst can facilitate conversion of glycerol to acetyl derivatives, which could potentially provide value to both renewable diesel producers, as well as wastewater treatment facilities.

2. Results

2.1. Preliminary Considerations Regarding Catalyst-Induced Reactivity and Composition

Glycerol acetylation proceeds through a well-known, multi-step mechanism (Figure 1). The first esterification reaction occurs using one of the hydroxylic functionalities bounded to either a terminal or internal carbon atom of glycerol. The acetylation of terminal groups is both statistically and chemically preferred due to less steric hindrance. As a consequence, the formation of 1-monoacetin is far more common than the formation of 2-monoacetin. The second acetylation occurs preferentially on the remaining terminal carbon of 1-monoacetin, again due to steric hindrance. According to Zhou et al. [54], the production of internal acetin is very slow and is the rate determining step in each acetylation reaction [31,55] causing a low yield of triacetin, 1,2-diacetin and 2-moacetin.

Catalysts currently used for glycerol acetylation are mainly solid acids, such as resins or inorganic salts [43], that promote a classical acid-mediated esterification. Nonetheless, Reddy et al. [56] reported the use of metal oxides as effective catalysts for glycerol acetylation. Considering the beneficial activity of metal oxides joined to acidic sites, the heterogeneous biosolids-based catalyst studied by Bartoli et al. [23] for the urea glycerolysis process is also a promising candidate for glycerol acetylation. The solid residue obtained through the thermal hydrolysis of biosolids showed very promising characteristics, such as a high concentration of acidic sites (up to 5.32 ± 0.03 mmol/g) and several transitional metals (Table 1), mainly as oxide derivatives [23], as shown by XRD analysis reported in Figure 2.

Table 1. The concentration of the main metal species in the solid heterogeneous catalyst acquired through the thermal hydrolysis of biosolids.

Catalyst	Concentration (mg/g)									
	Cr	Fe	Mn	Sr	Al	Cu	Zn	Pb	Ti	Ref.
Initial	0.4 ± 0.2	26.5 ± 1.3	0.58 ± 0.03	0.4 ± 0.1	26.9 ± 1.3	0.7 ± 0.1	1.1 ± 0.1	0.1 ± 0.1	3.2 ± 0.2	[23]
After 5 reactions	0.4 ± 0.2	28.7 ± 1.7	0.36 ± 0.02	0.4 ± 0.1	25.2 ± 1.6	0.2 ± 0.01	0.7 ± 0.1	0.2 ± 0.1	0.1 ± 0.1	Current study

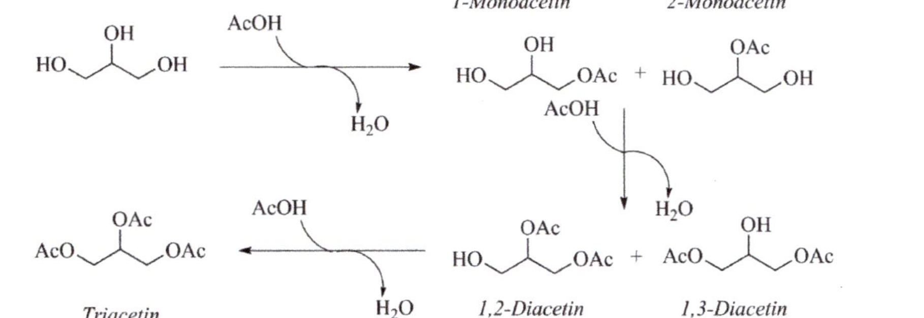

Figure 1. Esterification reaction pathways of glycerol to acetyl derivatives.

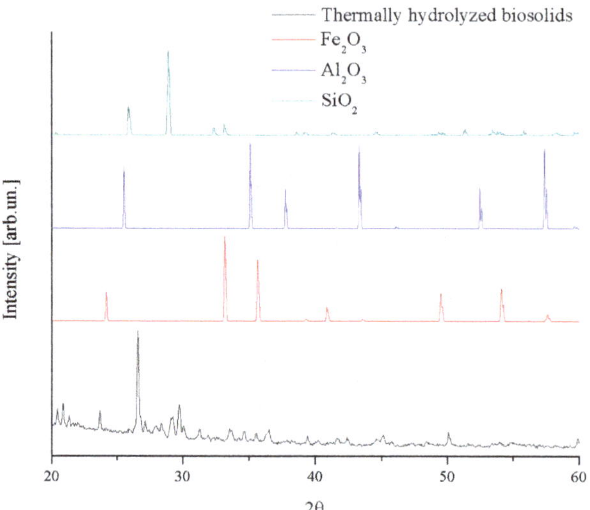

Figure 2. XRD pattern of thermally hydrolysed biosolids-based material in the range from 20 2θ to 60 2θ compared with pure phases of Fe$_2$O$_3$, Al$_2$O$_3$ and SiO$_2$ [57].

Due to the very complex composition of thermally hydrolysed biosolids-based material, XRD analysis can provide only qualitative data. The XRD pattern of thermally hydrolysed biosolids-based material using QualX software (IC-CNR, Bari, Italy) and the RRUFF™ Project database for pure materials showed the presence of silicon oxide, iron oxide and aluminum oxide (quality match of 94%, 89% and 79%, respectively) as amorphous or mixed phase materials. The additional peaks in the spectrum were reasonable due to residual oxides, halides, phosphate metals and organic residues that compose the complex mixture of biosolids-based catalysts.

Since the early years of 19th century, it has been known that metals and their oxides could improve the performance of catalysts during the esterification reactions [58]. Consequently, the metals mixture recovered from thermally-hydrolysed biosolids could potentially be used as a catalyst without further purification procedures.

2.2. Influence of Temperature on Glycerol Acetylation Conversion Rate and Selectivity

The influence of the reaction temperature on glycerol acetylation was investigated during a 4 h period using four different temperatures (60 °C, 80 °C, 100 °C, 120 °C) and a fixed catalyst/glycerol weight ratio (4 wt%) and acetic acid/glycerol molar ratio (6:1) as reported in Figure 3. According to two-way ANOVA tests, the temperature and reaction time both had significant relevance ($p < 0.05$) on the conversion values. At a temperature of 60 °C, conversions were generally low reaching a maximum of 32.1 ± 1.6% after 4 h. When the reaction temperature was increased to 80 °C, the conversion slightly increased to 41.1 ± 2.1% after 4 h, but at the 1 h or 2 h time points, there were no significant differences compared with the corresponding time points at 60 °C. After 1 h at 100 °C, the conversion value observed (8.9 ± 0.7%) was not statistically different from the 1 h reactions conducted at both 60 °C and 80 °C. However, an increase in reaction time at 100 °C led to improved conversion values, reaching a maximum of 63.2 ± 1.5% after 4 h. Furthermore, significantly higher conversion was achieved at all time points at 120 °C, reaching 27.7 ± 1.2% and 78.4 ± 0.2% after 1 h and 4 h, respectively.

According to two-way ANOVA tests ($p < 0.05$), selectivity was also affected by both temperature and reaction time (Table 2). At 60 °C, the use of a catalyst derived from thermally hydrolysed biosolids led to selective formation of monoacetins within the first 3 h of the reaction. However, diacetins were detected in appreciable amounts (11.4 ± 0.9%) after 4 h. Although the selectivity for the monoacetins at 60 °C was quite high, the overall yields (calculated as the product of selectivity and conversion) were quite low reaching a maximum of around 28% after 4 h. At 80 °C, selectivity for monoacetins was still extremely high (95.7 ± 0.6%) after 1 h, decreasing to 87.5 ± 1.0% after 4 h. At the same time, diacetins conversion increased from 3.6 ± 0.1 after 1 h to 11.3 ± 0.6% after 4 h. Moreover, traces of triacetin were detected in appreciable amounts after 4 h (1.2 ± 0.4%). A further increase in temperature to 100 °C dramatically reduced monoacetins selectivity to 53.3 ± 0.7% after 4 h, while the occurrence of diacetins reached 41.3 ± 0.4%. Under these conditions, triacetin was also detected at 5.4 ± 0.8%. After 4 h at 120 °C, the selectivity of monoacetins (45.5 ± 1.4%) and diacetins (46.6 ± 1.5%) were statistically similar ($p < 0.05$), while the selectivity for triacetin increased to 6.6 ± 0.8%. Generally, increments in diacetin and triacetin conversion were observed as a result of both increasing temperature, as well as longer reaction times.

Table 2. Effect of the temperature on the selectivity during glycerol acetylation catalysed by a biosolids-based catalyst. Reactions were carried out using a catalyst/glycerol ratio of 4% (*w/w*), and a molar ratio of acetic acid:glycerol of 6:1.

T [h]	T [°C]	Conversion [a] (%)	Selectivity [b] (%)		
			Monoacetins	Diacetins	Triacetin
1		8.3 ± 1.7	>99	Not detected	Not detected
2	60	20.0 ± 3.4	>99	Not detected	Not detected
3		26.8 ± 0.5	>99	Not detected	Not detected
4		32.1 ± 1.6	88.4 ± 0.8	11.4 ± 0.9	Not detected
1		8.9 ± 1.3	95.7 ± 0.6	3.6 ± 0.3	0.7 ± 0.2
2	80	22.8 ± 1.2	93.7 ± 0.5	6.1 ± 0.5	0.5 ± 0.2
3		29.3 ± 0.8	91.0 ± 0.7	8.6 ± 0.4	0.5 ± 0.3
4		41.1 ± 2.5	87.5 ± 1.0	11.3 ± 0.6	1.2 ± 0.4
1		8.9 ± 0.7	86.6 ± 0.7	12.5 ± 0.2	0.8 ± 0.2
2	100	37.3 ± 1.2	73.9 ± 0.3	25.3 ± 0.6	0.9 ± 0.3
3		51.1 ± 0.7	62.7 ± 0.9	33.5 ± 0.3	3.8 ± 0.9
4		63.2 ± 1.5	53.3 ± 0.7	41.3 ± 0.4	5.4 ± 0.8
1		27.7 ± 1.2	70.9 ± 1.9	23.9 ± 2.0	1.5 ± 0.1
2	120	51.1 ± 1.1	64.7 ± 0.5	33.1 ± 1.7	2.0 ± 0.7
3		66.1 ± 0.3	54.2 ± 1.4	40.8 ± 3.0	4.6 ± 0.1
4		78.4 ± 0.2	45.5 ± 1.4	46.6 ± 1.5	6.6 ± 0.8

[a] Conversion: $100 \times ((\text{initial mass of glycerol} - \text{final mass of glycerol})/(\text{initial mass of glycerol}))$; [b] *Selectivity* : $100 \times (area\ of\ product_i\ /\ \sum_i area\ product_i)$.

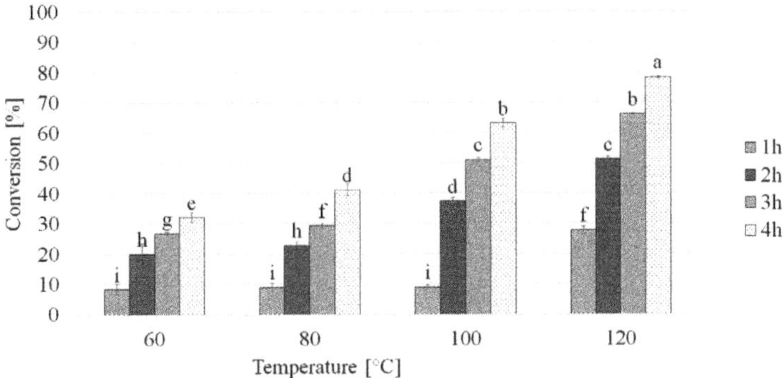

Figure 3. The effect of temperature on the conversion of glycerol to acetyl derivatives catalysed by the biosolids-based catalyst. Reactions were carried out using a catalyst:glycerol ratio of 4% (*w/w*), and a 6:1 molar ratio of acetic acid:glycerol. The conversion values were examined using four different temperatures: 60 °C, 80 °C, 100 °C and 120 °C. The averages ± standard deviations shown were calculated using values from three experiments. Data annotated with different letters are significantly different (at a 95% confidence level).

2.3. Influence of the Acetic Acid:Glycerol Ratio on Glycerol Acetylation Conversion Rate and Selectivity

The influence of the molar ratio of acetic acid and glycerol on glycerol acetylation using a biosolids-based catalyst was investigated using four different acetic acid:glycerol molar ratios (9:1, 6:1, 3:1 and 1:1) at 120 °C, and a catalyst/glycerol weight ratio of 4% as shown in Figure 4. According to two-way ANOVA tests, both reaction time and acetic acid/glycerol ratio had a significant impact on conversion. Despite this, there were no statistically differences ($p < 0.05$) in conversion after 1 h between the various acid:glycerol molar ratios tested, with an average conversion value around 27%. Using an acetic acid:glycerol molar ratio of 1:1, a slight increase in conversion from 26.0 ± 1.2% after 1 h to 40.8 ± 0.5% after 4 h was observed. A greater increase in conversion was observed using an acetic acid:glycerol ratio of 3:1, reaching 73.1 ± 0.7% after 4 h. Further increments in the acetic acid:glycerol molar ratio (6:1 and 9:1) led to increases in conversion of 78.4 ± 0.2% and 85.8 ± 1.2% after 4 h, respectively.

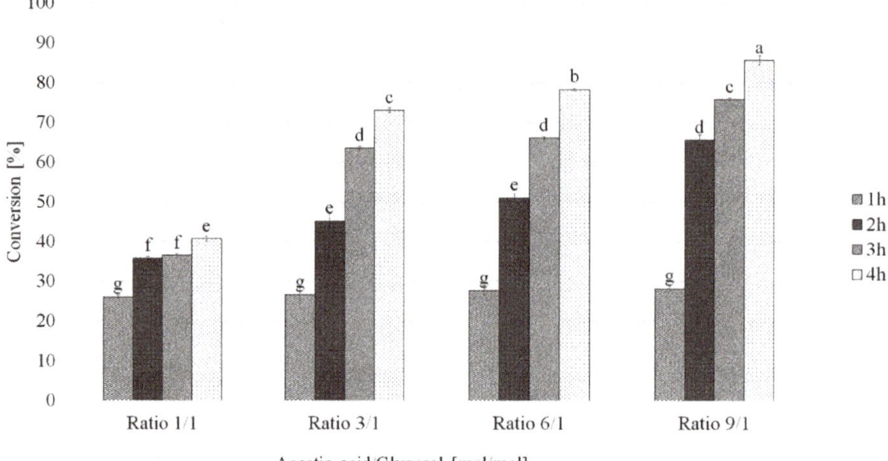

Figure 4. Effect of the acetic acid:glycerol molar ratio on the conversion of glycerol to acetyl derivatives catalysed by a biosolids-based catalyst. For the reactions shown above, the catalyst/glycerol ratio was maintained at 4 wt%, with a reaction temperature of 120 °C. To determinate the relationship between conversion and the acetic acid:glycerol ratios, four molar ratios were employed: 9:1, 6:1, 3:1 and 1:1. The errors bars represent the standard deviations from the averages calculated from triplicate experiments. Data annotated with different letters are significantly different (at a confidence level of 95%).

Selectivity (Table 3) was also affected by both temperature and acetic acid:glycerol molar ratio according to two-way ANOVA outputs ($p < 0.05$). Considering a reaction time of 1 h, the biosolids-based catalyst showed a monoacetins selectivity of 74.3 ± 3.2% using an acetic acid:glycerol molar ratio of 1:1. A decrease in monoacetins selectivity to around 70% was observed using molar ratios of acetic acid:glycerol of 3:1 and 6:1. A further increment in acetic acid:glycerol ratio (9:1) led to a monoacetins selectivity of 80.4 ± 3.8%. Monoacetins selectivity generally decreased with reaction time down to 43.8 ± 2.0% using an acetic acid:glycerol molar ratio of 1:1 after 4 h. A similar trend was observed for all other ratios tested. Conversely, diacetins selectivity increased with increasing acetic acid:glycerol molar ratios. Considering yields instead of the selectivity values, it is very clear that the acetic acid:glycerol molar ratio had a remarkable impact. Using acetic acid:glycerol at a molar ratio of 1:1, the yields of monoacetins and diacetins were 17.9 ± 1.0% and 20.0 ± 1.4% after 4 h, respectively. Increasing the acetic acid:glycerol molar ratio to 9:1 led to increased monoacetins and diacetins yields of 40.2 ± 2.4% and 41.5 ± 0.8%, respectively.

Table 3. Effect of the acetic acid:glycerol molar ratio on the selectivity during glycerol acetylation catalysed by a biosolids-based catalyst. For the reactions, the catalyst/glycerol weight ratio was maintained at 4 wt% and the reaction temperature used was 120 °C.

T [h]	Acetic Acid: Glycerol (mol/mol)	Conversion [a] (%)	Selectivity [b] (%)		
			Monoacetins	Diacetins	Triacetin
1		26.0 ± 0.5	74.3 ± 3.2	24.5 ± 1.8	1.2 ± 0.3
2	1	35.8 ± 0.3	59.7 ± 2.6	34.6 ± 1.1	4.1 ± 0.9
3		36.5 ± 0.5	51.7 ± 0.9	42.5 ± 2.6	5.5 ± 0.5
4		40.8 ± 0.7	43.8 ± 2.0	49.0 ± 2.8	6.3 ± 0.1
1		26.6 ± 0.3	70.9 ± 1.9	23.9 ± 2.0	1.5 ± 0.1
2	3	45.3 ± 1.9	64.7 ± 0.5	33.1 ± 1.7	2.0 ± 0.7
3		63.5 ± 0.6	54.2 ± 1.4	40.8 ± 3.0	4.6 ± 0.1
4		73.1 ± 0.7	45.5 ± 1.4	46.6 ± 1.5	6.6 ± 0.8
1		27.7 ± 0.7	70.9 ± 1.9	23.9 ± 2.0	1.5 ± 0.1
2	6	51.1 ± 1.1	64.7 ± 0.5	33.1 ± 1.7	2.0 ± 0.7
3		66.1 ± 0.3	54.2 ± 1.4	40.8 ± 3.0	4.6 ± 0.1
4		78.4 ± 0.2	45.5 ± 1.4	46.6 ± 1.5	6.6 ± 0.8
1		28.1 ± 0.9	80.4 ± 3.8	17.0 ± 0.5	0.9 ± 0.1
2	9	65.6 ± 1.2	62.1 ± 2.6	34.2 ± 1.5	2.4 ± 0.4
3		75.9 ± 0.5	49.7 ± 2.1	44.1 ± 0.7	3.9 ± 0.3
4		85.8 ± 1.2	46.9 ± 2.0	48.4 ± 0.3	6.3 ± 0.4

[a] Conversion: $100 \times$ ((initial mass of glycerol – final mass of glycerol)/(initial mass of glycerol)); [b] *Selectivity* : $100 \times (area\ of\ product_i / \sum_i area\ product_i)$.

2.4. Influence of Catalyst Loading on Glycerol Acetylation Conversion Rate and Selectivity

The effect of catalyst loading on acetylation of glycerol was studied using five different catalyst/glycerol weight ratios (0 wt%, 2 wt%, 4 wt%, 8 wt%, 16 wt%). The reactions were carried out for 4 h at 120 °C, with an acetic acid:glycerol molar ratio of 6:1 as shown in Figure 5. Two-way ANOVA testing ($p < 0.05$) confirmed that both catalyst loading and reaction time affected the conversion during acetylation of glycerol. The uncatalysed reaction in which no biosolids-based catalyst was employed proceeded very slowly, reaching a conversion of only 3.6 ± 0.4% after 1 h and 65.7 ± 0.4% after 4 h. Using the same conditions, a catalyst loading of 2 wt% substantially improved conversion, reaching 22.6 ± 0.7% after 1 h and 72.2 ± 1.5% of after 4 h. Further increments in catalyst loading induced significant increases in conversion, achieving 90.8 ± 2.3% after 4 h using a catalyst loading of 16 wt%. The effect of catalyst was particularly evident after 1h. In this experiment, a catalyst loading of 2–8% promoted conversions of ~22–27% while a catalyst loading of 16 wt% resulted in a conversion of 54.9 ± 1.1%.

The selectivity achieved during glycerol acetylation was also affected by catalyst loading and reaction time. According to the data shown in Table 4, the trends observed with regards to the selectivity of monoacetins and diacetins were comparable with other experiments. In general, a simultaneous increase in diacetins and a decrease in monoacetins was observed with increasing reaction time and catalyst loading. Triacetin selectivity followed the same trend as diacetins, albeit at much lower levels, with the exception of the experiments without the use of catalyst where no triacetin was detected. This suggests that a catalyst is necessary in order to promote the third esterification reaction, which is associated with a very low kinetic rate due to the high steric hindrance of diacetins. Thus, the catalyst obtained through thermal hydrolysis of biosolids positively affected the esterification of the internal hydroxylic functionality that is characterized by a reaction ΔG higher than terminal groups [32].

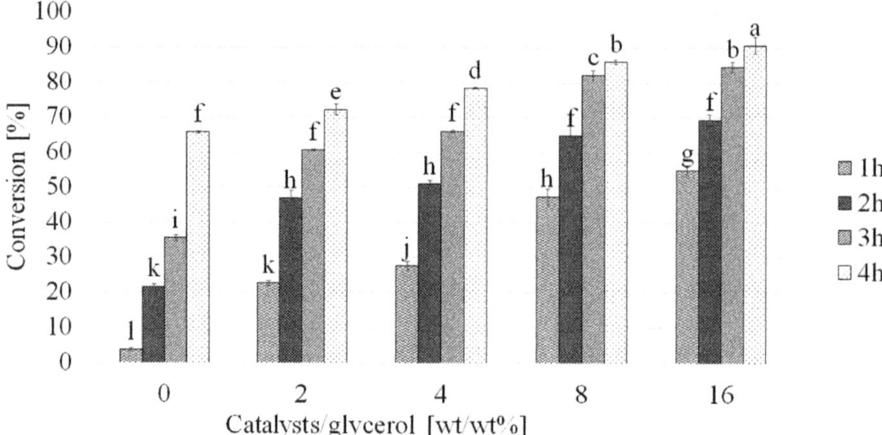

Figure 5. The effect of the catalyst loading on the conversion of glycerol to acetyl derivatives. Reactions were performed at 120 °C, with a 6:1 molar ratio (acetic acid:glycerol). For these experiments, different catalyst/glycerol weight ratios (0 wt%, 2 wt%, 4 wt%, 8 wt%, 16 wt%) were examined, using a biosolids-based catalyst. The data reported represent averages ± standard deviations of triplicate experiments. Values that are annotated with different letters are statistically different at a 95% confidence level.

Table 4. The effect of catalyst loading on selectivity during glycerol acetylation catalysed by the biosolids-based catalyst. Reactions were performed at 120 °C, with a 6:1 molar ratio of acetic acid:glycerol.

T [h]	Catalyst [wt%]	Conversion [a] (%)	Selectivity [b] (%)		
			Monoacetins	Diacetins	Triacetin
1		3.6 ± 0.4	65.2 ± 2.6	34.5 ± 2.6	Not detected
2	0	21.5 ± 0.9	56.6 ± 1.2	42.9 ± 2.8	Not detected
3		35.5 ± 0.9	49.0 ± 3.7	52.3 ± 1.9	Not detected
4		65.7 ± 0.4	44.4 ± 1.0	55.8 ± 2.6	Not detected
1		22.6 ± 0.3	74.3 ± 3.2	24.5 ± 3.0	1.2 ± 0.3
2	2	47.0 ± 2.0	59.7 ± 2.6	34.6 ± 1.5	4.1 ± 0.9
3		60.7 ± 0.3	51.7 ± 0.9	42.5 ± 1.8	5.5 ± 0.5
4		72.2 ± 1.5	43.8 ± 2.0	49.0 ± 1.1	6.3 ± 0.1
1		27.7 ± 1.2	70.9 ± 1.9	23.9 ± 2.0	1.5 ± 0.1
2	4	51.1 ± 1.1	64.7 ± 0.5	33.1 ± 1.7	2.0 ± 0.7
3		66.1 ± 0.3	54.2 ± 1.4	40.8 ± 3.0	4.6 ± 0.1
4		78.4 ± 0.2	45.5 ± 1.4	46.6 ± 1.5	6.6 ± 0.8
1		47.5 ± 2.2	78.6 ± 2.2	20.3 ± 2.3	1.0 ± 0.3
2	8	64.8 ± 3.1	65.2 ± 2.5	33.0 ± 1.9	2.5 ± 0.4
3		82.2 ± 1.3	54.1 ± 1.2	41.4 ± 1.1	4.6 ± 0.4
4		86.0 ± 0.5	46.1 ± 1.3	47.2 ± 0.3	6.8 ± 0.7
1		54.9 ± 1.1	80.4 ± 3.8	17.0 ± 0.5	0.9 ± 0.1
2	16	69.3 ± 1.6	62.1 ± 2.6	34.2 ± 1.5	2.4 ± 0.4
3		84.6 ± 1.3	49.7 ± 2.1	44.1 ± 0.7	3.9 ± 0.3
4		90.8 ± 2.3	46.9 ± 2.0	48.4 ± 0.3	6.3 ± 0.4

[a] Conversion: $100 \times$ ((initial mass of glycerol − final mass of glycerol)/(initial mass of glycerol)); [b] *Selectivity*: $100 \times (area\ of\ product_i\ /\ \sum_i area\ product_i)$.

2.5. Influence of Catalyst Recycling on Glycerol Acetylation Conversion Rate and Selectivity

To assess reusability of the catalyst generated through thermal hydrolysis of biosolids, the same catalyst sample was used for five consecutive reactions. The reactions were performed at a catalyst loading of 4 wt%, an acetic acid:glycerol molar ratio of 6:1, and at 120 °C for 4 h (Figure 6). According to a one-way ANOVA test, there were no significant differences ($p < 0.05$) in the conversion values assessed during the five catalytic runs with average values ~78%. Furthermore, comparable selectivity was also observed in the reaction products from the five catalytic runs (Table 5). Considering the metals concentrations assessed after the fifth catalytic run (Table 1), it reasonable to assume that the catalytic activity of thermally hydrolysed biosolids-based materials during glycerol acetylation was exploited by both metals and acidic sites. Acid sites remained untouched after the fifth catalytic runs with a concentration of 5.28 ± 0.05 mmol/g while key metals (Fe, Al) did not show any leaching. According to these observations, acetylation of glycerol was promoted by acidic sites, such as hydroxylic functionalities (mainly on silica as reported by Bartoli et al. [23]) and metal centres that can act as Lewis acids [59].

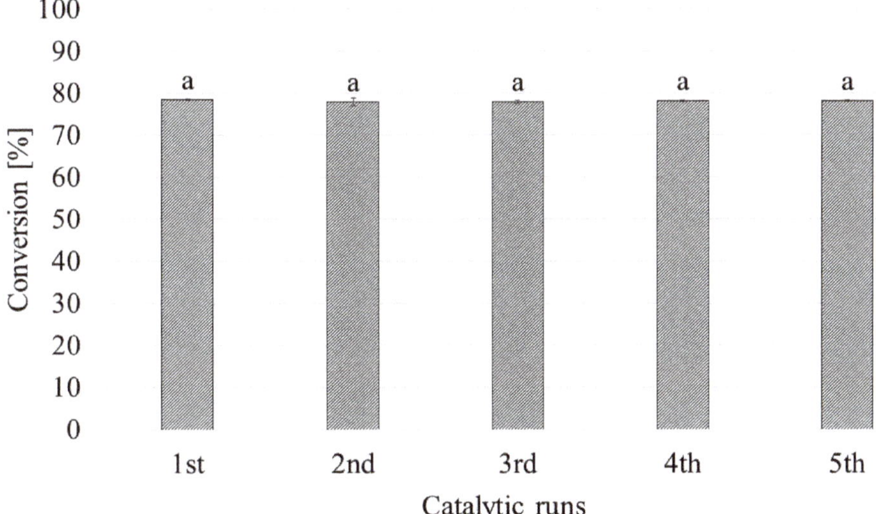

Figure 6. Conversion performance using recycled biosolids-based catalyst. The reactions were performed using a catalyst/glycerol ratio of 4 wt%, a temperature of 120 °C, an acetic acid:glycerol molar ratio of 6. These reactions were performed for 4 h. The values shown represent the average ± standard deviations of three independent experiments. Data annotated with different letters are significantly different (at a confidence level of 95%).

Table 5. Selectivity observed in glycerol acetylation reactions using recycled biosolids-based catalysts. Reactions were performed with a catalyst loading of 4 wt%, an acetic acid:glycerol molar ratio of 2:1, and at 120 °C for 4 h.

Catalytic Run	Conversion [a] (%)	Selectivity [b] (%)		
		Monoacetins	Diacetins	Triacetin
1st	78.4 ± 0.2	45.5 ± 1.4	46.6 ± 1.5	6.6 ± 0.8
2nd	77.9 ± 0.9	44.9 ± 0.5	45.9 ± 1.1	6.3 ± 0.6
3rd	77.9 ± 0.4	44.6 ± 0.3	46.3 ± 1.3	6.7 ± 0.4
4th	78.1 ± 0.2	46.6 ± 0.8	45.8 ± 0.8	6.0 ± 0.9
5th	78.1 ± 0.3	42.0 ± 0.7	46.2 ± 1.1	6.9 ± 0.5

[a] Conversion: $100 \times ((\text{initial mass of glycerol} - \text{final mass of glycerol})/(\text{initial mass of glycerol}))$; [b] *Selectivity* : $100 \times (area\ of\ product_i / \sum_i area\ product_i)$.

2.6. Considerations on Catalytic Performance of Thermally Hydrolysed Biosolids and Available Catalysts

The catalytic performance of the biosolids-based catalyst is not as high as other reported materials, such as superacid resins. Dosuna-Rodríguez et al. [40] compared several acidic resins reaching a maximum conversion of up to 95% after 200 min with a selectivity for monoacetins and diacetins of up to 81% and 5%, respectively, using 0.8 wt% of Dowex-2®. Similar and even better results were achieved only using a high loading of thermally hydrolysed biosolids catalyst (from 8 wt% to 16 wt%). Other acid catalyst such as ionic liquids outperformed the biosolids-based catalyst, reaching conversions of up to 99% and selectivity for triacetin up to 47% [60]. Furthermore, nanostructured sulphated zirconia led to higher yields of triacetin (up to 40%) with low catalyst loading [61]. A similar behaviour was also observed using acidic zeolites as reported by many authors [62–64]. Nonetheless, thermally-hydrolysed biosolids catalyst operates at lower temperature compared with heteropolyacidic catalytic systems reaching the same performances [65]. Furthermore, as the biosolids-based catalyst is generated from a negative value material, the cost of this catalyst may offer significant advantages in commercial applications.

3. Materials and Methods

3.1. Materials

Tetrahydrofuran (HPLC grade, >99.9%) and methyl nonadecanoate (used as internal standard for gas chromatography) were purchased from Sigma-Aldrich (St. Louis, MO, USA). Acetic acid (>98%) and glycerol (98%) were purchased from Fisher Chemical (Fair Lawn, NJ, USA). Gases (Air, N_2, H_2, He) were purchased from Praxair (Praxair Inc., Edmonton, AB, Canada). The biosolids-based catalyst was obtained through thermal hydrolysis (121 °C, 1 h, >15 psi) of digested biosolids (~3.5% dry solids; pH $\approx$ 9) as previously described [23,53] The solid material was collected through centrifugation, then dried at 105 °C to constant weight in a drying oven. The dried solids were then ground with a mortar and pestle, and then passed through a 200 mesh nylon cloth. The material passing through the filter was used as the biosolid-based catalyst.

3.2. Methods

3.2.1. Catalytic Acetylation of Glycerol

Glycerol (5 g) was placed into a one neck balloon with acetic acid (acetic acid:glycerol molar ratios of 9:1, 6:1, 3:1, and 1:1) and the biosolids-based catalyst (catalyst/glycerol weight ratio of 16 wt%, 8 wt%, 4 wt%, 2 wt%, and 0 wt%), which was then connected to a condenser. The reaction mixtures were stirred and heated at different temperatures (60 °C, 80 °C, 100 °C and 120 °C), with sampling every hour for a total of 4 h. After, the reaction mixture was cooled to room temperature and the catalyst was recovered through centrifugation (7155× *g* for 10 min), washed with acetone (5 mL three times), dried overnight at 105 °C, and then analysed as described below. Each test was replicated three times.

3.2.2. Sample Analysis

The reaction mixture was analysed as previously reported [23]. Briefly, 100 μL of the crude reaction mixture were diluted with 900 μL of tetrahydrofuran and 10 mg of methyl nonadecanoate were added as the internal standard. The solution was analysed using a gas chromatograph (6890N, Agilent Technologies, Fort Worth, TX, USA) equipped with an autosampler (Agilent 7683 series; Agilent Technologies, Fort Worth, TX, USA), a flame ionization detector (FID) and a mass spectrometer (Agilent 5975B inert XL EI/CI MSD; Agilent Technologies, Fort Worth, TX, USA) used for the quantification and identification of each peaks, respectively. Analyses were carried out by injecting 1 μL of the samples into a DB-5 column (100 m × 250 μm × 0.25 μm; Agilent Technologies, Fort Worth, TX) according to the methodology described previously [66]. Analysis were carried out using helium as the carrier gas (flow rate of 1 mL/min), an injector temperature of 310 °C and FID temperature of 320 °C, respectively. The starting oven temperature was set at 35 °C, which was held for 0.1 min, increased to 280 °C at 10 °C/min, and then held for 5.4 min to reach a total time of 30 min. Mass spectrometry outputs were analysed and matched to those of the National Institute of Standards and Testing (NIST) mass spectral library. The concentration of unreacted glycerol was evaluated using a five point calibration curve (m = 0.185 ± 0.02; R^2 = 0.994) established through the use of methyl nonadecanoate as the internal standard.

Analysis of the main inorganics in the catalyst was performed using a Thermo iCAP 6000 series Inductively Coupled Plasma-Optical Emission Spectrometer (ICP-OES; Thermo Fisher Scientific, Madison, WI, USA) at the Natural Resources Analytical Laboratory (Department of Chemistry, University of Alberta). 50 mg of each sample were digested dissolved in 5 mL of trace-metal grade HNO_3 and further diluted with Milli-Q water to 25 mL. An internal standard solution containing Yttrium is used to correct for matrix effects in each sample. Final analyte values are reported after this correction factor has been applied. Analyses were carried out before and after the five catalytic runs. Analysis of acid residual groups on the surface of the catalyst was carried out according to methodology reported by Bartoli et al. [23], which was based on a modified Boehm's test [67].

XRD analysis was carried out using a D8 Discover X-ray diffractometer (Bruker, Madison, WI, USA) employing Cu Ka (k = 1.5418 Å) radiation. Diffractograms were acquired in the 2θ range from 10.0° to 90.0°, applying a step size of 0.50°. Inorganic species were detected using freeware software QualX software. Only species with a quality match of 80% or higher were reported.

3.2.3. Statistical Analysis

One-way and two-way ANOVA tests and *t*-tests with a significance level of 0.05 ($p < 0.05$) were carried out using Excel™ software (Microsoft Corp., Redmond, WA, USA) and the "Data analysis" tool. Statistical analysis was performed using triplicate data, with the means and standard deviations reported.

Five grams (5 g) was placed into a one neck balloon with acetic acid (acetic acid:glycerol molar ratios of 9:1, 6:1, 3:1, and 1:1) and the biosolids-based catalyst (catalyst/glycerol weight ratio of 16 wt%, 8 wt%, 4 wt%, 2 wt% and 0 wt%), which was then connected to a condenser. The reaction mixtures were stirred and heated at different temperatures (60 °C, 80 °C, 100 °C and 120 °C), with sampling every hour for a total of 4 h. After, the reaction mixture was cooled to room temperature and the catalyst was recovered through centrifugation (7155× *g* for 10 min), washed with acetone (5 mL three times), dried overnight at 105 °C, and then analysed as described below. Each test was replicated three times.

4. Conclusions

In this work, the effectiveness of a heterogeneous catalyst obtained from thermally hydrolysed biosolids in promoting glycerol acetylation was demonstrated. Glycerol conversion was studied using various temperatures, acetic acid:glycerol molar ratios, catalyst loading and reaction times. The results highlight the ability of the biosolids-based catalyst to promote a rapid conversion of glycerol to a mixture of monoacetins, diacetins, and triacetin. The maximum conversion obtained was 90.8 ± 2.3% after 4 h at 120 °C using a catalyst loading of 16 wt% and an acetic acid:glycerol molar ratio of 6:1. Furthermore, under the same conditions, an extremely high selectivity of monoacetins were achieved after only 1 h. Considering the catalytic activity and negative cost of the material recovered from biosolids, a waste stream from wastewater management processes, catalysts recovered from thermally hydrolysed biosolids is a promising tool for the glycerol acetylation process.

Author Contributions: Conceptualization: M.B. and D.C.B.; methodology: M.B.; formal analysis: M.B.; investigation: M.B. and C.Z.; data curation: M.B.; writing—original draft preparation: M.B.; writing—review and editing: M.C. and D.C.B.; supervision: D.C.B.; project administration: M.C. and D.C.B.; funding acquisition: M.C. and D.C.B. All authors have read and agreed to the published version of the manuscript.

Funding: The work described in this research was achieved through generous financial support from the Natural Sciences and Engineering Research Council of Canada (NSERC) (grant number RGPIN 298352-20130, Mitacs Canada (grant number MI MA IT05367), BioFuelNet Canada (grant number NCEBFNC 6F), and our collaborators from Forge Hydrocarbons Inc. who also hosted Mattia Bartoli and Chengyong Zhu as interns.

Acknowledgments: The work described in this manuscript was supported by BioFuelNet Canada (grant number NCEBFNC 6F), Mitacs Canada (grant number MI MA IT05367), and the Natural Sciences and Engineering Research Council of Canada (NSERC) (grant numbers RGPIN 298352-2013 and RGPIN-2019-04184). The authors wish to thank Forge Hydrocarbons Inc. who hosted Mattia Bartoli and Chengyong Zhu as Mitacs Interns. The authors are also thankful for the support provided throughout the project by the Natural Resources Analytical Laboratory (Chemistry Department, University of Alberta), NanoFab Centre (University of Alberta), the City of Edmonton, EPCOR Water Services Inc., and Suez.

Conflicts of Interest: The funders had no role in the design of the study, in the collection, analyses, or interpretation of data, or in the writing of the manuscript. Forge Hydrocarbons Inc. reviewed the manuscript prior to submission strictly to ensure that there was no release of confidential information.

References

1. Gasparatos, A.; Doll, C.N.H.; Esteban, M.; Ahmed, A.; Olang, T.A. Renewable Energy and Biodiversity: Implications for Transitioning to A Green Economy. *Renew. Sustain. Energy Rev.* **2017**, *70*, 161–184. [CrossRef]

2. Mahmudul, H.; Hagos, F.; Mamat, R.; Adam, A.A.; Ishak, W.; Alenezi, R. Production, Characterization and Performance of Biodiesel as An Alternative Fuel in Diesel Engines—A Review. *Renew. Sustain. Energy Rev.* **2017**, *72*, 497–509. [CrossRef]

3. Mardhiah, H.H.; Ong, H.C.; Masjuki, H.; Lim, S.; Lee, H. A Review on Latest Developments and Future Prospects of Heterogeneous Catalyst in Biodiesel Production from Non-Edible Oils. *Renew. Sustain. Energy Rev.* **2017**, *67*, 1225–1236. [CrossRef]

4. Espinosa-Gonzalez, I.; Parashar, A.; Chae, M.; Bressler, D.C. Cultivation of Oleaginous Yeast Using Aqueous Fractions Derived from Hydrothermal Pretreatments of Biomass. *Bioresour. Technol.* **2014**, *170*, 413–420. [CrossRef] [PubMed]

5. Asomaning, J.; Mussone, P.; Bressler, D.C. Two-Stage Thermal Conversion of Inedible Lipid Feedstocks to Renewable Chemicals and Fuels. *Bioresour. Technol.* **2014**, *158*, 55–62. [CrossRef]

6. Espinosa-Gonzalez, I.; Asomaning, J.; Mussone, P.; Bressler, D.C. Two-Step Thermal Conversion of Oleaginous Microalgae into Renewable Hydrocarbons. *Bioresour. Technol.* **2014**, *158*, 91–97. [CrossRef]

7. OCED. *OECD-FAO Agricultural Outlook 2018–2027*; OECD Publishing: Paris, France, 2018.

8. Monteiro, M.R.; Kugelmeier, C.L.; Pinheiro, R.S.; Batalha, M.O.; Da Silva César, A. Glycerol fom Biodiesel Production: Technological Paths for Sustainability. *Renew. Sustain. Energy Rev.* **2018**, *88*, 109–122. [CrossRef]

9. França, R.G.; Souza, P.A.; Lima, E.R.; Costa, A.L. An Extended Techno-Economic Analysis of the Utilization of Glycerol as An Alternative Feedstock for Methanol Production. *Clean Technol. Environ. Policy* **2017**, *19*, 1855–1865. [CrossRef]

10. Ardi, M.S.; Aroua, M.K.; Hashim, N.A. Progress, Prospect and Challenges in Glycerol Purification Process: A Review. *Renew. Sustain. Energy Rev.* **2015**, *42*, 1164–1173. [CrossRef]

11. Ciriminna, R.; Pina, C.D.; Rossi, M.; Pagliaro, M. Understanding the Glycerol Market. *Eur. J. Lipid Sci. Technol.* **2014**, *116*, 1432–1439. [CrossRef]

12. Bartoli, M.; Rosi, L.; Frediani, M. Introductory Chapter: A Brief Insight about Glycerol. In *Glycerine Production And Transformation—An Innovative Platform For Sustainable Biorefinery And Energy*; Intechopen: London, UK, 2019.

13. Dunphy, P.J.; Meyers, A.J.; Rigg, R.T. Cosmetic Water-In-Oil Emulsion Lipstick Comprising A Phospholipid and Glycerol Fatty Acid Esters Emulsifying System. U.S. Patent 5,085,856, 2 April 1992.

14. Suh, Y.; Kil, D.; Chung, K.; Abdullayev, E.; Lvov, Y.; Mongayt, D. Natural Nanocontainer for the Controlled Delivery of Glycerol as A Moisturizing Agent. *J. Nanosci. Nanotechnol.* **2011**, *11*, 661–665. [CrossRef] [PubMed]

15. Tamarkin, D.; Besonov, A. Glycerol Ethers Vehicle and Pharmaceutical Compositions Thereof. U.S. Patent 20090130029A1, 21 May 2009.

16. Van Cleef, E.H.C.B.; Ezequiel, J.M.B.; D'aurea, A.P.; Fávaro, V.R.; Sancanari, J.B.D. Crude Glycerin in Diets for Feedlot Nellore Cattle. *Rev. Bras. Zootec.* **2014**, *43*, 86–91. [CrossRef]

17. Meale, S.; Chaves, A.; Ding, S.; Bush, R.; Mcallister, T. Effects of Crude Glycerin Supplementation on Wool Production, Feeding Behavior, and Body Condition of Merino Ewes. *J. Anim. Sci.* **2013**, *91*, 878–885. [CrossRef] [PubMed]
18. Razali, N.; Abdullah, A.Z. Production of Lactic Acid from Glycerol Via Chemical Conversion Using Solid Catalyst: A Review. *Appl. Catal. A Gen.* **2017**, *543*, 234–246. [CrossRef]
19. Oberhauser, W.; Evangelisti, C.; Jumde, R.P.; Psaro, R.; Vizza, F.; Bevilacqua, M.; Filippi, J.; Machado, B.F.; Serp, P. Platinum on Carbonaceous Supports for Glycerol Hydrogenolysis: Support Effect. *J. Catal.* **2015**, *325*, 111–117. [CrossRef]
20. Magar, S.; Kamble, S.; Mohanraj, G.T.; Jana, S.K.; Rode, C. Solid-Acid-Catalyzed Etherification of Glycerol to Potential Fuel Additives. *Energy Fuels* **2017**, *31*, 12272–12277. [CrossRef]
21. Shukla, K.; Srivastava, V.C. Synthesis of Organic Carbonates from Alcoholysis of Urea: A Review. *Catal. Rev.* **2017**, *59*, 1–43. [CrossRef]
22. Oliverio, M.; Costanzo, P.; Nardi, M.; Calandruccio, C.; Salerno, R.; Procopio, A. Tunable Microwave-Assisted Method for the Solvent-Free and Catalyst-Free Peracetylation of Natural Products. *Beilstein J. Org. Chem.* **2016**, *12*, 2222–2233. [CrossRef]
23. Bartoli, M.; Zhu, C.; Chae, M.; Bressler, D. Value-Added Products from Urea Glycerolysis Using A Heterogeneous Biosolids-Based Catalyst. *Catalysts* **2018**, *8*, 373. [CrossRef]
24. Amaral, P.F.F.; Ferreira, T.F.; Fontes, G.C.; Coelho, M.A.Z. Glycerol Valorization: New Biotechnological Routes. *Food Bioprod. Process.* **2009**, *87*, 179–186. [CrossRef]
25. Costanzo, P.; Calandruccio, C.; Di Gioia, M.L.; Nardi, M.; Oliverio, M.; Procopio, A. First Multicomponent Reaction Exploiting Glycerol Carbonate Synthesis. *J. Clean. Prod.* **2018**, *202*, 504–509. [CrossRef]
26. Mufrodi, Z.; Rochmadi, R.; Sutijan, S.; Budiman, A. Synthesis Acetylation of Glycerol Using Batch Reactor and Continuous Reactive Distillation Column. *Eng. J.* **2014**, *18*, 29–40. [CrossRef]
27. Rastegari, H.; Ghaziaskar, H.S. From Glycerol as the by-Product Of Biodiesel Production to Value-Added Monoacetin by Continuous and Selective Esterification in Acetic Acid. *J. Ind. Eng. Chem.* **2015**, *21*, 856–861. [CrossRef]
28. Odibi, C.; Babaie, M.; Zare, A.; Nabi, M.N.; Bodisco, T.A.; Brown, R.J. Exergy Analysis of A Diesel Engine with Waste Cooking Biodiesel and Triacetin. *Energy Convers. Manag.* **2019**, *198*. [CrossRef]
29. Nda-Umar, U.I.; Ramli, I.; Taufiq-Yap, Y.H.; Muhamad, E.N. An Overview of Recent Research in the Conversion of Glycerol into Biofuels, Fuel Additives and Other Bio-Based Chemicals. *Catalysts* **2019**, *9*, 15. [CrossRef]
30. Gama, N.; Santos, R.; Godinho, B.; Silva, R.; Ferreira, A. Triacetin as A Secondary Pvc Plasticizer. *J. Polym. Environ.* **2019**, *27*, 1294–1301. [CrossRef]
31. Okoye, P.; Abdullah, A.; Hameed, B. A Review on Recent Developments and Progress in the Kinetics and Deactivation of Catalytic Acetylation of Glycerol—A Byproduct of Biodiesel. *Renew. Sustain. Energy Rev.* **2017**, *74*, 387–401. [CrossRef]
32. Liao, X.; Zhu, Y.; Wang, S.-G.; Chen, H.; Li, Y. Theoretical Elucidation of Acetylating Glycerol with Acetic Acid and Acetic Anhydride. *Appl. Catal. B Environ.* **2010**, *94*, 64–70. [CrossRef]
33. Melero, J.A.; Van Grieken, R.; Morales, G.; Paniagua, M. Acidic Mesoporous Silica for the Acetylation of Glycerol: Synthesis of Bioadditives to Petrol Fuel. *Energy Fuels* **2007**, *21*, 1782–1791. [CrossRef]
34. Herrada-Vidales, J.A.; García-González, J.M.; Martínez-Palou, R.; Guzmán-Pantoja, J. Integral Process for Obtaining Acetins from Crude Glycerol and Their Effect on the Octane Index. *Chem. Eng. Commun.* **2019**, *207*, 1–11. [CrossRef]
35. Gorji, Y.M.; Ghaziaskar, H.S. Optimization of Solketalacetin Synthesis as A Green Fuel Additive from Ketalization of Monoacetin with Acetone. *Ind. Eng. Chem. Res.* **2016**, *55*, 6904–6910. [CrossRef]
36. Lachman, L.; Drubulis, A. Factors Influencing the Properties of Films Used for Tablet Coating I. Effects of Plasticizers on the Water Vapor Transmission Of Cellulose Acetate Phthalate Films. *J. Pharm. Sci.* **1964**, *53*, 639–643. [CrossRef] [PubMed]
37. Weber, F. Pharmaceutical Composition. U.S. Patent US7074412B2, 11 July 2006.
38. Knowles, M.; Pridgen, H.S. Solvents for Fat and Oil Antioxidants. U.S. Patent 2944908A, 12 July 1960.
39. Deline, G.D. Homogeneous, Free-Flowing Liquid Black Pepper Oleoresin Composition. European Patent 0137082A1, 11 October 1983.

40. Dosuna-Rodríguez, I.; Gaigneaux, E.M. Glycerol Acetylation Catalysed by Ion Exchange Resins. *Catal. Today* **2012**, *195*, 14–21. [CrossRef]

41. Caballero, K.V.; Guerrero-Amaya, H.; Baldovino-Medrano, V.G. Revisiting Glycerol Esterification with Acetic Acid over Amberlyst-35 Via Statistically Designed Experiments: Overcoming Transport Limitations. *Chem. Eng. Sci.* **2019**. [CrossRef]

42. De Abreu Dessimoni, A.L.; De Oliveira Pereira, L.; Penido, E.S.; Lima Abreu Veiga, T.R.; De Barros Fernandes, R.V.; Teixeira, M.L.; Bonésio, M.D.R.; Bianchi, M.L. Characterization of Catalysts for Glycerol Ester Production with Various Acetylating Agents. *Anal. Lett.* **2017**. [CrossRef]

43. Gonçalves, V.L.C.; Pinto, B.P.; Silva, J.C.; Mota, C.J.A. Acetylation of Glycerol Catalyzed by Different Solid Acids. *Catal. Today* **2008**, *133–135*, 673–677.

44. Dill, L.P.; Kochepka, D.M.; Melinski, A.; Wypych, F.; Cordeiro, C.S. Microwave-Irradiated Acetylation of Glycerol Catalyzed by Acid Activated Clays. *React. Kinet. Mech. Catal.* **2019**, *127*, 1–14. [CrossRef]

45. Okoye, P.; Abdullah, A.; Hameed, B. Synthesis of Oxygenated Fuel Additives Via Glycerol Esterification with Acetic Acid over Bio-Derived Carbon Catalyst. *Fuel* **2017**, *209*, 538–544. [CrossRef]

46. Okoye, P.U.; Hameed, B.H. Review On Recent Progress in Catalytic Carboxylation and Acetylation of Glycerol as A Byproduct of Biodiesel Production. *Renew. Sustain. Energy Rev.* **2016**, *53*, 558–574. [CrossRef]

47. Silva, L.N.; Gonçalves, V.L.C.; Mota, C.J.A. Catalytic Acetylation of Glycerol with Acetic Anhydride. *Catal. Commun.* **2010**, *11*, 1036–1039. [CrossRef]

48. Cunningham, J.K.; Liu, L.-M.; Callaghan, R.C. Essential ("Precursor") Chemical Control for Heroin: Impact of Acetic Anhydride Regulation on Us Heroin Availability. *Drug Alcohol Depend.* **2013**, *133*, 520–528. [CrossRef] [PubMed]

49. Len, C.; Luque, R. Continuous Flow Transformations of Glycerol to Valuable Products: An Overview. *Sustain. Chem. Process.* **2014**, *2*, 1. [CrossRef]

50. Mufrodi, Z.; Budiman, A. Continuous Process of Reactive Distillation to Produce Bio-Additive Triacetin from Glycerol. *Mod. Appl. Sci.* **2013**, *7*, 70. [CrossRef]

51. Costa, I.C.; Itabaiana, I., Jr.; Flores, M.C.; Lourenço, A.C.; Leite, S.G.; De, M.E.; Miranda, L.S.; Leal, I.C.; De Souza, R.O. Biocatalyzed Acetins Production under Continuous-Flow Conditions: Valorization of Glycerol Derived from Biodiesel Industry. *J. Flow Chem.* **2013**, *3*, 41–45. [CrossRef]

52. Varma, R.S.; Len, C. Glycerol Valorization under Continuous Flow Conditions-Recent Advances. *Curr. Opin. Green Sustain. Chem.* **2018**, *15*, 83–90. [CrossRef]

53. Chae, M.; Xia, L.; Zhu, C.; Bressler, D.C. Accelerating Settling Rates of Biosolids Lagoons through Thermal Hydrolysis. *J. Environ. Manag.* **2018**, *220*, 227–232. [CrossRef]

54. Zhou, L.; Nguyen, T.-H.; Adesina, A.A. The Acetylation of Glycerol over Amberlyst-15: Kinetic and Product Distribution. *Fuel Process. Technol.* **2012**, *104*, 310–318. [CrossRef]

55. Patel, A.; Singh, S. A Green and Sustainable Approach for Esterification of Glycerol Using 12-Tungstophosphoric Acid Anchored to Different Supports: Kinetics and Effect of Support. *Fuel* **2014**, *118*, 358–364. [CrossRef]

56. Reddy, P.S.; Sudarsanam, P.; Raju, G.; Reddy, B.M. Selective Acetylation of Glycerol over Ceo2–M and So42–/Ceo2–M (M = Zro2 And Al2o3) Catalysts for Synthesis of Bioadditives. *J. Ind. Eng. Chem.* **2012**, *18*, 648–654. [CrossRef]

57. Arizona, U.O. Rruff™ Project. Available online: Http://Rruff.Info/ (accessed on 21 October 2019).

58. Jaeger, A.O.; Selden Co, Assignee. Catalytic Esterification. U.S. Patent 1,819,818, 18 August 1931.

59. Greenwood, N.N.; Earnshaw, A. *Chemistry of the Elements*; Elsevier: Amsterdam, The Netherlands, 2012.

60. Keogh, J.; Tiwari, M.S.; Manyar, H. Esterification of Glycerol with Acetic Acid Using Nitrogen-Based Brønsted-Acidic Ionic Liquids. *Ind. Eng. Chem. Res.* **2019**, *58*, 17235–17243. [CrossRef]

61. Testa, M.L.; La Parola, V.; Mesrar, F.; Ouanji, F.; Kacimi, M.; Ziyad, M.; Liotta, L.F. Use of Zirconium Phosphate-Sulphate as Acid Catalyst for Synthesis of Glycerol-Based Fuel Additives. *Catalysts* **2019**, *9*, 148. [CrossRef]

62. Liu, J.; Wang, Z.; Sun, Y.; Jian, R.; Jian, P.; Wang, D. Selective Synthesis of Triacetin fom Glycerol Catalyzed by Hzsm-5/Mcm-41 Micro/Mesoporous Molecular Sieve. *Chin. J. Chem. Eng.* **2019**, *27*, 1073–1078. [CrossRef]

63. Bandyopadhyay, M.; Tsunoji, N.; Bandyopadhyay, R.; Sano, T. Comparison of Sulfonic Acid Loaded Mesoporous Silica in Transesterification of Triacetin. *React. Kinet. Mech. Catal.* **2019**, *126*, 167–179. [CrossRef]

64. Marwan, M.; Indarti, E.; Darmadi, D.; Rinaldi, W.; Hamzah, D.; Rinaldi, T. Production of Triacetin by Microwave Assisted Esterification of Glycerol Using Activated Natural Zeolite. *Bull. Chem. React. Eng. Catal.* **2019**, *14*, 672. [CrossRef]
65. Huang, M.-Y.; Han, X.-X.; Hung, C.-T.; Lin, J.-C.; Wu, P.-H.; Wu, J.-C.; Liu, S.-B. Heteropolyacid-Based Ionic Liquids as Efficient Homogeneous Catalysts for Acetylation of Glycerol. *J. Catal.* **2014**, *320*, 42–51. [CrossRef]
66. Omidghane, M.; Jenab, E.; Chae, M.; Bressler, D.C. Production of renewable hydrocarbons by thermal cracking of oleic acid in the presence of water. *Energ. Fuel.* **2017**, *31*, 9446–9454. [CrossRef]
67. Boehm, H. Acidic and Basic Properties of Hydroxylated Metal Oxide Surfaces. *Discuss. Faraday Soc.* **1971**, *52*, 264–275. [CrossRef]

© 2019 by the authors. Licensee MDPI, Basel, Switzerland. This article is an open access article distributed under the terms and conditions of the Creative Commons Attribution (CC BY) license (http://creativecommons.org/licenses/by/4.0/).

Article

Influence of Base-Catalyzed Organosolv Fractionation of Larch Wood Sawdust on Fraction Yields and Lignin Properties

Markus Hochegger [1], Gregor Trimmel [2], Betty Cottyn-Boitte [3], Laurent Cézard [3], Amel Majira [3], Sigurd Schober [1] and Martin Mittelbach [1,*]

[1] Institute of Chemistry, NAWI Graz, Central Lab Biobased Materials, University of Graz, 8010 Graz, Austria; markus.hochegger@uni-graz.at (M.H.); si.schober@uni-graz.at (S.S.)

[2] Institute for Chemistry and Technology of Materials (ICTM), NAWI Graz, Graz University of Technology, 8010 Graz, Austria; gregor.trimmel@tugraz.at

[3] Institut Jean-Pierre Bourgin, INRA, AgroParisTech, CNRS, Université Paris-Saclay, 78000 Versailles, France; betty.cottyn@inra.fr (B.C.-B.); laurent.cezard@inra.fr (L.C.); amel.majira@versailles.inra.fr (A.M.)

*   Correspondence: martin.mittelbach@uni-graz.at; Tel.: +43316380-5353

Received: 30 October 2019; Accepted: 25 November 2019; Published: 27 November 2019

Abstract: Lignocellulose-based biorefineries are considered to play a crucial role in reducing fossil-fuel dependency. As of now, the fractionation is still the most difficult step of the whole process. The objective of this study is to investigate the potential of a base-catalyzed organosolv process as a fractionation technique for European larch sawdust. A solvent system comprising methanol, water, sodium hydroxide as catalyst, and anthraquinone as co-catalyst is tested. The influence of three independent process variables, temperature (443–446 K), catalyst loading (20–30% w/w), and alcohol-to-water ratio (30–70% v/v), is studied. The process conditions were determined using a fractional factorial experiment. One star point (443 K, 30% v/v MeOH, 30% w/w NaOH) resulted in the most promising results, with a cellulose recovery of 89%, delignification efficiency of 91%, pure lignin yield of 82%, residual carbohydrate content of 2.98% w/w, and an ash content of 1.24% w/w. The isolated lignin fractions show promising glass transition temperatures ($\geq$424 K) with high thermal stabilities and preferential O/C and H/C ratios. This, together with high contents of phenolic hydroxyl ($\geq$1.83 mmol/g) and carboxyl groups ($\geq$0.52 mmol/g), indicates a high valorization potential. Additionally, Bjorkman lignin was isolated, and two reference Kraft cooks and a comparison to three acid-catalyzed organosolv fractionations were conducted.

Keywords: lignin; organosolv; fractionation; European larch; characterization

1. Introduction

Interest in renewable resources to supply energy and products significantly increased over the last 20 years. The driving factors of this trend are the reduction of fossil-fuel dependence, sustainability, mitigation of greenhouse gas emissions, and the ever-increasing global energy demand. Lignocellulosic biorefineries will play a key role in achieving these goals, as they are able to generate chemicals, fuels, and energy from non-feed/food biomass such as primary wood processing (e.g., wood chips and sawdust) [1,2].

The pretreatment step of lignocellulosic biomass is one of the crucial steps of biomass valorization, as well as one of the most cost-intensive, accounting for up to 40% of the total processing costs [1]. A plethora of biological, chemical, physical, and thermal fractionation methods, as well as combinations thereof, were developed to efficiently separate biomass into its main components, cellulose, hemicellulose, and lignin, as well as extractives [1]. For softwoods, sulfite, sulfate, and soda pulping are available on a commercial scale; alternative techniques include steam explosion and (auto)catalyzed organosolv pretreatment [3].

The first studies on the organosolv process were published in 1931 by Kleinert and Tayanthal, who found that wood could be delignified using aqueous ethanol at elevated temperature and pressure [4]. Over the years, different solvent systems were tested, including alcohols, ketones, lactones, organic acids, phenols, amines, and various catalysts including acids, bases, and neutral salts [5]. There are two major advantages of organosolv over conventional sulfate and sulfite pulping processes: production of high amounts of valuable side streams/byproducts, such as sulfur-free lignin, and ease of solvent recovery via distillation. The latter is unfortunately only true for simple solvent systems, such as alcohol–water [6].

Recent studies in the field of alcohol based organosolv fractionation of softwoods were mainly performed under acid catalysis. Løhre et al. investigated a two-step semi-continuous process comprising a sulfuric acid-catalyzed organosolv fractionation followed by lignin solvolysis. They used spruce as feedstock and primarily focused on fractionation performance and oil yield [7]. Rossberg et al. investigated the impact of different processing conditions of a sulfuric acid-catalyzed ethanol organosolv process on lignin properties, using spruce, and focusing on downstream processability of the lignin fraction [8]. Nitsos et al. also used spruce to study the effect of sulfuric acid-catalyzed ethanol organosolv pretreatment on fractionation performance and resulting lignin properties, with focus on process performance [9]. Imlauer-Vedoya et al. studied a combined steam explosion and uncatalyzed organosolv process for the fractionation of radiata pine. However, they were only interested in the pulp fraction [10]. Lesar et al. investigated the uncatalyzed ethanol organosolv fractionation of spruce and mixed softwood. They also focused their research on the hydrolyzability of the pulp fraction [11].

A two-stage base-catalyzed organosolv process, named Organocell, was developed in the 1970s in Germany, as an alternative to the Kraft process [12]. In the first step, hemicellulose and part of the lignin fraction are dissolved at elevated temperatures, 443–463 K, in a 1:1 methanol–water mixture. In the second step, the methanol content is reduced (30% v/v) and the chips are cooked with up to 30% NaOH on a dry wood basis. Additionally, 0.1%–0.2% anthraquinone (AQ) is added as a redox co-catalyst, to reduce pulp loss and increase delignification efficiency [13]. As the initial step, AQ oxidizes a reducing end of a carbohydrate chain (to form an aldonic acid group), limiting carbohydrate degradation in the process. The in situ formed anthrahydroquinone species then reduces a lignin quinone methide structure, cleaving a β-aryl ether bond in the process, significantly improving delignification efficiency and reducing lignin condensation reactions [14]. Furthermore, terminal oxidation of primary alcohols to aldehydes leads to extensive lignin fragmentation via retro aldol reactions [13]. The process was also run as a single alkaline step, due to economic considerations [6].

However, to the best of the authors' knowledge, no detailed investigation of fractionation performance and direct comparison to other processes was published; the only available literature to date dealing with the lignin fraction of an Organocell-type pretreatment process is a four-part study by Lindner and Wegener published from 1988 to 1990 [15–18]. In these studies, the lignin fractions from both pulping stages were isolated and analyzed. However, they only investigated the reaction time dependence of these properties and kept all other parameters constant. The isolated lignin fractions had promising properties for valorization, such as low carbohydrate content, relatively low molecular weight, and high content of functional groups; therefore, the authors decided to further investigate this fractionation process.

The aim of this work was to evaluate the potential of a one-step Organocell process using European larch sawdust. The feedstock was chosen due to the high volume of larch timber on the European wood market [19]. The valorization potential of the pulp fraction and the isolated lignin fraction is of major interest. The usability of the lignin fraction mainly depends on the physicochemical properties, as conventional pulping processes, namely, sulfate and sulfite processes, result in technical lignins of low purity that contain sulfur and are highly degraded [20], which reduce the valorization potential significantly. Thus, a high fractionation performance combined with a high-quality lignin fraction would significantly increase the attractiveness of the base-catalyzed methanol organosolv process for softwood fractionation. The efficiency of the fractionation was investigated using a fractional factorial design on reaction temperature, catalyst loading, and methanol-to-water ratio. The conditions were derived from literature reporting on the Organocell process [12,15–18]. The results were compared to two lab-scale Kraft processes and three previously reported acid-catalyzed organosolv fractionations [21]. The properties of the resulting lignin fractions were analyzed and compared using ultimate, thermal, ^{31}P-NMR, thioacidolysis GC/MS, and HPLC–SEC analysis.

2. Results and Discussion

In this publication, the feasibility of a base-catalyzed organosolv treatment with focus on usability of the lignin fraction is explored. In the first part, the fractionation performance is investigated, including purity of the resulting fractions and comparison to two lab-scale Kraft and three acid-catalyzed ethanol organosolv processes from a previous study [21]. In the second part, the physicochemical properties of the isolated lignin fractions are investigated and compared.

2.1. Feedstock Composition

The chemical composition of European larch sawdust was determined according to the NREL/TP-510-42618 protocol published by Sluiter et al. [22]. In a previous publication by the authors [21], the composition was determined using a different technique. However, due to the distinctly higher throughput of the protocol described by Sluiter et al. [22], it was analyzed again for a better comparison to the pulp fractions (see Table 1).

Table 1. Composition of European larch, reported as mean (n = 3) ± standard deviation.

	Cellulose	Hemicellulose [a]	Total Lignin [b]	Extractives [c]	Ash [21]
Content % w/w	42.3 ± 0.4	29.6 ± 0.2	26.4 ± 0.1	2.70 ± 0.09	0.17 ± 0.02

[a] 1.17% ± 0.01% arabinan, 5.77% ± 0.07% xylan, 5.54% ± 0.07% galactan, and 17.0% ± 0.2% glucomannan; [b] 26.07% ± 0.02% Klason lignin and 0.3% ± 0.1% acid-soluble lignin; [c] acetone extractives.

In comparison, Scots pine (*Pinus sylvestris*) contained a slightly higher amount of cellulose and hemicellulose with 40.0% and 28.5% respectively. However, the lignin and extractive contents were both slightly decreased, with 27.7% and 3.5%, respectively. White spruce (*Picea glauca*) contained a lower amount of cellulose and extractives, with 39.5% and 2.1%, respectively. Both the hemicellulose and the lignin content were elevated in comparison, with 30.6% and 27.7%, respectively [23].

2.2. Interpretation of Results from Experimental Design

2.2.1. Pulp Fraction

Firstly, preliminary fractionation experiments on alcohol-to-water ratio, type of alcohol used, and concentration of base catalyst were performed in a limited set of combinations (see Supplementary Materials). Based on that, the fractional factorial design of experiment was set up. The quantitative results from the 2^{3-1} fractional factorial design are summarized in Table 2. Statistical analyses of the data from the three center point replicates (CP$_{1-3}$) showed that the response to the base-catalyzed organosolv fractionation (BCOS) was reproducible, when comparing lignin and pulp yields, as well as the composition of the respective fractions. The validity of the optimization was limited due to the small number of experiments, as chemical pulping is a rather complex system. Moreover, several responses showed significant curvature (see Supplementary Materials), indicating that a more sophisticated design of experiments, such as response surface methodology, should be used for a proper optimization. However, it was deemed sufficient for an exploratory analysis.

Table 2. Mass flows for the pulp fraction of the base-catalyzed organosolv fractionation.

	Parameters [a]			Pulp Fraction [b]					
	T	*S*	*C*	Total	KL [c]	ASL	TL	Cellulose	Hemicellulose
Sawdust				100	26.07	0.35	26.42	42.34	29.56
SP$_1$ (−/+/−)	443	70	20	54.00	4.28	0.35	4.63	37.61	11.76
SP$_2$ (−/−/+)	443	30	30	52.46	2.12	0.37	2.49	37.73	12.25
SP$_3$ (+/−/−)	463	30	20	44.69	0.35	0.20	0.55	35.12	9.02
SP$_4$ (+/+/+)	463	70	30	34.42	0.62	0.19	0.81	26.50	7.10
CP$_1$ (0/0/0)	453	50	25	45.79	0.96	0.26	1.22	34.07	10.50
CP$_2$ (0/0/0)	453	50	25	46.19	0.88	0.31	1.20	34.51	10.74
CP$_3$ (0/0/0)	453	50	25	46.12	0.92	0.26	1.18	34.27	10.67
CP$_{1-3}$				*46.0*	*0.92*	*0.28*	*1.20*	*34.3*	*10.6*
Avg ± SD				*0.2*	*0.03*	*0.02*	*0.02*	*0.2*	*0.1*

[a] *T*, cooking temperature (K); *S*, solvent ratio (% *v/v* methanol); *C*, catalyst concentration (% oven-dried wood (odw) NaOH). [b] All data are yields of components (g) per 100 g (oven-dried weight) untreated larch sawdust. [c] *KL*, Klason lignin; *ASL*, acid-soluble lignin; *TL*, total lignin.

Using the Design-Expert 12 software, the major effects of three factors, temperature (X_1), methanol-to-water ratio (X_2), and NaOH concentration (X_3) on pulp and cellulose yield, as well as residual lignin content, were investigated. These three responses were chosen as they are among the most critical results for an efficient fractionation. The Shapiro–Wilk test was used to test for normality. ANOVA and multiple linear regression analysis were used to determine if the effects were either positive or negative [24] (see Supplementary Materials). Only the residual lignin content did not follow a normal distribution ($\alpha < 0.05$). Except for the methanol (MeOH)/H$_2$O ratio in the case of residual lignin content, all factors had a negative effect. Pareto charts were drawn to visualize the magnitude and importance of the effects. Additionally, the linear correlation coefficients of the three coded factors on each response were plotted, to visualize the significance of the respective effects (see Figures 1 and 2). As can be seen in both figures, all three factors had a significant influence on the response. Temperature, as already indicated by the preliminary experiments (see Supplementary Materials), indeed had the greatest impact, especially for cellulose and lignin yield. Interestingly, catalyst loading and MeOH/H$_2$O ratio had, on average, similar magnitudes of effect over all responses. These data clearly show that a delicate balance of catalyst loading, MeOH/H$_2$O ratio, and temperature has to be found for optimal fractionation performance, since the effect of both delignification and cellulose degradation is positive, albeit with different magnitudes. However, as the effect of the MeOH/H$_2$O ratio is inverse for delignification and cellulose degradation, a low MeOH content is to be preferred for maximum fractionation performance.

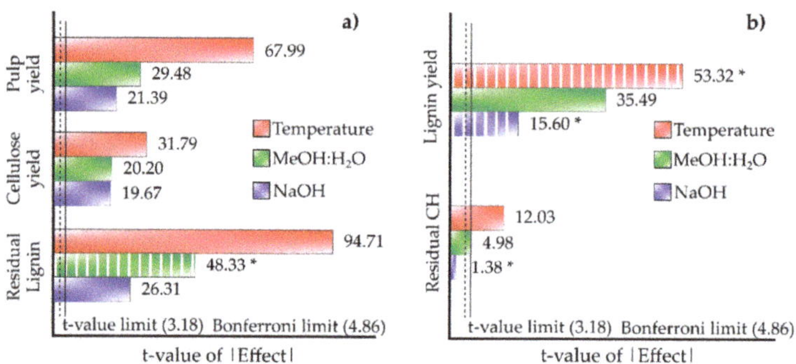

Figure 1. Pareto charts showing the effect of temperature, methanol (MeOH)/H$_2$O ratio, and NaOH concentration on the respective responses. (**a**) Responses of the pulp fraction; (**b**) responses of the lignin fraction. On the *x*-axis, the square roots of the *F*-values from the ANOVA analysis are plotted. Factors denoted with an asterisk and dashed lines are positive, while all others are negative. The *t*-value limit (line of significance) is denoted by a dashed line, whereas the Bonferroni limit (absolute significance) by a solid line.

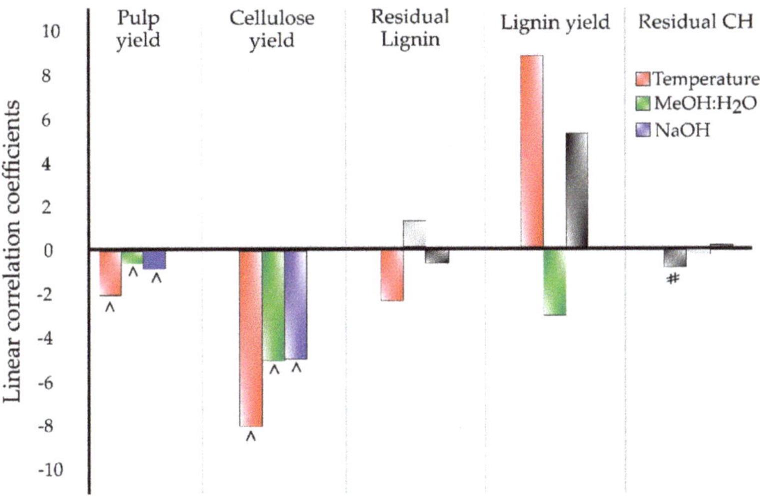

Figure 2. Effects of the three factors, temperature, MeOH/H_2O ratio, and NaOH concentration, on the responses of pulp, cellulose, and lignin yield, as well as residual lignin and residual carbohydrate (CH) content. The responses are increased/decreased by positive/negative correlation coefficients, respectively. Colored bars have a significant correlation at $p = 0.05$, bars with ^ have a significant correlation at $p = 0.01$. # denotes a significant effect at $p = 0.10$. The linear correlation coefficients are derived from multiple linear regression analysis (see Supplementary Materials).

The lowest amount of residual lignin in the pulp fraction was achieved under SP_3 conditions (463 K, 30% v/v MeOH, and 20% w/w NaOH) with 1.23% Klason lignin (2.07% of initial wood content) (see Table 2). The pulp and cellulose yields of 44.69% oven-dried wood (odw) and 82.96% initial wood (i.w.), respectively, were comparable to center point conditions but were significantly lower compared to the reactions performed at 443 K (SP_1, SP_2) (see Table 2). Pulping run SP_4 was given the maximum values of all variables. These maximum conditions significantly decreased the pulp and cellulose recovery, resulting in the lowest yields of this study, with 46.74% i.w. and 62.59% i.w., respectively. Additionally, the delignification efficiency was reduced compared to SP_3, making these experimental conditions unfeasible.

Although being less efficient than SP_3 in terms of raw fractionation performance, the authors decided that the conditions of SP_2 resulted in the most preferential results. Cellulose recovery was excellent at 89%, with a delignification efficiency of 91%. It was deduced from a publication by van Tran [25] that, after bleaching via oxygen delignification to the same residual lignin content as SP_3, the expected pulp yield would still be significantly higher, with an estimated 68% i.w. versus 61% i.w. It also would not significantly impair the physical properties of the pulp fraction according to Yang et al. [26].

As can be seen in all tested experimental conditions (see Table 2), the presence of anthraquinone hindered efficient hemicellulose removal. This was expected, since anthraquinone (AQ) protects the reducing sugar ends groups via oxidation to aldonic acid groups [14]. For a proper hemicellulose removal, the authors suggest either a pre- or a post-pulping hydrolysis step at elevated temperatures. The former concept was used in the two-step Organocell process, the basis for the investigations in this study [12]. However, recently, it was demonstrated by Borrega et al. [27] that a post-pulping hydrolysis step increases both cellulose yield and purity in comparison to pre-pulping hydrolysis in traditional Kraft pulps. Possible high-value applications for the pulp fraction, due to the high cellulose and low lignin content, after hemicellulose removal and possible bleaching, include cellulose nanofibers [28,29] and dissolving pulp preparation [6].

Higher fractionation efficiency in comparison to catalyzed ethanol organosolv pulping

Both Kraft processes were performed at a lower liquid-to-solid ratio of 3.5 versus 8. Therefore, the solid loading was increased to 40 g oven-dried wood, to reach the solvent level required for adequate mixing in the stirred batch reactor.

The low-severity Kraft process (Kraft (−)) was the least efficient fractionation process tested in this study, resulting in a delignification efficiency of just 78.95% (see Table 3). The conditions were chosen, as reaction temperature (440 K vs. 443 K), residence time (58 min vs. 60 min), and effective alkali charge (19% odw vs. 20% odw) were almost identical to SP_1. Compared to that, the pulp and cellulose yields were similar, albeit slightly lower. Hemicellulose recovery was distinctly decreased for the low-severity Kraft process, which was expected due to the protective properties of anthraquinone, which is detrimental for an efficient hemicellulose removal. As expected, pulp, cellulose, and hemicellulose yields, as well as residual lignin content, were significantly lower for the high-severity Kraft process (see Table 3). The residual lignin content was comparable to that of SP_2 (8.61% i.w. vs. 9.36% i.w.); however, the cellulose and hemicellulose yields were significantly decreased, closer to that of SP_3, albeit slightly lower. Comparing the Kraft cooks of this study to the results by Hörhammer et al. [30] on Siberian larch, significantly higher pulp yields were achieved for the low-severity pulping run, with 52.1% *w/w* vs. 43.2% *w/w*. This also coincided with a higher kappa number, 50.3 vs. 67.1, which was calculated using a previously published approximation (% Klason lignin = 0.159 × kappa number) [6]. The kappa number is an indicator for bleachability and degree of delignification for pulp samples and is based on the reaction of a 0.1 N potassium permanganate solution with lignin [6]. This trend can be also seen in the high-severity process, albeit distinctly less pronounced, with 42.8% *w/w* vs. 40.2% *w/w* and 33.4 vs. 24.2, respectively. These results indicate a significantly poorer fractionation performance of the Kraft processes in this study, which can be attributed to lower humidity (40% *w/w* vs. 21% *w/w*), poorer mixing, and difference in feedstock composition.

Table 3. Mass flows for the Kraft and acid-catalyzed organosolv fractionations.

	Pulp Fraction [a]					Lignin Fraction					
	Total	KL [b]	ASL	Cell	Hcell	Total	KL	ASL	TL	CH	Ash
Sawdust	100	26.07	0.35	42.34	29.56	-	-	-	-	-	-
Kraft (−)	52.08	4.84	0.72	36.86	9.66	23.27	19.14	0.35	19.50	0.51	0.56
Kraft (+)	42.83	1.98	0.29	33.05	7.51	26.90	22.76	0.55	23.31	0.46	0.68
EOS (−) [c]	47.79	8.45	0.18	36.98	2.23	16.36	15.48	0.07	15.55	0.09	>0.05
EOS (0)	38.56	4.91	0.15	32.84	0.67	19.42	18.12	0.15	18.27	0.08	>0.05
EOS (+)	29.74	5.31	0.12	24.05	0.27	20.83	19.52	0.24	19.78	0.18	>0.05

[a] Data are yields of components (g) per 100 g (oven-dried weight) untreated larch sawdust. [b] *EOS*, ethanol organosolv; *KL*, Klason lignin; *ASL*, acid-soluble lignin; *Cell*, cellulose; *Hcell*, hemicellulose; *TL*, total lignin; *CH*, residual carbohydrates. [c] Sulfuric acid loading, reaction time, and liquid-to-solid ratio were set to 1.10% odw, 30 min, and 7 *v/w*, respectively; reaction temperatures were 440 K (−), 450 K (0), and 460 K (+), respectively.

Higher fractionation efficiency in comparison to catalyzed ethanol organosolv pulping

To be able to better assess the fractionation performance of the base-catalyzed organosolv process, three pulp and corresponding lignin samples were included that were isolated in a previous study on acid-catalyzed ethanol organosolv fractionation of the same feedstock [21].

The most distinct difference between the results of the acid-catalyzed organosolv fractionation and both base-catalyzed processes was obviously hemicellulose yield, as hemicellulose is significantly more susceptible toward acid hydrolysis [31]. Even in low-severity conditions, the hemicellulose content of the pulp fraction was just 7.52% of initial wood content, compared to 24.02% i.w. for the most severe base-catalyzed organosolv process (SP_4) (see Table 3). On the downside, with increasing temperature (440–460 K), the cellulose yield decreased significantly, from 87.33% i.w. to 56.80% i.w., dropping to levels lower than that of SP_4 (62.59% i.w.). At the same time, the residual lignin content did not decrease accordingly, with a minimal content of 19.02% i.w., only being able to surpass the least efficient base-catalyzed process in terms of fractionation performance (Kraft low, 21.05% i.w.). Additionally, at 460 K, pseudo-lignin formation likely occurred, increasing the Klason lignin content. Pseudo-lignins are aromatic compounds that are hypothesized to form from carbohydrate degradation products, such as 5-hydroxymethylfurfural (C6 sugars) and furfural (C5 sugars) via a series of rearrangement, oxidation, polymerization, and condensation reactions [32,33].

2.2.2. Lignin Fraction

Lignin isolation yields must also be considered when deciding the optimum process conditions of a lignocellulosic biomass fractionation process. The pure lignin yields ranged from 64.7% (SP_1) to 88.0% (SP_3) of initial wood content (i.w.) (see Table 4). As expected, the maximum yield coincided with the lowest residual lignin content of the pulp fraction (SP_3) and vice versa (SP_1). The main factor affecting lignin yield was temperature, having a positive effect according to ANOVA analysis (see Figures 1 and 2). The methanol-to-water ratio had the second biggest effect, correlating negatively to lignin yield. The average of lignin purity, including all tested reaction conditions, was 93.7% ± 0.2% w/w. The t-value of effect for all three factors (X_1–X_3) failed to reach the t-value limit, suggesting their independence (not shown). Impurities of the lignin fractions included not only residual carbohydrates and ash but also residual water, as well as water/diluted methanol-insoluble wood extractives/resins and degradation products. The latter two only reduce apparent purity if they either decompose or evaporate at 378 K; they contribute to the acid-insoluble lignin fraction otherwise. The content of residual carbohydrates was low for each lignin fraction, ranging from 2.98% w/w (SP_2) to 0.76% w/w (SP_4). Interestingly, temperature was the only factor having a significant effect on the residual carbohydrate content of the lignin fraction, correlating negatively to it (see Figures 1 and 2). The t-value of the effect of MeOH/H_2O ratio barely exceeded the Bonferroni limit, and that of NaOH concentration did not even reach the t-value limit. The ash content of all lignin fractions was low as well, ranging from just 0.53% w/w to 1.24% w/w. The ash content was only determined by the initial base loading and the washing efficiency during work-up.

Table 4. Mass flows for the lignin fractions of the base-catalyzed organosolv fractionations.

	Lignin Fraction [a]						Aqueous Fraction
	Total	KL	ASL	TL	CH	Ash	ASL
SP_1 (−/+/−)	18.44	16.78	0.56	17.34	0.40	0.10	4.56
SP_2 (−/−/+)	21.53	19.57	0.62	20.19	0.64	0.27	3.94
SP_3 (+/−/−)	24.82	22.38	0.87	23.24	0.31	0.19	3.61
SP_4 (+/+/+)	23.08	21.07	0.52	21.59	0.18	0.28	7.06
CP_1 (0/0/0)	22.28	20.01	0.65	20.66	0.28	0.23	5.63
CP_2 (0/0/0)	22.16	20.24	0.58	20.82	0.22	0.24	5.62
CP_3 (0/0/0)	22.57	20.42	0.67	21.10	0.25	0.25	5.62
CP_{1-3}	22.3	20.2	0.64	20.9	0.25	0.24	5.62
$Avg ± SD$	0.2	0.1	0.05	0.2	0.03	0.01	0.01

[a] Data are yields of components (g) per 100 g (oven-dried weight) untreated larch sawdust.

The mass balances for lignin for the base-catalyzed organosolv fractionations are shown in Table 5. Note that the lignin recoveries were all over 100%, except for the two experiments performed at lower temperatures (443 K). This can be attributed to the more excessive carbohydrate degradation at higher temperatures that can lead to overestimation of the acid-soluble lignin content of the aqueous fraction. Additionally, the presence of 2.70% w/w acetone extractives in European larch wood further complicated the exact determination, as they can influence both Klason lignin and acid-soluble lignin content [34].

Table 5. Lignin recovery for the base-catalyzed organosolv fractionations.

	Pulp Fraction [a]		Lignin Fraction		Aqueous Fraction	Lignin Recovery
	KL	**ASL**	**KL**	**ASL**	**ASL**	**%**
SP_1 (−/+/−)	4.28	0.35	16.78	0.56	4.56	100.4
SP_2 (−/−/+)	2.12	0.37	19.57	0.62	3.94	100.8
SP_3 (+/−/−)	0.35	0.20	22.38	0.87	3.61	103.8
SP_4 (+/+/+)	0.62	0.19	21.07	0.52	7.06	111.5
CP_1 (0/0/0)	0.96	0.26	20.01	0.65	5.63	104.1
CP_2 (0/0/0)	0.88	0.31	20.24	0.58	5.62	104.6
CP_3 (0/0/0)	0.92	0.26	20.42	0.67	5.62	105.6
CP_{1-3}	*0.92*	*0.28*	*20.2*	*0.64*	*5.62*	*104.8*
Avg ± SD	*0.03*	*0.02*	*0.1*	*0.05*	*0.01*	*0.5*

[a] Data are yields of components (g) per 100 g (oven-dried weight) untreated larch sawdust

Higher purity compared to Kraft pulping

The pure lignin yields of both Kraft processes were comparatively high, reaching up to 88.24% i.w., with Kraft (+) being the highest in this study, on par with SP_3 (87.98% i.w.). The high yields despite lower delignification efficiency can be attributed to the decreased solubility of lignin in pure acidified water compared to acidified 20% aqueous methanol for the base-catalyzed organosolv fractionations. The purity of both lignin fractions was significantly lower compared to the base-catalyzed organosolv processes, with 83.78% *w/w* (−) and 86.67% *w/w* (+). The content of residual carbohydrates was low, with 2.21% *w/w* and 1.71% *w/w*, which was similar to the BCOS. The ash content was distinctly higher for both sulfate processes, with 2.21% *w/w* and 2.41% *w/w* on lignin content. This is to be expected, as (co-)catalyst concentration was between 1.7 and 3.2 times higher based on pulping liquor.

Higher yields compared to catalyzed ethanol organosolv pulping

As expected, lignin yields were lower for the acid-catalyzed process (61.92%–78.84% i.w.), due to the lower delignification efficiency. However, lignin purity was slightly higher in all three runs (94.08%–95.6%), with both lower ash and residual carbohydrate content. The former can be explained by the significantly decreased catalyst loading, whereas the latter due can be explained by the enhanced carbohydrate fragmentation under acidic conditions.

2.3. Influence of Pulping Conditions on Lignin Properties

As the second part of this study, the physicochemical properties of the isolated lignin fractions were investigated. They are a key factor for both lignin downstream valorization and feasibility of the whole fractionation process.

2.3.1. Ultimate Analysis

The influence of pulping conditions on the elemental composition was determined via ultimate analysis, and the results are shown in Table 6. Both molar hydrogen-to-carbon and oxygen-to-carbon molar ratios can be used as indicators for degrees of aromaticity and of oxygenation, respectively.

Table 6. Results from ultimate and HPLC–SEC analysis.

	C % w/w	H % w/w	O % w/w	S % w/w	H/C mol/mol	O/C mol/mol	M_n [a] kg/mol	M_w kg/mol
MWL	60.78	5.71	33.51	- [b]	1.04	0.50	1.76	2.61
SP$_1$ (−/+/−)	66.13	5.78	28.09	-	1.05	0.32	1.89	2.94
SP$_2$ (−/−/+)	65.32	5.78	28.90	-	1.06	0.33	1.80	2.88
SP$_3$ (+/−/−)	66.75	5.77	27.48	-	1.04	0.31	1.68	2.77
SP$_4$ (+/+/+)	67.27	5.85	26.88	-	1.04	0.30	1.75	2.56
CP$_1$ (0/0/0)	66.43	5.79	27.78	-	1.05	0.31	1.65	2.58
CP$_2$ (0/0/0)	66.36	5.78	27.86	-	1.05	0.31	1.62	2.52
CP$_3$ (0/0/0)	66.27	5.82	27.91	-	1.05	0.32	1.62	2.55
CP$_{1-3}$	*66.35*	*5.80*	*27.85*	*-*	*1.05*	*0.31*	*1.63*	*2.55*
Avg ± SD	*0.08*	*0.02*	*0.07*	*-*	*<0.01*	*<0.01*	*0.02*	*0.03*
Kraft (−)	61.78	5.33	30.23	2.66	0.99	0.44	1.81	3.05
Kraft (+)	61.57	5.44	30.26	2.73	0.97	0.43	1.78	2.92
EOS (−)	66.42	6.22	27.36	-	1.12	0.31	1.06	2.50
EOS (0)	66.83	5.86	27.32	-	1.05	0.31	1.12	2.40
EOS (+)	67.18	6.04	26.78	-	1.08	0.30	0.97	2.02

[a] M_n, number average molecular weight; M_w, weight average molecular weight. [b] Below 0.05%.

All samples were free of nitrogen and, except for both Kraft samples, free of sulfur, which could not be effectively removed with the milled wood lignin (MWL) purification method. Of the tested base-catalyzed organosolv conditions, SP$_4$ produced the lignin fraction with the lowest H/C and O/C ratios, indicating both severe lignin side-chain degradation and condensation. However, it was recently mentioned by Rossberg et al. [8] that a low H/C and O/C ratio is desirable for special application such as carbon-fiber production, for which Kraft and soda lignins are frequently used. The Kraft lignin samples isolated in this study had the highest degree of both oxygenation and aromaticity. However, the difference of aromaticity was comparably small. This, together with the high purity, makes the base-catalyzed organosolv samples promising feedstocks for carbon-fiber production. The previously reported acid-catalyzed samples (ethanol organosolv, EOS) [21] had a lower aromaticity with a similar degree of oxygenation. The former was expected, especially for EOS (−), since the fractionation conditions were distinctly less severe.

2.3.2. Molecular Weight Analysis

The average molecular weights of the isolated lignin fractions were investigated via size exclusion chromatography (SEC) (see Table 6, Figure 3b). Due to the lack of proper lignin standards, this technique can only provide an estimation on relative molecular mass. Several attempts for a standardization of molecular weight analysis were made previously [35,36]. The objective of this investigation was to use the molecular weights (MWs) as an indicator for downstream processability, especially solubility.

As can be seen in Table 6, MWs were not significantly impacted by the pulping conditions. The lowest MWs were achieved under center point conditions, with 1.63 kg/mol and 2.55 kg/mol, respectively, which was just 16% lower than the maximum (SP$_1$). The M_w range for the acid/base-catalyzed organosolv samples was in agreement with data found in the literature, which also reported lower values for organosolv compared to commercial soda and Kraft processes [8,37–39]. In this study, this was only true for the average molecular weight (M_n) of the acid-catalyzed organosolv samples, whereas the difference in average molecular weight (M_w) was distinctly less pronounced. Interestingly, the heterogeneity (M_w/M_n) of the acid-catalyzed organosolv lignins was higher (2.1–2.4) than for the other lignin fractions (1.5–1.7). This might be attributed to the milder fractionation conditions, as less extensive fragmentation of β-O-4 bonds occurred, with similar observations being made by others [40,41]. All heterogeneities, however, were distinctly lower compared to literature values [38,39,42]. Each lignin fraction, regardless of fractionation technique, was still fully soluble in 1,4-dioxane, DMSO, THF, and ethanol/dichloroethane (1:2).

2.3.3. Quantitative ^{31}P-NMR analysis

^{31}P-NMR analysis can be used to distinguish and quantify functional groups of lignin fractions after phosphorylation [43]. In this study, the content of aliphatic and phenolic hydroxyl groups, as well as carboxylic acid groups, was investigated (see Figure 3a). Cyclohexanol was used as an internal standard, despite possible underestimation of hydroxyl groups as reported by Balakshin et al. [44], due to its high stability under the used reaction conditions.

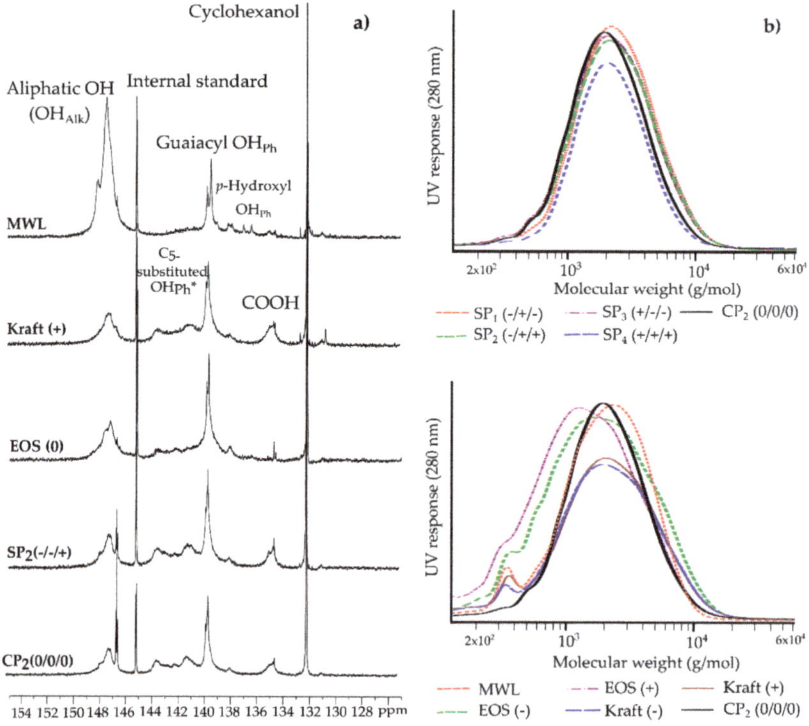

Figure 3. (a) ^{31}P-NMR spectra of selected lignin samples. The integrals of the C5-substituted OH$_{Ph}$ are added to the integrals of the guaiacyl OH$_{Ph}$ to yield the total phenolic hydroxyl groups (OH$_{Ph.}$). (b) Molecular weight distribution of selected lignin samples. Note that CP$_2$ is displayed in both figures to increase comparability.

A distinct decrease in aliphatic hydroxyl groups can be seen in Table 7, when comparing the organosolv and Kraft lignins to the milled wood lignin sample. This indicates that the OH_{alk} groups were either cleaved off or chemically modified during the pulping process. The presence of anthraquinone in the base-catalyzed organosolv process favors the oxidation of primary hydroxyl groups to aldehydes, leading to either retro aldol reaction or subsequent oxidation to carboxylic acids. AQ also increases the oxidation of secondary alcohols to ketones, further decreasing OH_{alk} [13]. As can be seen in Table 7, the OH_{alk} content for the base-catalyzed organosolv samples was mainly determined by reaction temperature, reaching a minimum of 1.18 mmol/g at SP$_3$. In roughly comparable pulping conditions (SP$_1$/Kraft (−)) in terms of temperature, time, and effective alkali charge, the Kraft sample contained a slightly lower amount of OH_{alk}, indicating similar to slightly higher lignin fragmentation in the given conditions.

Table 7. Results from ^{31}P-NMR-, thioacidolysis GC/MS, and thermal analysis.

	^{31}P-NMR Analysis						Thermal Analysis	
	OH_{Alk} mmol/g	OH_{Ph} mmol/g	OH_{total} mmol/g	Ph/Alk ratio	COOH mmol/g	β-O-4 % (μmol/g)	T_g K	$T_{5\%}$ K
MWL[a]	3.67	1.84	5.51	0.5	0.11	100 (947)	432	563
SP$_1$ (−/+/−)	1.54	3.47	5.01	2.3	0.43	8.2 (78)	433	496
SP$_2$ (−/−/+)	1.52	3.44	4.96	2.3	0.52	4.1 (39)	433	528
SP$_3$ (+/−/−)	1.18	3.60	4.78	3.1	0.51	2.6 (25)	429	567
SP$_4$ (+/+/+)	1.38	3.83	5.21	2.8	0.47	2.7 (26)	426	578
CP$_1$ (0/0/0)	1.51	3.68	5.19	2.4	0.50	4.6 (44)	424	497
CP$_2$ (0/0/0)	1.54	3.69	5.23	2.4	0.50	5.0 (47)	425	512
CP$_3$ (0/0/0)	1.59	3.78	5.37	2.4	0.52	4.8 (45)	425	483
CP$_{1-3}$	*1.55*	*3.72*	*5.27*	*2.4*	*0.51*	*4.8 (45)*	*425*	*497*
Avg ± SD	*0.04*	*0.06*	*0.09*	*>0.1*	*0.01*	*0.2 (2)*	*1*	*14*
Kraft (−)	1.48	3.56	5.04	2.41	0.61	7.4 (70)	-[b]	473
Kraft (+)	1.44	4.01	5.45	2.78	0.68	3.7 (35)	-[b]	412
EOS (−)	1.92	1.78	3.70	0.9	0.09	45.8 (443)	393	512
EOS (0)	1.61	2.35	3.96	1.5	0.07	36.8 (365)	402	516
EOS (+)	1.37	2.5	3.87	1.8	0.09	24.6 (233)	401	471

[a] *MWL*, milled wood lignin; OH_{Alk}, aliphatic hydroxyl; OH_{Ph}, phenolic hydroxyl; *COOH*, carboxylic acid; β-*O*-4, content of residual β-O-4 linkages relative to *MWL*; T_g, glass transition temperature, $T_{5\%}$, temperature at 5% weight loss. [b] No T_g detectable.

The content of phenolic hydroxyl groups (OH_{Ph}) of all isolated lignins was distinctly higher than that of the milled wood lignin sample, due to excessive breaking of ether bonds during the different fractionation processes [31]. Looking at the data shown in Tables 6 and 7, a clear indirect correlation between molecular weight and free phenolic hydroxyl groups can be identified for both catalyzed organosolv processes. Among those, the highest content of OH_{Ph}, 3.83 mmol/g (SP$_4$), did not coincide with the lowest M$_W$, indicating a higher degree of lignin condensation. This theory is supported by the fact that both catalyst loading and temperature were the highest for this pulping run. However, the lignin fractions from acid-catalyzed organosolv fractionations displayed significantly lower amounts of OH_{Ph}. Overall, the Kraft (+) sample had the maximum number of OH_{Ph} groups, which was expected due to the high concentration of Na$_2$S and NaOH, leading to excessive lignin fragmentation [20]. When comparing molecular weights and phenolic hydroxyl contents of both Kraft lignins to the base-catalyzed organosolv samples, a lower degree of recondensation at a similar level of fragmentation was indicated for the latter, due to lower MW and reduced OH_{ph} content. This observation was also supported by a higher H/C ratio (see above), making them more promising feedstocks for downstream valorization. The content of phenolic hydroxyl groups for the base-catalyzed organosolv samples from European larch was high compared to other softwood organosolv lignins found in literature. The reported values ranged from 2.6 to 3.5 mmol/g [8], from 2.44 to 3.64 mmol/g [9], and 2.99 mmol/g [42] for spruce, 2.8 mmol/g for lodgepole pine [45], and 2.81 mmol/g for loblolly pine [46].

A distinct increase in carboxylic acid groups can be found for the lignin fractions of the tested Kraft and base-catalyzed organosolv processes. This can be mainly attributed to oxidation of the primary alcohols in the aliphatic side chains [13,20]. All three acid-catalyzed lignin fractions displayed a distinctly lower *COOH* content, similar to that of the milled wood lignin sample, due to the lower process severity [31]. High amounts of phenolic and carboxyl hydroxyl groups, together with comparably low average molecular weight, make the base-catalyzed organosolv samples promising additives for cellulose nanofiber–starch composite films, which were recently investigated by Zhao et al. [47]. The same properties make them also usable as adhesives for wood panel production, as discussed by Koumba-Yoya et al. [48].

2.3.4. Thioacidolysis GC/MS Analysis

As this study serves as a basis for further investigations in this field, thioacidolysis GC/MS was used to quantify the fragmentation of the lignin fraction during the different pulping processes. In this method, the aryl ether linkages in the lignin samples were selectively cleaved, derivatized, and quantified via GC/MS. A high content of releasable monomers indicates less lignin degradation and consequently more benign fractionation conditions.

The purified MWL sample of European larch contained a total of 947 µmol/g of releasable monomers, 845 µmol/g G-units, 82 µmol/g vanillic units, and 20 µmol/g H-units, respectively (see Table 6). The former two originated from guaiacyl monomers, while the second was mostly formed via lignin side-chain oxidation that can even occur during mild MWL extraction [31]. H-units are the thioacidolysis derivates of *p*-hydroxyphenyl monomers, which are uncommon in the heartwood lignin of softwood species, but common in bark lignin [49].

As determined by the Design-Expert 12 software, all three factors (temperature, methanol content, NaOH loading) had a significant impact on the number of residual β-O-4 linkages, with MeOH/H_2O ratio having a positive effect and the other two having a negative effect (not shown). All lignin fractions from both Kraft and base-catalyzed organosolv processes contained less than 10% of the initial β-O-4 linkages, demonstrating significant lignin fragmentation (see above). Overall, G-units made up 69–88% of the total releasable monomers with the rest being vanillic units (see Supplementary Materials). The lowest values for the base-catalyzed samples were achieved under SP_3 and SP_4 conditions (2.6/2.7 µmol/g), which coincided with the highest amount of OH_{Ph} (SP_4) and lowest amount of OH_{alk} (SP_3). This trend was also true for the Kraft samples and the acid-catalyzed organosolv lignins. The *EOS* retained a significantly higher content of β-O-4 linkages, ranging from 24.6% to 45.8%, as described previously [21]. Therefore, only the acid-catalyzed organosolv lignins are suitable feedstocks for downstream depolymerization. However, the base-catalyzed organosolv lignins are a suitable feedstock for slow pyrolysis, due to the absence of sulfur and the small amount of impurities, such as ash and carbohydrates, as well as low O/C ratio, and higher hydrophobicity compared to soda and Kraft lignins [50].

2.3.5. Thermal Analysis

As this study aimed to evaluate the valorization potential of the lignin fraction, the thermal processability was investigated, which is crucial for applications such as melt pressing and melt spinning [8]. The thermal properties reflect the molecular features of lignin including rigidity and flexibility that are influenced by inter-/intramolecular hydrogen bonds, degree of cross-linking, and aromaticity, as well as molecular weight distribution [51]. Therefore, the glass transition temperature (T_g) and thermal stability of the lignin fractions were determined via differential scanning calorimetry (DSC) and thermogravimetric analysis (TGA), respectively.

The acid-catalyzed organosolv lignins had the lowest T_gs (see Table 7, Figure 4), with *EOS* (−) being the minimum at 393 K. This coincided with the highest content of residual β-O-4 linkages, lowest amount of free phenolic hydroxyl groups, lowest average molecular weight, and lowest aromaticity (highest H/C ratio). As expected, the glass transition temperatures of the base-catalyzed organosolv samples were higher in comparison, ranging from 424 K to 433 K. No glass transition temperature was detected for both Kraft lignins. This can be attributed to the presence of sulfur (as –SH) and other impurities, high aromaticity, M_w and OH_{Ph}, extensive cross-linking, and a low number of ether linkages [51]. The acid-catalyzed organosolv lignins from European larch exhibited slightly lower T_g values compared to values reported recently by Rossberg et al. for spruce organosolv lignins, with 393–402 K versus ~405 K [8] at distinctly lower average molecular weights, 2.0–2.5 kg/mol versus 6.2 kg/mol [8]. The thermal stability of the samples was investigated by determining the 5% mass loss temperature via TGA. The values ranged from 412 K to 578 K; the comparably low values of the Kraft fraction can be explained due to the presence of impurities such as sulfur (see Figure 5). The better performance of base-catalyzed organosolv lignins (SP_2–SP_4) compared to the acid-catalyzed lignins originates from the higher degree of condensation and aromaticity of the former [51]. The $T_{5\%}$ temperatures were comparable to literature values for organosolv and Kraft lignins of spruce, with 516 K [42] and 536 K [42]/544 K [47], respectively. Note that water/solvent residue from incomplete drying during lignin purification can negatively impact $T_{5\%}$. The high differences between T_g and $T_{5\%}$ of up to 152 K (SP$_4$) indicate good workability, as materials are not instantly ductile or flowable at the respective T_g but need to be heated further. These findings further demonstrate the promising properties of the base-catalyzed lignin fraction for applications such as additives for composite films and biopolymers [47], feedstock for carbon fiber [8], and adhesive for wood panel production [48].

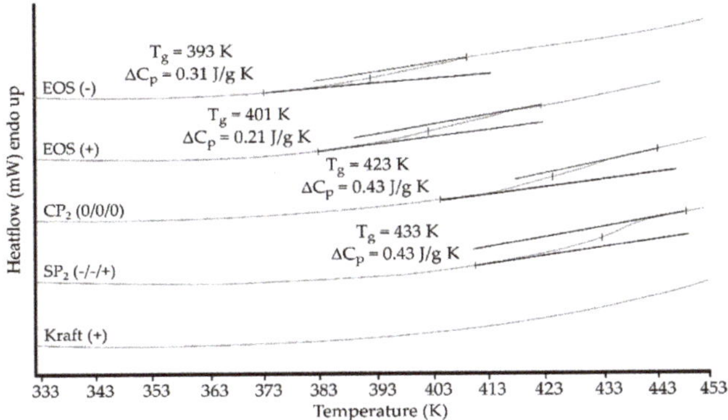

Figure 4. Differential scanning calorimetry (DSC) curves for selected lignin samples. ΔCP denotes the change in heat capacity caused by the glass transition.

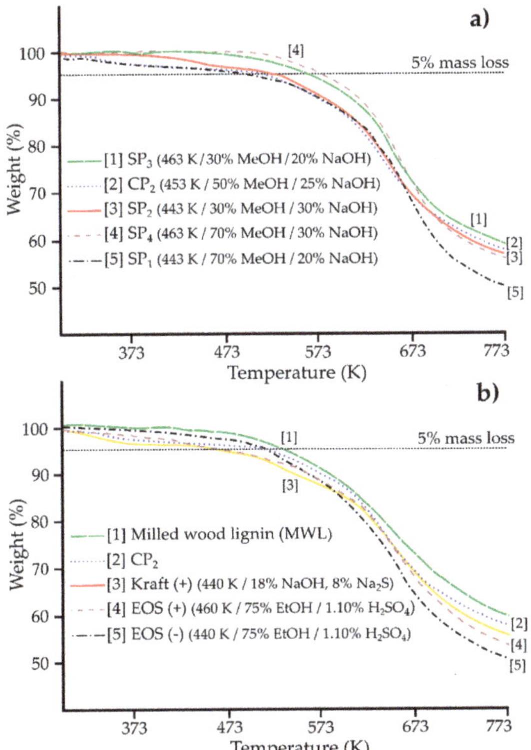

Figure 5. Thermogravimetric analysis (TGA) curves for selected lignin samples. (**a**) Results for the base-catalyzed organosolv lignins. (**b**) Results for the milled wood lignin sample, one Kraft, two acid-catalyzed organosolv, and one base-catalyzed organosolv lignin sample again for comparison.

3. Materials and Methods

European sawdust was used as feedstock in this study; the origin, storage conditions, and sampling were described previously [21]. Selected lignin and pulp fractions from this previously reported acid-catalyzed organosolv fractionation using the same feedstock were also analyzed and compared.

3.1. Base-Catalyzed Organosolv Fractionation

The reactions were performed in a stirred batch reactor with an internal volume of 0.5 L (Series 4575A HP/HT, Parr Instrument Company, St. Moline, IL, USA), which was described previously in detail [21]. The used process conditions were based on the Organocell process [12]. As a first step, preliminary fractionation experiments on alcohol-to-water ratio, type of alcohol used, and concentration of base catalyst were performed using a limited set of combinations (see Supplementary Materials). Based on that, a fractional factorial design of experiment was used to investigate the dependence of sugar composition, delignification, lignin yield, and properties on the experimental conditions. The three variables of the 2^{3-1} experimental design were reaction temperature (X_1), solvent ratio (X_2), and catalyst concentration (X_3). To reduce the required number of experiments, only a balanced half of the possible variable combinations was used. Four star point ($+/-$) plus three center point experiments (0) were performed; the latter were carried out to check for curvature and determine reproducibility. The residence time was kept constant at 1 h at the reaction temperature with a heating rate of 8–10 K/min, whereas mixing was achieved using a turbine type impeller at 300 rpm. Furthermore, the loading of the co-catalyst, as well as the solid-to-liquid ratio, was kept constant at 0.1% (oven-dried wood, odw) and 8 v/w, respectively (see Table 8).

Table 8. Experimental variables for the base catalyzed organosolv process.

Variable	Assignment	($-$)	(0)	(+)
X_1	Temperature (K)	443	453	463
X_2	Methanol (MeOH)/H_2O (v/v)	30:70	50:50	70:30
X_3	NaOH (w/w odw [a])	20	25	30

[a] odw, oven-dried wood.

In detail, for each run, a fresh methanol (99.9%, Fischer Scientific, Schwerte, Germany)/water mixture was prepared. The required amount of NaOH (≥99%, VRW Chemicals, Germany) was then dissolved in the solution and left to cool down to room temperature. Then, 250 mL of this solution was combined with 32 g of sawdust (odw) and 32 mg of anthraquinone (97%, Sigma Aldrich, Steinheim, Germany) as co-catalyst in the reactor. After the desired reaction time elapsed, the ceramic heater was removed, and the pulping liquor was cooled down via internal cooling loop to 323 K at an average rate of 6 K/min. The pulping liquor was then filtered in a Büchner funnel with a Whatman No. 5 filter paper to separate the cellulose-rich pulp fraction from the lignin- and hemicellulose-containing liquid fraction. The pulp was washed three times with 65 mL of warm (323 K) aqueous MeOH with identical solvent ratio to the pulping liquor, followed by three washes with warm (323 K) water (65 mL each), which was combined with the initial filtrate. The washed pulp fraction was then transferred to a 500-mL round-bottom flask and freeze-dried for 40 h. To ease the consequent lignin precipitation, the methanol content in the combined liquor was diluted to roughly 20% v/v via water addition. Then, concentrated HCl (37%, Merck, Darmstadt, Germany) was added dropwise under constant stirring until a pH of 3.5 was reached, which was monitored with a pH meter (Metrohm 691, Metrohm, Herisau, Switzerland). To be able to isolate the lignin fraction, the resulting suspension was centrifuged at 12,000× g for 20 min, and the supernatant liquid was decanted, filtered, and stored for analysis. The lignin fraction was resuspended in deionized H_2O, thoroughly mixed, ultrasonicated of 5 min, and centrifuged, which was repeated once, before freeze-drying for 40 h in a 250 mL round-bottom flask (see Figure 6).

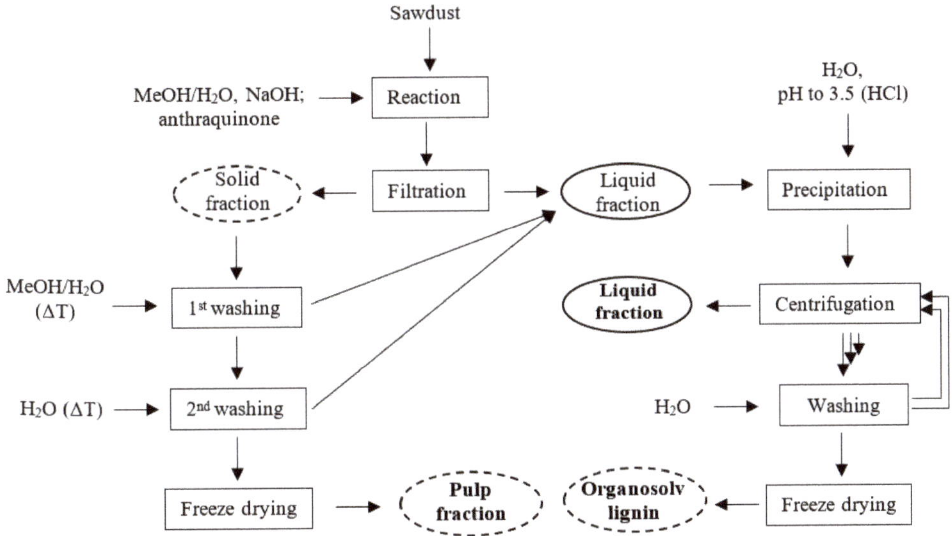

Figure 6. Reaction scheme for the base-catalyzed organosolv fractionation including the work-up procedure. Product fractions are written in bold, and dashed lines indicate solid fractions, whereas solid lines indicate liquid fractions (modified from a previous study by the authors [21]).

3.2. Data Analysis

To determine the significant factors and their influence on process yield and pulp and lignin characteristics, Design-Expert software for Windows (version 12, 2019, Stat-Ease Inc, Minneapolis, MN, USA) was used. The Shapiro–Wilk test was used to for normality, while ANOVA analysis and multiple linear regression were used to determine the effects of the factors on the responses [24].

3.3. Lab-Scale Kraft Pulping

As reference for the fractionation processes, two different lab-scale Kraft cooks were performed, based on conditions published by Hörhammer et al. [30]. A low-severity process and a high-severity process were chosen, according to both H-factor and catalyst/co-catalyst loading. The H-factor combines both temperature and time into a single factor; however, it does not account for catalyst dosage [52]. For both cooks, the liquid-to-solid ratio, reaction temperature, heating time, and stirring speed were kept constant at 3.5 *v/w*, 440 K, 15 min and 300 rpm, respectively. Then, 40 g of sawdust on a dry wood basis was loaded into the reactor with 140 mL of deionized H$_2$O, containing the desired amount of NaOH and Na$_2$S (≥60%, Sigma Aldrich, Steinheim, Germany) (see Table 9). The work-up procedure was performed similarly to the base-catalyzed organosolv processes (see Figure 6), with some modifications. Firstly, the pulp fraction was washed with five times with 65 mL of warm (323 K) water. Secondly, the lignin fraction was washed with water acidified to pH 3.5, using acetic acid (≥99%, Merck, Darmstadt, Germany), to remove excess sulfur as H$_2$S.

Table 9. Experimental conditions of the low- and high-severity Kraft cooks (liquid-to-wood ratio 3.5, heat-up/cooldown time of 15 min, 440 K).

Parameter	Low	High
Reaction time, min	58	134
H-factor, h	850	1650
NaOH, % (odw)	15.0	17.7
Na$_2$S, % (odw)	5.76	7.61

3.4. Crude Milled Wood Lignin (MWLc) Preparation

Firstly, 20 g of European larch sawdust was milled to a fraction <60 mesh and extracted with acetone in a Soxhlet apparatus for 16h at ≥6 cycles per hour with 300 mL of solvent prior to MWL preparation. The crude milled wood lignin was prepared based on a method described by Zinovyev et al. [53]. In detail, 4 g of extracted sawdust was milled in a 250-mL ZrO_2 bowl, with 34 ZrO_2 balls (Ø 5 mm) in a planetary ball mill (Pulverisette 6, Fritsch, Idar-Oberstein, Germany). The milling was performed at 500 rpm with 30 min of milling followed by 15 min of pause, to avoid excess heating, for a total of 15 cycles. The whole procedure was performed three times. Then, 11.2 g of fine wood meal was subjected to a fourfold aqueous dioxane (96% *v/v*) (≥99.5%, Sigma Aldrich, Steinheim, Germany) extraction. For this, 2.81 g of wood meal was mixed with 39.3 mL of solvent in separate 50-mL centrifuge tubes (Greiner, Sigma Aldrich, Steinheim, Germany) and shaken constantly for 24 h in the absence of light. Afterward, the suspensions were centrifuged at 3000× *g* for 10 min (Centrifuge 5702, Eppendorf, Wesseling-Berzdorf, Germany) to easily decant the supernatant solution, which was combined in a single flask and stored at 4 °C. This procedure was repeated four times, and the combined solution was filtered in a Büchner funnel with Whatman No. 5 filter paper (Sigma Aldrich, Vienna, Austria), and the solvent removed under vacuum at 35 °C. The solid matter was resuspended in 100 mL of water, ultrasonicated for 15 min, and freeze-dried for 40 h to remove traces of both dioxane and water. The yield of the crude milled wood lignin was 0.78 g, corresponding to 26% of the total lignin content.

3.5. Milled Wood Lignin Purification

To remove residual carbohydrates, which impede the quantitative determination of structural features, a purification according to Björkman et al. [54] was performed. In detail, the crude MWL was firstly dissolved in 20 mL/g 90% *v/v* acetic acid (≥99%, Merck, Darmstadt, Germany) and then added dropwise to 230 mL/g DI water under constant stirring. The suspension was then centrifuged at 12,000× *g* for 20 min (CR 22 N, Hitachi, Tokyo, Japan), the supernatant solution was discarded, and the solid matter was freeze-dried for 48 h. As the last step, the dried lignin phase was dissolved in a 20 mL/g 2:1 mixture of 1,2-dichloroethane (≥99.8%, Sigma Aldrich, Steinheim, Germany) and ethanol (≥99.5%, Merck, Darmstadt, Germany), before being filtered and precipitated in 230 mL/g petrol ether (extra pure, Carl ROTH, Karlsruhe, Germany). The insoluble fraction was separated via filtration, washed three times with 20 mL of fresh petrol ether, and dried in vacuo overnight to yield the purified milled wood lignin (MWL).

3.6. Lignin Purification

To remove residual carbohydrates and extractives, catalyst, and cocatalyst, which all interfere with follow-up lignin analyses, approximately 3.5 g of the raw lignin fractions from the base-catalyzed organosolv and Kraft fractionations were purified according to milled wood lignin protocol (see above).

3.7. Analysis of Chemical Composition

The composition of the feedstock, pulp, and lignin fractions was analyzed according to Sluiter et al. [22] using a two-step hydrolysis method. In detail, as the first step, the pulp sample was hydrolyzed in 72% *w/w* H_2SO_4, at an acid-to-pulp ratio of 10 mL/g, at 303 K, for 1 h. The resulting suspension was subjected to the second hydrolysis in 4% *w/w* H_2SO_4, with an acid-to-pulp ratio of 300 mL/g, in an autoclave at 394 ± 1 K, for 1 h. After filtration using a glass frit (porosity 4), the acid-insoluble (Klason) lignin was determined gravimetrically, while the content of acid-soluble lignin (ASL) was determined by measuring the absorbance at 205 nm (spectrophotometer Shimadzu UV-2550), using 110 L/(g·cm) as the extinction coefficient. The monosaccharides were determined via high-performance anion exchange chromatography (HPAEC–PAD) in a Dionex ICS-3000 system, equipped with a CarboPac

PA20 column. The hemicellulose and cellulose content of European larch sawdust and the resulting pulp fractions were calculated according to the Janson formulas [55].

3.8. Thermal Analysis

To determine thermal stability and processability of the purified lignin fractions, thermal analyses were performed. Thermogravimetric analysis (TGA) was carried out on a STA 449 C Jupiter thermal analyzer (NETZCH, Selb, Germany). Here, 2–3 mg of sample was placed into an aluminum pan crucible, and the temperature range was set to 293 to 823 K at a heating rate of 10 K/min with a 15-min holding time at maximum temperature. Helium was used as the purge gas with a flow rate of 50 cm^3/min. Differential scanning calorimetry (DSC) was performed on a DSC 8500 (Perkin Elmer Rodgau, Germany) with a temperature range from 273 to 453 K at a heating rate of 20 K/min. The analysis was carried out under nitrogen atmosphere, with a flow rate of 20 cm^3/min and 8–10 mg per sample. Glass transition temperatures (T_g) from the second heating run were interpreted as the midpoint of change in heat capacity

3.9. Further Analysis Techniques

The methods below were described in detail in a previous publication [21] and, thus, are only briefly mentioned here.

The phenolic (OH_{Ph}) and aliphatic hydroxyl (OH_{Alk}) groups, as well as carboxyl groups (COOH), were determined via ^{31}P-NMR analysis on a Bruker Biospin Avance III 400 MHz spectrometer (Bruker, Billerica, MA, USA) using cyclohexanol as the internal standard and chromium(III) acetylacetonate as the relaxation agent.

The number of residual β-O-4 linkages was determined via thioacidolysis GC/MS on a Saturn 2100 (Varian, Agilent, Palo Alto, CA, USA) equipped with a poly(dimethylsiloxane) column (30 m × 0.25 mm; PB-1, Supelco, Sigma Aldrich, St. Louis, MO, USA) with heneicosane (C21H44, Fluka Analytical, Darmstadt USA) as the internal standard.

The molecular weight of the lignin fractions was estimated via HPLC–SEC–UV using a styrene–divinylbenzene PLgel column (5 mm, 100 Å, 600 mm length, 7.5 mm inner diameter, Agilent, Polymer Laboratories, Palo Alto, CA) with a photodiode array detector (Dionex Ultimate 3000 UV/VIS detector) and propylene oxide standards (Igepal, Sigma Aldrich Steinheim, Germany).

4. Conclusions

In this study, for the first time, a comprehensive analysis of fractionation performance and the influence of pulping conditions on lignin properties of a NaOH/anthraquinone-catalyzed organosolv fractionation of softwood was discussed. This study should also serve as the step stone for future valorization of European larch sawdust, and it demonstrated that lignins with promising physicochemical properties can be isolated.

The most promising fractionation conditions, 443 K, 30% *v/v* MeOH, 30% *w/w* NaOH, achieved a cellulose recovery of 89% and a delignification efficiency of 91%. The pure lignin yield was 82% with a residual carbohydrate content of 2.98% *w/w* and 1.24% *w/w* ash. The high purity and high aromaticity, on the one hand, make the lignin fractions good feedstocks for carbon fibers or slow pyrolysis. Furthermore, the high contents of phenolic hydroxyl and carboxyl groups, together with low molecular weight and promising thermal properties (low T_g with high $T_{5\%}$), indicate good usability as feedstock for biopolymer synthesis and as an additive in composite films and in wood panels.

Further investigations on larch wood sawdust and the isolated lignin fractions are planned to fully assess the potential of the feedstock and its fractionation products. As a next step, a green fractionation technique, γ-valerolactone pulping, will be investigated and compared. Additional analysis techniques are also planned such as zeta potential, two-dimensional (2D) HSQC NMR, and quantitative ^{13}C-NMR analysis.

Supplementary Materials: The following are available online at http://www.mdpi.com/2073-4344/9/12/996/s1: Table S1: A detailed description of the preliminary experiments; Table S2: The results of Shapiro–Wilk test for normality; Tables S3–S7: ANOVA analysis for the responses of the fractional factorial design; Tables S8–S12: Results from multilinear regression analyses of the responses; Table S13: Detailed results from the thioacidolysis GC/MS experiments.

Author Contributions: M.H. designed the experiments and wrote the manuscript; M.H., G.T., B.C.-B., L.C., and A.M. performed the experimental analyses; B.C.-B., G.T., M.M., and S.S. critically reviewed the manuscript; M.M. and S.S. provided supervision throughout the study.

Funding: This research received no external funding.

Acknowledgments: The authors would like to thank Herbert Sixta for his suggestions, as well as Quang Lê Hui and Rita Hatakka for their expertise in determining the feedstock composition; as the authors also thank Josefine Hobisch for her help with DSC/TGA analyses. Furthermore, the authors acknowledge the Open Access Funding by the University of Graz.

Conflicts of Interest: The authors declare no conflict of interest.

References

1. *Biomass Fractionation Technologies for a Lignocellulosic Feedstock Based Biorefinery*; Mussatto, S.I. (Ed.) Elsevier: Amsterdam, The Netherlands, 2016; ISBN 9780128023235.

2. Rinaldi, R.; Jastrzebski, R.; Clough, M.T.; Ralph, J.; Kennema, M.; Bruijnincx, P.C.A.; Weckhuysen, B.M. Paving the way for lignin valorisation: Recent advances in bioengineering, biorefining and catalysis. *Angew. Chem. Int. Ed. Engl.* **2016**, *55*, 8164–8215. [CrossRef] [PubMed]

3. Nitsos, C.; Rova, U.; Christakopoulos, P. Organosolv fractionation of softwood biomass for biofuel and biorefinery applications. *Energies* **2018**, *11*, 50. [CrossRef]

4. Kleinert, T.; Tayenthal, K.v. Über neuere Versuche zur Trennung von Cellulose und Inkrusten verschiedener Hölzer. *Angew. Chem.* **1931**, *44*, 788–791. [CrossRef]

5. Rodríguez, A.; Espinosa, E.; Domínguez-Robles, J.; Sánchez, R.; Bascón, I.; Rosal, A. Different solvents for organosolv pulping. In *Pulp and Paper Processing*; Kazi, S.N., Ed.; InTechOpen: London, United Kingdom, 2018; ISBN 978-1-78923-847-1.

6. Sixta, H. *Handbook of Pulp*; Wiley-VCH: Weinheim, Germany, 2006; ISBN 9783527309979.

7. Løhre, C.; Kleinert, M.; Barth, T. Organosolv extraction of softwood combined with lignin-to-liquid-solvolysis as a semi-continuous percolation reactor. *Biomass Bioenergy* **2017**, *99*, 147–155. [CrossRef]

8. Rossberg, C.; Janzon, R.; Saake, B.; Leschinsky, M. Effect of process parameters in pilot scale operation on properties of organosolv lignin. *Bioresources* **2019**, *14*. [CrossRef]

9. Nitsos, C.; Stoklosa, R.; Karnaouri, A.; Vörös, D.; Lange, H.; Hodge, D.; Crestini, C.; Rova, U.; Christakopoulos, P. Isolation and characterization of organosolv and alkaline lignins from hardwood and softwood biomass. *ACS Sustain. Chem. Eng.* **2016**, *4*, 5181–5193. [CrossRef]

10. Imlauer-Vedoya, C.M.; Vergara-Alarcón, P.; Area, M.C.; Revilla, E.; Felissia, F.E.; Villar, J.C. Fractionation of Pinus radiata wood by combination of steam explosion and organosolv delignification. *Maderas. Cienc. Tecnol.* **2019**, *21*, 587–598. [CrossRef]

11. Lesar, B.; Humar, M.; Hora, G.; Hachmeister, P.; Schmiedl, D.; Pindel, E.; Siika-aho, M.; Liitiä, T. Utilization of recycled wood in biorefineries: Preliminary results of steam explosion and ethanol/water organosolv pulping without a catalyst. *Eur. J. Wood Prod.* **2016**, *74*, 711–723. [CrossRef]

12. Baumeister, M.; Edel, E. Äthanol-Wasser-Aufschluss. *Das Papier* **1980**, *34*, V9–V18.

13. Schröter, M.C. Possible lignin reaction in the organocell pulping process. *Tappi J.* **1991**, *73*, 197–200.

14. Fullerton, T.J. Soda-anthraquinone pulping. The advantages of using oxygen-free conditions. *Tappi J.* **1979**, *62*, 55–57.

15. Lindner, A.; Wegener, G. Characterization of lignins from organosolv pulping according to the organocell process. Part 1. Elemental analysis, nonlignin portions and functional groups. *J. Wood Chem. Technol.* **1988**, *8*, 323–340. [CrossRef]

16. Lindner, A.; Wegener, G. Characterization of lignins from organosolv pulping according to the organocell process. Part 2. Residual lignins. *J. Wood Chem. Technol.* **1989**, *9*, 443–465. [CrossRef]

17. Lindner, A.; Wegener, G. Characterization of lignins from organosolv pulping according to the organocell process. Part 3. Permanganate oxidation and thioacidolysis. *J. Wood Chem. Technol.* **1990**, *10*, 331–350. [CrossRef]

18. Lindner, A.; Wegener, G. Characterization of lignins from organosolv pulping according to the organocell process. Part 4. Molecular weight determination and investigation of fractions isolated by GPC. *J. Wood Chem. Technol.* **1990**, *10*, 351–363. [CrossRef]

19. Da Ronch, F.; Caudullo, G.; Tinner, W.; de Rigo, D. Larix decidua and other larches in Europe: Distribution, habitat, usage and threats. In *European Atlas of Forest Tree Species*, 2016th ed.; San-Miguel-Ayanz, J., de Rigo, D., Caudullo, G., Durrant, T.H., Mauri, A., Eds.; Publication Office of the European Union: Luxembourg, 2016; ISBN 978-92-79-36740-3.

20. Crestini, C.; Lange, H.; Sette, M.; Argyropoulos, D.S. On the structure of softwood Kraft lignin. *Green Chem.* **2017**, *19*, 4104–4121. [CrossRef]

21. Hochegger, M.; Cottyn-Boitte, B.; Cézard, L.; Schober, S.; Mittelbach, M. Influence of ethanol organosolv pulping conditions on physicochemical lignin properties of European larch. *Int. J. Chem. Eng.* **2019**, *2019*, 1–10. [CrossRef]

22. Sluiter, A.; Hames, B.; Ruiz, R.; Scarlata, C.; Sluiter, J.; Templeton, D.; Crocker, D. Determination of Structural Carbohydrates and Lignin in Biomass. Technical Report NREL/TP-510-42618; 2012. Available online: https://www.nrel.gov/docs/gen/fy13/42618.pdf (accessed on 26 November 2019).

23. Sjöström, E. *Wood Chemistry. Fundamentals and Applications*, 2nd ed.; Academic Press: San Diego, CA, USA, 1993; ISBN 9780126474817.

24. Carlson, R.; Carlson, J.E. *Design and Optimization in Organic Synthesis*, 2nd ed.; Elsevier: Amsterdam, The Netherlands, 2005; ISBN 9780080455273.

25. Tran, A. Effect of cooking temperature on Kraft pulping of hardwood. *Tappi J.* **2002**, *1*, 13–19.

26. Yang, R.; Lucia, L.; Ragauskas, A.; Jameel, H. Oxygen delignification chemistry and its impact on pulp fibers. *J. Wood Chem. Technol.* **2003**, *23*, 13–29. [CrossRef]

27. Borrega, M.; Larsson, P.T.; Ahvenainen, P.; Ceccherini, S.; Maloney, T.; Rautkari, L.; Sixta, H. Birch wood pre-hydrolysis vs pulp post-hydrolysis for the production of xylan-based compounds and cellulose for viscose application. *Carbohydr. Polym.* **2018**, *190*, 212–221. [CrossRef]

28. Ehman, N.; Tarrés, Q.; Delgado-Aguilar, M.; Vallejos, M.; Felissia, F.; Area, M.; Mutjé, P. From pine sawdust to cellulose nanofibres. *Cell Chem. Technol.* **2016**, *50*, 361–367.

29. Liu, J.; Korpinen, R.; Mikkonen, K.S.; Willför, S.; Xu, C. Nanofibrillated cellulose originated from birch sawdust after sequential extractions: A promising polymeric material from waste to films. *Cellulose* **2014**, *21*, 2587–2598. [CrossRef]

30. Hörhammer, H. A Larch Biorefinery Producing Pulp and Lactic Acid. Ph.D. Thesis, Aalto University, Espoo, Finland, 2014.

31. McDonough, T.J. The chemistry of organosolv delignification. *Tappi J.* **1993**, *76*, 186–193.

32. Hu, F.; Jung, S.; Ragauskas, A. Pseudo-lignin formation and its impact on enzymatic hydrolysis. *Bioresour. Technol.* **2012**, *117*, 7–12. [CrossRef] [PubMed]

33. Shinde, S.D.; Meng, X.; Kumar, R.; Ragauskas, A.J. Recent advances in understanding the pseudo-lignin formation in a lignocellulosic biorefinery. *Green Chem.* **2018**, *20*, 2192–2205. [CrossRef]

34. Burkhardt, S.; Kumar, L.; Chandra, R.; Saddler, J. How effective are traditional methods of compositional analysis in providing an accurate material balance for a range of softwood derived residues? *Biotechnol. Biofuels* **2013**, *6*, 90. [CrossRef]

35. Baumberger, S.; Abaecherli, A.; Fasching, M.; Gellerstedt, G.; Gosselink, R.; Hortling, B.; Li, J.; Saake, B.; de Jong, E. Molar mass determination of lignins by size-exclusion chromatography: Towards standardisation of the method. *Holzforschung* **2007**, *61*, 205. [CrossRef]

36. Gosselink, R.J.A.; Abächerli, A.; Semke, H.; Malherbe, R.; Käuper, P.; Nadif, A.; van Dam, J.E.G. Analytical protocols for characterisation of sulphur-free lignin. *Ind. Crops Prod.* **2004**, *19*, 271–281. [CrossRef]

37. Bauer, S.; Sorek, H.; Mitchell, V.D.; Ibáñez, A.B.; Wemmer, D.E. Characterization of miscanthus giganteus lignin isolated by ethanol organosolv process under reflux condition. *J. Agric. Food Chem.* **2012**, *60*, 8203–8212. [CrossRef]

38. Constant, S.; Wienk, H.L.J.; Frissen, A.E.; Peinder, P.D.; Boelens, R.; van Es, D.S.; Grisel, R.J.H.; Weckhuysen, B.M.; Huijgen, W.J.J.; Gosselink, R.J.A.; et al. New insights into the structure and composition of technical lignins: A comparative characterisation study. *Green Chem.* **2016**, *18*, 2651–2665. [CrossRef]
39. Zhu, M.-Q.; Wen, J.-L.; Su, Y.-Q.; Wei, Q.; Sun, R.-C. Effect of structural changes of lignin during the autohydrolysis and organosolv pretreatment on Eucommia ulmoides Oliver for an effective enzymatic hydrolysis. *Bioresour. Technol.* **2015**, *185*, 378–385. [CrossRef]
40. Gilarranz, M.A.; Rodríguez, F.; Oliet, M. Lignin behavior during the autocatalyzed methanol pulping of eucalyptus globulus changes in molecular weight and functionality. *Holzforschung* **2000**, *54*, 373–380. [CrossRef]
41. Pan, X.; Kadla, J.F.; Ehara, K.; Gilkes, N.; Saddler, J.N. Organosolv ethanol lignin from hybrid poplar as a radical scavenger: Relationship between lignin structure, extraction conditions, and antioxidant activity. *J. Agric. Food Chem.* **2006**, *54*, 5806–5813. [CrossRef] [PubMed]
42. Gordobil, O.; Moriana, R.; Zhang, L.; Labidi, J.; Sevastyanova, O. Assesment of technical lignins for uses in biofuels and biomaterials: Structure-related properties, proximate analysis and chemical modification. *Ind. Crops Prod.* **2016**, *83*, 155–165. [CrossRef]
43. Argyropoulos, D.S. 31P NMR in wood chemistry: A review of recent progress. *Res. Chem. Intermed.* **1995**, *21*, 373. [CrossRef]
44. Balakshin, M.; Capanema, E. On the quantification of lignin hydroxyl groups with 31P and 13C NMR spectroscopy. *J. Wood Chem. Technol.* **2015**, *35*, 220–237. [CrossRef]
45. Yang, X.; Li, N.; Lin, X.; Pan, X.; Zhou, Y. Selective cleavage of the aryl ether bonds in lignin for depolymerization by acidic lithium bromide molten salt hydrate under mild conditions. *J. Agric. Food Chem.* **2016**, *64*, 8379–8387. [CrossRef]
46. Lai, C.; Tu, M.; Shi, Z.; Zheng, K.; Olmos, L.G.; Yu, S. Contrasting effects of hardwood and softwood organosolv lignins on enzymatic hydrolysis of lignocellulose. *Bioresour. Technol.* **2014**, *163*, 320–327. [CrossRef]
47. Zhao, Y.; Tagami, A.; Dobele, G.; Lindström, M.E.; Sevastyanova, O. The impact of lignin structural diversity on performance of cellulose nanofiber (CNF)-starch composite films. *Polymers* **2019**, *11*, 538. [CrossRef]
48. Koumba-Yoya, G.; Stevanovic, T. Study of organosolv lignins as adhesives in wood panel production. *Polymers* **2017**, *9*, 46. [CrossRef]
49. Huang, F.; Singh, P.M.; Ragauskas, A.J. Characterization of milled wood lignin (MWL) in loblolly pine stem wood, residue, and bark. *J. Agric. Food Chem.* **2011**, *59*, 12910–12916. [CrossRef] [PubMed]
50. Toloue Farrokh, N.; Suopajärvi, H.; Mattila, O.; Umeki, K.; Phounglamcheik, A.; Romar, H.; Sulasalmi, P.; Fabritius, T. Slow pyrolysis of by-product lignin from wood-based ethanol production—A detailed analysis of the produced chars. *Energy* **2018**, *164*, 112–123. [CrossRef]
51. Kubo, S.; Kadla, J.F. Poly(ethylene oxide)/organosolv lignin blends: Relationship between thermal properties, chemical structure, and blend behavior. *Macromolecules* **2004**, *37*, 6904–6911. [CrossRef]
52. Vroom, K.E. The H factor: A means of expressing cooking times and temperatures as a single variable. *Pulp Paper Mag. Can.* **1957**, *58*, 228–231.
53. Zinovyev, G.; Sumerskii, I.; Rosenau, T.; Balakshin, M.; Potthast, A. Ball milling's effect on pine milled wood lignin's structure and molar mass. *Molecules* **2018**, *23*, 2223. [CrossRef]
54. Björkman, A. Studies on finely divided wood. Part 1. Extraction of lignin with neutral solvents. *Sven Papperstidn* **1956**, *59*, 477–485.
55. Janson, J. Calculation of the polysaccharide composition of wood and pulp. *Paperi Ja Puu* **1970**, *52*, 323–329.

© 2019 by the authors. Licensee MDPI, Basel, Switzerland. This article is an open access article distributed under the terms and conditions of the Creative Commons Attribution (CC BY) license (http://creativecommons.org/licenses/by/4.0/).

Article

Ca-based Catalysts for the Production of High-Quality Bio-Oils from the Catalytic Co-Pyrolysis of Grape Seeds and Waste Tyres

Olga Sanahuja-Parejo, Alberto Veses, José Manuel López, Ramón Murillo, María Soledad Callén and Tomás García *

Instituto de Carboquímica (ICB-CSIC), C/ Miguel Luesma Castán, 50018 Zaragoza, Spain;
osanahuja@icb.csic.es (O.S.-P.); a.veses@icb.csic.es (A.V.); jmlopez@icb.csic.es (J.M.L.);
ramon.murillo@csic.es (R.M.); marisol@icb.csic.es (M.S.C.)
* Correspondence: tomas@icb.csic.es

Received: 31 October 2019; Accepted: 22 November 2019; Published: 26 November 2019

Abstract: The catalytic co-pyrolysis of grape seeds and waste tyres for the production of high-quality bio-oils was studied in a pilot-scale Auger reactor using different low-cost Ca-based catalysts. All the products of the process (solid, liquid, and gas) were comprehensively analysed. The results demonstrate that this upgrading strategy is suitable for the production of better-quality bio-oils with major potential for use as drop-in fuels. Although very good results were obtained regardless of the nature of the Ca-based catalyst, the best results were achieved using a high-purity CaO obtained from the calcination of natural limestone at 900 °C. Specifically, by adding 20 wt% waste tyres and using a feedstock to CaO mass ratio of 2:1, a practically deoxygenated bio-oil (0.5 wt% of oxygen content) was obtained with a significant heating value of 41.7 MJ/kg, confirming its potential for use in energy applications. The total basicity of the catalyst and the presence of a pure CaO crystalline phase with marginal impurities seem to be key parameters facilitating the prevalence of aromatisation and hydrodeoxygenation routes over the de-acidification and deoxygenation of the vapours through ketonisation and esterification reactions, leading to a highly aromatic biofuel. In addition, owing to the CO_2-capture effect inherent to these catalysts, a more environmentally friendly gas product was produced, comprising H_2 and CH_4 as the main components.

Keywords: co-pyrolysis; biomass; waste tyres; bio-oils; Ca-based catalyst; auger reactor

1. Introduction

The bio-oil produced from lignocellulosic biomass pyrolysis has outstanding potential as a renewable source of green energy and chemicals [1]. It should also be mentioned that several companies have implemented the commercial stages for bio-oil production [2]. However, an attractive market for this liquid product is not yet available [3], and it is still necessary to facilitate the standardisation and commercialisation of these liquid bio-oils [4]. Interestingly, the first insights into the application of this liquid as a drop-in fuel have shown great potential, increasing its market opportunities [5,6]. However, the complex nature of the bio-oil poses challenges for its competitive incorporation into the current energy market, particularly given its excessive content in reactive oxygen compounds, which affect both its handling and storage, and therefore, limiting its value for consideration as a drop-in fuel in current energy infrastructure. To overcome these limitations, upgrading methods such as hydrodeoxygenation, catalytic cracking, fast catalytic pyrolysis, hydrogenation, and molecular distillation are considered the most promising strategies to reach this aim [7]. Interestingly, due to their potential economic advantages and moderate operating conditions, catalytic processes focused on the in-situ cracking of this bio-oil are emerging as a potentially effective method to achieve this [8].

Zeolites are the main type of catalysts studied for this purpose [9,10]. In general terms, the main drawback of using these high-cost materials lies in catalyst deactivation, essentially the result of zeolite coking. Although zeolite regeneration may be carried out by char combustion in an integrated process [11,12], the high combustion temperatures required (above 700 °C), together with the presence of steam (approximately 10% in volume), would restrict their use in industrial processes due to their long-time deactivation by partial dealumination. Hence, the search for low-cost catalysts [13] must be considered as a promising economically viable strategy to compete in the current energy market. Basic catalysts such as MgO or calcium-based materials are emerging as attractive and relatively economical catalysts with which to carry out catalytic pyrolysis [13–15]. In this regard, Kalogiannis et al. [16] evaluated the fast catalytic pyrolysis of commercial lignocellulosic biomass at a representative scale using a circulating fluidised-bed facility with MgO catalysts and commercial zeolites. These authors found that ketonisation and aldol condensation reactions were promoted by the basic sites of MgO, resulting in a hydrogen-rich bio-oil being produced. In another interesting study conducted by our research group in a pilot auger reactor plant, calcined calcite and calcined dolomite were evaluated as catalysts for use in a self-sustained energy system. The oxygen fraction and acidic character of the bio-oil were decreased in both cases, while pH and calorific values were increased. Unfortunately, these bio-oils still require a further upgrading step if they are to be contemplated as a real substitute for fossil fuels or even as a drop-in fuel.

New alternative strategies such as the incorporation of polymer residues into conventional pyrolysis processes have also been considered in recent years. Detailed information on the latest advances in this process can be found in different exhaustive reviews [17–19]. In particular, the addition of waste tyres [20–22] and different plastic waste such as polystyrene [23], polyethylene [24], and polypropylene [25] to lignocellulosic biomass was studied as a potential route to improving the properties of the bio-oil, both by decreasing oxygen content and increasing heating value [26]. In this case, waste plastics acted as hydrogen donors by promoting oligomerisation, cycling, and hydrodeoxygenation reactions. Again, there is general consensus, mainly owing to its high viscosity, water content, oxygen content, acidic nature, solid, and ash content, and non-homogenous nature, that a further upgrading step is still necessary to facilitate the incorporation of this bio-oil obtained by co-pyrolysis processes into the energy market, especially for use as drop-in fuels.

Thus, a combined strategy involving the catalytic co-pyrolysis of lignocellulosic biomass with polymer waste has emerged as a very promising solution for the production of drop-in fuels in a simple, one-step process. Recent progress in this strategy can be found in different extensive reviews [27–29]. The catalysts successfully studied for this purpose were mainly zeolites (HY and HZSM-5), mesoporous materials (MCM-41 and SBA-15), alumina, spent Fluid Catalytic Cracking (FCC) catalysts, and a number of minerals, such as bentonite. Interestingly, our research group was also able to demonstrate the potential of this process to facilitate the production of drop-in fuels in a laboratory-scale fixed-bed reactor. Particularly, the catalytic co-pyrolysis of grape seeds (GSs), as a form of agricultural waste, and waste tyres (WTs) was successfully carried out using calcined limestone as the catalyst [30]. Thus, this combined strategy made it feasible to produce a deoxygenated and aromatic-rich drop-in fuel by applying one simple stage. More specifically, the positive effects attached to CaO were not only responsible for promoting the dehydration and dehydrogenation reactions of acids and phenols, but also for the production of a more environmentally friendly gas because of the in-situ CO_2-capture effect associated with CaO carbonation. Although the results of catalytic co-pyrolysis are very encouraging, most of the work in this field has been carried out in micro and lab-scale reactors [31–37], and the demonstration of this process on a larger scale is still limited [38]. Therefore, the study of this upgrading strategy to a larger scale (Technology Readiness Levels, TRL ≥ (5)) seems necessary as a first step towards the implementation of the catalytic co-pyrolysis processes of lignocellulosic biomass with waste plastics in future bio-refineries. In this sense, the use of auger reactors at TRL-5 have shown to be a robust and flexible option for pyrolysis processes [15,39–42]. However, further experimental research is required to verify the proposed approaches in a relevant environment. Specifically, to the best of our knowledge, there is no published data on catalytic co-pyrolysis using both of these feedstocks (grape seeds and waste tyres) on a relevant scale that could contribute to validating the potential of these upgraded bio-oils.

Thus, the aim of this work is to demonstrate the potential of the catalytic co-pyrolysis of waste tyres and grape seeds for the production of improved bio-oils in a relevant environment (TRL-5). For this purpose, the experiments were conducted in an Auger reactor pilot plant operating at atmospheric pressure. GSs and WTs (up to 20 wt%) were used as feedstock. Four Ca-based materials obtained from different companies were studied as low-cost commercial catalysts.

2. Results

2.1. Characterisation of Catalysts

The purity and crystallinity of each catalyst was verified by X-ray diffraction (XRD), see Figure 1. It is worth mentioning that while the Ca-based-1, Ca-based-2, and Ca-based-3 catalysts were obtained from limestone calcination at 900 °C, the Ca-based-4 catalyst was obtained from dolomite calcination at the same temperature. Accordingly, it can be observed that the Ca-based-1, Ca-based-2, and Ca-based-3 catalysts mainly presented the characteristic diffraction peaks of calcium oxide. These peaks were situated at 2θ 33°, 38°, 53°, 63°, and 68°, corresponding to the crystallographic planes (111), (200), (202), (311), and (222), respectively [43,44]. Diffraction patterns also showed diffraction peaks related to $Ca(OH)_2$, which could be associated with the fact that water was absorbed from the environment during handling [45], which is more significant in the case of the Ca-based-3 catalyst. The presence of magnesium calcite cannot be ruled out in this sample. Finally, CaO and MgO diffraction peaks were identified for the Ca-based-4 catalyst, in line with the use of dolomite as its raw mineral. Diffraction peaks could also be observed at 2θ 23°, 29°, 40°, and 48°, corresponding to the structure of magnesium calcite and highlighting that 900 °C was not a high enough temperature to ensure the total decomposition of this phase during the pre-calcination step. Finally, $Ca(OH)_2$ was also identified as a minor phase. It is worth mentioning that while these hydroxides were again transformed into the metal oxide under pyrolysis conditions, the opposite was true in the case of magnesium calcite.

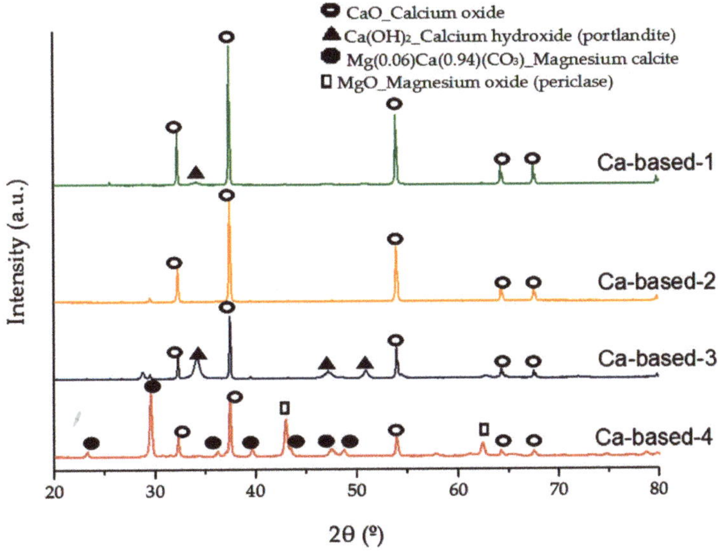

Figure 1. XRD patterns of the different Ca-based catalysts.

Table 1 shows the elemental composition of the different Ca-based catalysts as determined by inductively coupled plasma-optical emission spectrometry (ICP-OES). As expected, the main elements were Ca and Mg. The remaining elements found in the catalysts were identified in quantities lower than 1.0 wt% (Al, Fe, K, S, and Si). As could be expected, the highest MgO content was identified in the Ca-based-4 catalyst (47.6 wt%), while the remaining catalysts showed lower MgO content, between 0.33-0.86 wt%. The high purity of the Ca-based-1 catalyst should be pointed out since CaO was present in a very high percentage, 95 wt%. For the different catalysts, Ca content in the order from higher to lower was as follows: Ca-based-1 > Ca-based-2 > Ca-based-3 > Ca-based-4. Table 1 also shows that all the Ca-based catalysts displayed quite a low specific surface area, with this being slightly higher in the case of the catalyst obtained from dolomite. It is also worth highlighting that the average pore diameter of the Ca-based-4 catalyst ranged between 2 and 50 nm, which corresponded to a mesoporous structure according to the International Union of Pure and Applied Chemistry (IUPAC) classification. The remaining catalysts (Ca-based-1, Ca-based-2, Ca-based-3) had an average pore diameter >50 nm, which indicated a macroporous structure. With regards to porosity values, no significant differences were observed since values remained in the range of 42–53% for all the catalysts.

Table 1. ICP-OES and the structural characterisation of the Ca-based catalysts. BET and D_p determined by N_2 adsorption. Porosity was determined by Hg porosimetry.

	CaO (wt%)	MgO (wt%)	Al_2O_3 (wt%)	Fe_2O_3 (wt%)	K_2O (wt%)	SiO_2 (wt%)	BET (m^2/g)	D_p (nm)	Porosity (%)
Ca-based-1	95.1 ± 0.7	0.86 ± 0.03	0.14 ± 0.01	0.29 ± 0.02	0.02 ± 0.00	0.51 ± 0.02	< 2	520	50
Ca-based-2	89.8 ± 0.6	0.78 ± 0.03	0.14 ± 0.01	0.10 ± 0.01	0.15 ± 0.01	1.09 ± 0.04	5	428	42
Ca-based-3	79.8 ± 0.4	0.33 ± 0.02	0.14 ± 0.01	0.07 ± 0.00	0.04 ± 0.00	0.60 ± 0.00	6	192	53
Ca-based-4	47.6 ± 0.2	33.2 ± 0.10	0.08 ± 0.00	–	0.02 ± 0.00	0.24 ± 0.01	12	48	50

The acidity of the Ca-based catalysts was studied with the TPD-NH_3 technique, using a mass spectrometer as a detector, and values of desorbed $\mu molNH_3$/g were calculated by dividing the area of the peak between the mass of the treated sample. Those values were included in Table 2.

Table 2. Amount of CO_2 and NH_3 desorbed from the TPD-CO_2 and TPD-NH_3 profiles of the different catalysts.

	Ca-based-1	Ca-based-2	Ca-based-3	Ca-based-4
$molNH_3$/g (T Peak (°C))	0.2 (630)	0.1 (650)	0.3 (690)	0.8 (750)
$mmolCO_2$/g (T Peak (°C))	0.04 (550)	0.10 (562)	0.10 (592)	0.11 (598)

Figure 2 shows a quantitative analysis of the desorbed NH_3 based on the fragment $m/z = 15$ for the different Ca-based catalysts. Values were quite small (ranging between 0.1 and 0.8 $\mu molNH_3$/g), in line with the marginal acidity also observed by other researchers using similar catalysts [13]. The Ca-based-4 catalyst was the solid with the highest amount of acid sites, suggesting that these strong acid sites could be linked to the presence of magnesium calcite. Accordingly, a slightly lower acidity can be seen for the Ca-based-3 catalyst, while the Ca-based-1 and Ca-based-2 catalysts presented the lowest values.

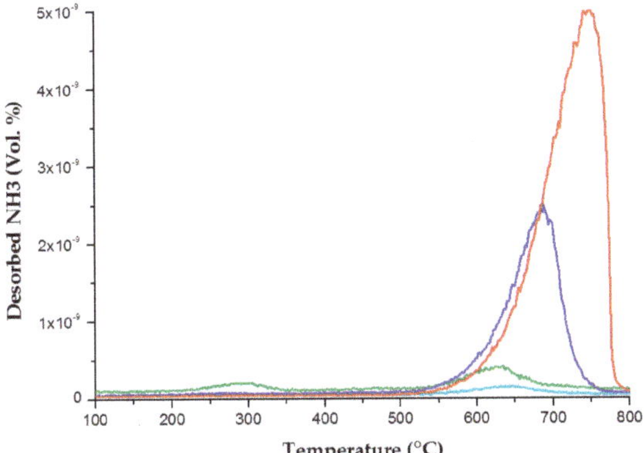

Figure 2. Quantitative analysis of the desorbed NH_3 referenced to the fragment $m/z = 15$. Ca-based-1 (green), Ca-based-2 (cyan), Ca-based-3 (blue), and Ca-based-4 (red).

TPD-CO_2 technique was performed to know the basicity of the Ca-based catalysts. These basic sites can play an important role as active centres in such specific deoxygenation reactions [46] as esterification and ketonisation. As an example, and for simplification, the TPD-CO_2 profile of the Ca-based-1 catalyst was plotted in Figure 3. Similar strength of the basicity sites was observed for all the catalysts showing CO_2 desorption temperatures of 550–600 °C. More importantly, the Ca-based-1 catalyst showed the lowest amount of evolved CO_2, 0.04 mmolCO_2/g, whereas comparable higher values of 0.10–0.11 mmolCO_2/g were found for the other materials (see Table 2). The lower specific surface area observed in the Ca-based-1 catalyst could explain its lower total basicity [13,47].

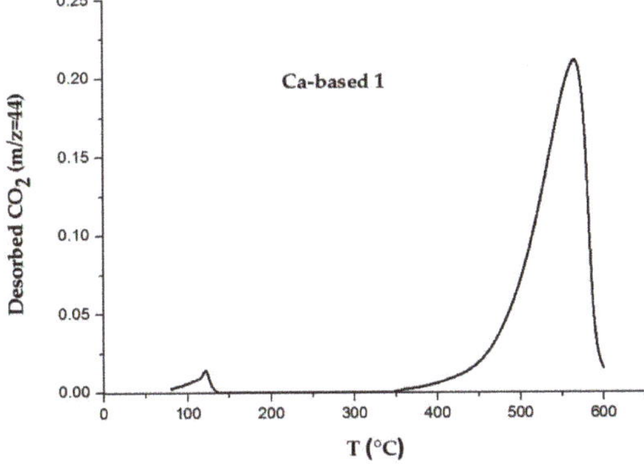

Figure 3. TPD-CO_2 profile of the Ca-based-1 catalyst (fragment $m/z = 44$).

2.2. Product Yields

To evaluate the importance of the addition of WTs on the distribution of the products, different mixtures of GSs and WTs (10 and 20 wt%) were treated in an auger pilot plant. Yields obtained from the pyrolysis of both raw materials are also shown in Table 3 for comparative purposes. Thus, the liquid yield derived from GSs was 42.5 wt%, while the solid and gas yields were 31.2 and 23.2 wt%, respectively. On the other hand, the pyrolysis of WTs generated a lower liquid yield (37.8 wt%) and a comparable gas fraction (23.5 wt%), leading to a higher solid yield (37.5 wt%). These values could be considered in line with other work related to WT pyrolysis in auger reactors [40]. According to these values, the addition of WTs to GSs follows an expected trend in the distribution of products. Thus, the liquid yield decreased with respect to GSs. Correspondingly, the solid yields were higher, and gas yields were close to those obtained in biomass pyrolysis and WTs pyrolysis.

Table 3. Product yields (liquid, solid, and non-condensable gas) and liquid phase distribution (organic and aqueous phases) obtained after the pyrolysis and co-pyrolysis experiments carried out in the Auger pilot plant. Catalytic co-pyrolysis experiments were carried out, adding 20 wt% of waste tyres. Values were determined as the average of two runs.

	Liquid (wt%)		Solid (wt%)			Gas	Balance	Phase Distribution (wt%)	
	Org.	Aq.	Char	Coke	CO_2			Org.	Aq.
GSs	16.0 ± 0.8	26.5 ± 1.3	31.2 ± 1.6	0	0	23.2 ± 1.2	96.9	37.7	62.3
WTs	37.8 ± 1.9	0.0 ± 0.0	37.5 ± 1.8	0	0	23.5 ± 1.2	98.8	100.0	0.0
GSs/WTs (90/10)	12.8 ± 0.6	20.2 ± 1.0	36.7 ± 1.2	0	0	28.9 ± 1.5	98.5	38.9	61.1
GSs/WTs (80/20)	17.4 ± 0.8	18.2 ± 0.9	35.4 ± 1.8	0	0	24.2 ± 1.2	95.1	48.8	51.2
Ca-based-1	15.7 ± 0.8	23.0 ± 1.2	34.6 ± 1.7	0.37	1.8	20.5 ± 1.0	95.2	40.6	59.4
Ca-based-2	15.8 ± 0.8	21.4 ± 1.1	37.7 ± 1.9	0.39	1.6	19.7 ± 1.0	96.6	42.4	57.6
Ca-based-3	15.7 ± 0.8	20.6 ± 1.0	37.2 ± 1.8	0.41	2.4	20.1 ± 1.0	96.4	43.2	56.8
Ca-based-4	16.8 ± 0.9	20.0 ± 1.0	37.6 ± 1.9	0.41	2.6	18.7 ± 0.9	96.2	45.6	54.4

With regard to GS pyrolysis, the liquid fraction resulting from the non-catalytic co-pyrolysis process was composed of two phases (aqueous and organic), which were easily separated by centrifugation. While the organic phase is the most interesting fraction for use in fuel and energy applications [48], the possibilities for valorising the aqueous phase are still limited, given that the production of chemical products such as H_2, ethanol or even light acids is the option attracting more interest. Accordingly, the main target of this study was focused on improving the yield and fuel properties of the organic phase. Interestingly, the yield to this fraction only increased with the highest WT percentage (20 wt%), reaching values up to 10% higher, approximately. This behaviour was also observed in previous co-pyrolysis work where the addition of polymer waste to the biomass pyrolysis process enhanced the yield to the organic fraction [18,27].

Several catalytic tests using the Ca-based-1 catalyst were carried out using GSs and WTs on their own. Since no significant improvement was evidenced, the resulting information was only included as supporting information (Tables S1 and S2). Subsequently, the catalytic co-pyrolysis of GSs and WTs using low-cost Ca-based materials was studied at TRL-5. We would point out here that a ratio (GSs + WTs) to catalyst of 2 was selected, since we had previously observed that higher catalyst ratios excessively promoted cracking reactions, leading to the formation of undesired heavy condensable organic compounds [49,50]. On the other hand, the use of lower catalyst proportions in the feedstock hardly affected process performance compared to the non-catalytic test. Table 3 compiles the result obtained for the different GS and WT catalytic co-pyrolysis tests. Liquid yields are observed to range between 36.8 and 38.0 wt%, slightly higher in comparison with non-catalytic experiments. Interestingly, organic fraction yields were only slightly lower in comparison with the non-catalytic tests, remaining in the same range of values for all the catalytic experiments. This fact could be explained by the promotion of dehydration reactions by Ca-based materials, eventually increasing the yield to the aqueous fraction. This tendency has also been observed by other researchers [51]. For these catalytic tests, the solid yield was separated into char, CO_2 (as carbonate), and coke yields (by TGA analysis). As can be observed in Table 3, the effect of CaO carbonation is evident after the experiments, while remarkably low values of coke yield (0.37–0.41 wt%) were achieved. Slightly higher carbonation was observed in those Ca-based materials where Magnesium calcite phase was observed by XRD since this phase could be decarbonated under TGA conditions. In addition, gas yields were slightly lower after the catalytic experiments in comparison with the non-catalytic tests.

2.3. Upgraded Bio-Oil: Analysis of the Organic Fraction Properties

Since the main target in this work was to improve the quality of the organic fraction, this product was comprehensively characterised. Table 4 summarises the main properties of this fraction. First, it can be seen that the organic phase obtained from GS pyrolysis had a remarkably high water content (5.09

wt%), whereas the water content of the WT pyrolytic oil was practically negligible. Correspondingly, the water content of the organic fraction was drastically reduced by adding WTs to the feedstock, especially at 20 wt% (1.07 wt%). Focusing on the catalytic experiments (see Table 4), the exceptional performance of Ca-based catalysts can be highlighted, with the water content in the organic fraction decreasing to values lower than 1 wt%. Other important physical properties of the organic phase as a liquid fuel, such as density and viscosity, were also determined [52]. While the organic fraction from GS pyrolysis is a viscous liquid (87.1 mPa.s) with higher density (1.30 g/mL) than commercial fuels [53,54], organic WTs pyrolytic oil can be considered similar to fossil fuel-derived oils, showing a viscosity of 3.2 mPa.s and a density of 0.87 g/mL. In this regard, positive effects on these physical properties can be directly obtained through to the addition of WT to the feedstock. Accordingly, viscosity and density values were reduced down to 21.3–16.3 mPa.s and 1.15–1.11 g/mL, respectively, depending on the WT percentage in the feedstock. These findings were in line with the results found by Cao et al. [55], which showed that there was an improvement in these physical properties when the co-pyrolysis of pine sawdust and waste tyres was carried out in a fixed-bed reactor. Significantly, these positive effects were further improved when Ca-based catalysts were incorporated into the process [56]. Thus, a remarkable reduction in the parameters of viscosity and density was achieved, leading to values lower than 6 mPa.s and 1 g/mL, respectively, for all the catalysts. Specifically, the Ca-based-1 catalyst produced the organic fraction with the lowest water content (0.62 wt%) and the best physical properties as a fuel (υ = 3.5 mPa.s; ρ = 0.91 (g/mL), which were seen as very remarkable achievements.

Table 4. Organic fraction properties after the pyrolysis, co-pyrolysis, and catalytic co-pyrolysis processes. Deox.: deoxygenation; LHV: lower heating value.

	Properties				Elemental Analysis (wt%)						
	H2O (wt%)	pH	υ (mPa.s)	ρ (g/mL)	C	N	H	S	O	Deox. (%)	LHV (MJ/kg)
GSs	5.09 ± 0.7	8.0	87.1 ± 2.6	1.30	77.4 ± 0.8	2.5 ± 0.2	8.3 ± 0.2	0.2 ± 0.02	11.6 ± 0.6	–	34.6 ± 1.0
WTs	<0.3 ± 0.0	8.3	3.2 ± 0.2	0.87	89.9 ± 1.0	1.1 ± 0.1	7.6 ± 0.1	0.7 ± 0.03	0.7 ± 0.1	–	42.4 ± 1.3
GSs/WTs (90/10)	4.83 ± 0.6	9.6	21.3 ± 1.2	1.15	80.6 ± 0.9	2.8 ± 0.2	8.3 ± 0.2	0.6 ± 0.03	7.8 ± 0.5	32.8	36.0 ± 1.4
GSs/WTs (80/20)	1.07 ± 0.1	9.5	16.3 ± 0.8	1.11	83.6 ± 0.9	2.6 ± 0.2	9.5 ± 0.2	0.4 ± 0.02	3.9 ± 0.2	66.4	38.78 ± 1.2
Ca-based-1	0.62 ± 0.1	9.1	3.5 ± 0.2	0.91	87.0 ± 0.8	2.5 ± 0.1	9.6 ± 0.2	0.4 ± 0.02	0.5 ± 0.1	95.7	40.7 ± 1.5
Ca-based-2	0.70 ± 0.1	8.9	4.7 ± 0.2	0.95	85.5 ± 0.7	2.7 ± 0.2	10.2 ± 0.3	0.6 ± 0.03	1.1 ± 0.2 0.3	90.5	40.7 ±1.5
Ca-based-3	0.90 ± 0.1	9.1	5.5 ± 0.3	0.97	85.0 ± 0.7	2.7 ± 0.2	9.9 ± 0.2	0.5 ± 0.02	1.9 ± 0.3	83.6	40.3 ±1.4
Ca-based-4	0.76 ± 0.1	9.1	5.6 ± 0.3	0.98	84.9 ± 0.4	2.8 ± 0.2	10.0 ± 0.1	0.6 ± 0.03	1.8 ± 0.2	84.5	40.1 ± 1.4

Another important parameter to measure the quality of the liquid is acidity. In this case, acidity was measured by pH determination. Interestingly, the GS-derived organic fraction presented a pH value of 8, preventing the presence of the corrosion issues associated with the handling and storage of the highly acidic bio-oils usually obtained from lignocellulosic biomass (pH 3–4). This value was in the range of other pyrolytic oils obtained from GSs, probably due to the presence of a remarkable amount of basic N-derived compounds in the feedstock [26,30]. The pH value for WT-derived oil was quite similar (8.3). It is worth highlighting that both co-pyrolysis and catalytic co-pyrolysis bio-oils only entailed a slight increase in pH values, while still preserving an appropriate value (8.9–9.6) for handling and storage purposes.

In relation to the elemental analysis, it could be also seen that the bio-oil derived from the pyrolysis of GSs contained 11.6 wt% of oxygen. Although it could be still considered a remarkable oxygen content for its further direct utilisation as a drop-in fuel, this value was quite low when compared to those found in bio-oils produced from the pyrolysis of other types of lignocellulosic biomass [14,57,58]. This result confirmed that those lignocellulosic agricultural wastes were exceptional alternatives to be used in the pyrolysis process. On the contrary, pyrolytic WT oil presented insignificant amounts of oxygen (<1 wt%). Obviously, the incorporation of WT led to a better-quality bio-oil in relation to lower oxygen content. It should be highlighted that this improvement was always better than that expected according to the rule of mixtures. Specifically, it was possible to reduce the oxygen content down to 3.9 wt% by adding up to 20 wt% of WTs. The effect of catalyst addition on the deoxygenation process was more apparent. Oxygen contents in the upgraded bio-oils were always lower than 2 wt%, reaching the lowest value of 0.5 wt% with the Ca-based-1 catalyst. In this sense, although the formation of H_2O soluble compounds cannot totally be ruled out, retention of CO_2 and H_2O seems to be enhanced by these catalysts. At this point, it should be pointed out that the elemental composition of the aqueous fraction was also determined (Table S3, supplementary data). The values of carbon content determined were quite similar and were in the range of 4.14–6.53 wt%, where small differences can be attributed to experimental error. Thus, these results suggest that a great effect of the catalysts is produced in improving the organic fraction since no additional organic components seem to be dissolved in the aqueous fraction.

Hence, a practically deoxygenated organic fraction was produced (values of deoxygenation in range of 83.6–95.7%), entailing significant lower heating values (LHVs) of about 40–41 MJ/kg. Although all the catalysts led to very good results, again, it is worth highlighting the performance of the Ca-based-1 catalyst, which produced the best organic oil in terms of lower oxygen content and LHV.

2.4. Non-Condensable Gas Composition

Table 5 summarises the non-condensable gas composition after pyrolysis, co-pyrolysis, and catalytic co-pyrolysis reactions. It can be observed that, in the case of GS pyrolysis, the composition of this gas was rich in CO_2 (10.5 g/100 g feedstock), and also in CO (3.1 g/100 g feedstock) and CH_4 (5.5 g/100 g feedstock). The remaining C_2-C_4 hydrocarbons were present as marginal content. In the case of pyrolysis of WTs, hydrogen and methane were the main gases, since CO and CO_2 were produced in lower amounts, as could be expected by the low oxygen content present in WTs. When WTs were added to the feed, there was a decrease in CO_2 in the gas stream and a slight increase in CH_4 and H_2. As a consequence, a more valuable non-condensable gas in terms of LHV was produced as the WT ratio in the feedstock was increased from 10 to 20 wt%, 28.5, and 30.3 MJ/Nm^3, respectively. By adding the catalyst to the GS/WT mixture, significant positive differences were seen in the composition of the gas. First, it can be highlighted the meaningful increase regarding the production of hydrogen and methane, and the significant reduction in the CO_2 emitted. This reduction in CO_2 can be explained assuming that CaO is a CO_2 sorbent. This reaction favours the production of hydrogen caused by the enhancement of the water gas shift reaction, which in turn affects the reaction of methane reforming [15]. Accordingly, a direct relationship can be found between the amount of CO_2 in the gas fraction and the CaO content in the catalysts, with the Ca-based-1 and Ca-based-4 catalysts leading to the lowest and highest amounts of CO_2 in the gas fraction, respectively. The poor performance of the Ca-based-4 catalyst was not surprising since it is well known that MgO in Ca-based sorbents does not significantly contribute to CO_2 capture at such temperature [15]. Additionally, some Magnesium calcite phase was detected, which cannot contribute to this process. Another positive effect of incorporating Ca-based catalysts was found in the increase in heating value. Thus, by means of the addition of these types of solids, it was possible to enhance heating values up to 40 MJ/Nm^3, approximately. Interestingly, emissions of H_2S in the catalytic co-pyrolysis processes were comparable to those obtained during the GS pyrolysis, which may be associated with the desulphuration capacity of Ca-based materials.

Table 5. Lower heating value (LHV) (MJ/Nm^3) and non-condensable gas composition (g/100 g feedstock) produced after the pyrolysis, co-pyrolysis, and catalytic co-pyrolysis processes.

Experiment	H_2	CO_2	CO	CH_4	C_2H_6	C_2H_4	C_3-C_4	H_2S	LHV (MJ/Nm^3)
GSs	0.21 ± 0.01	10.5 ± 0.60	3.1 ± 0.16	5.5 ± 0.31	0.8 ± 0.05	1.1 ± 0.06	1.9 ± 0.10	0.02	24.1 ± 1.2
WTs	0.34 ± 0.03	0.65 ± 0.06	0.21 ± 0.02	10.2 ± 0.60	2.1 ± 0.16	1.7 ± 0.14	8.1 ± 0.65	0.25	53.6 ± 2.0
GSs/WTs(90/10)	0.21 ± 0.02	7.9 ± 0.40	2.5 ± 0.13	5.3 ± 0.32	0.9 ± 0.10	1.1 ±.0.09	2.5 ± 0.25	0.04	28.5 ± 1.4
GSs/WTs(80/20)	0.22 ± 0.02	7.07 ± 0.40	2.2 ± 0.12	5.5 ± 0.35	0.9 ± 0.10	1.1 ± 0.09	2.7 ± 0.35	0.04	30.3 ± 1.5
Ca-based-1	0.31 ± 0.03	2.42 ± 0.12	3.2 ± 0.16	7.3 ± 0.45	1.5 ± 0.12	1.2 ± 0.12	4.3 ± 0.40	0.00	39.3 ± 1.8
Ca-based-2	0.33 ± 0.03	1.7 ± 0.09	3.0 ± 0.15	7.1 ± 0.40	1.3 ± 0.11	1.1 ± 0.10	4.2 ± 0.35	0.00	39.8 ± 1.8
Ca-based-3	0.33 ± 0.02	1.9 ± 0.10	3.3 ± 0.17	7.3 ± 0.40	1.4 ± 0.14	1.1 ± 0.09	4.5 ± 0.42	0.02	39.9 ± 1.9
Ca-based-4	0.19 ± 0.01	5.34 ± 0.35	3.2 ± 0.16	6.4 ± 0.38	1.3 ± 0.12	1.0 ± 0.11	4.1 ± 0.34	0.02	37.0 ± 1.7

2.5. GC-MS

A semi-quantitative identification of the compounds (relative area percentage, r.a.%) found in the organic phase was performed using gas chromatography-mass spectrometry (GC-MS). Table 6 summarises the chemical composition of the organic fractions derived from the pyrolysis, co-pyrolysis, and catalytic co-pyrolysis processes. The chemical composition of the organic fraction obtained after the GS pyrolysis was composed mainly of oxygenated compounds (phenols, esters, fatty acids, aromatics, ketones, cyclic hydrocarbons, olefins, and other oxygenated compounds) [59,60]. It should be pointed out that these compounds were already identified in previous work as the main oxygenated compounds in the organic fraction of the bio-oil obtained from the pyrolysis of grape seeds [24,30]. Interestingly, a noteworthy amount of aromatic compounds was also identified (31.6%). On the other hand, the organic fraction of the pyrolysis liquid of WTs presented a prevailing fraction of aromatics (more than 90 r.a.%), followed by a significant proportion of cyclic hydrocarbons (6.2 r.a.%). The more relevant aromatic components of the liquid fraction were also identified, which were benzene and benzene-derived compounds (toluene, xylene, ethyl-benzene, and styrene) [61]. A list of the main identified components can be found in the supporting information in Tables S4 and S5. It can be observed in Table 6 that in the case of GS/WT co-pyrolysis, a considerable increase in cyclic and aromatic hydrocarbon production was achieved in comparison with GS pyrolysis. A remarkable reduction in oxygenated compounds, mainly esters but also phenols and fatty acids, was observed at the same time. These tendencies may be associated with the promotion of hydrodeoxygenation reactions due to the H_2-donor effect of the WTs [3]. For more information on the compounds identified after the co-pyrolysis process, see Table S6, supplementary data.

Table 6. Chemical composition of bio-oils derived from the pyrolysis, co-pyrolysis, and catalytic co-pyrolysis processes as the percentage of the relative area (r.a.%). HC: Hydrocarbons; OC: Oxygenated Compounds. Catalytic co-pyrolysis experiments were carried out, adding 20 wt% of waste tyres.

Experiment	Aromatics	Olefins	Linear HC	Cyclic HC	Phenols	Esters	Ketones	Fatty Acids	Others OC
GSs	31.6 ± 2.5	7.6 ± 0.6		–	18.8 ± 1.5	28.7 ± 2.3	0.8 ± 0.3	7.4 ± 0.3	5.1 ± 0.2
WTs	91.8 ± 2.5	–		8.2 ± 0.5	–	–	–	–	–
GSs/WTs (90/10)	63.3 ± 2.6	2.9 ± 0.3	0.4 ± 0.2	23.0 ± 1.2	6.4 ± 0.8	0.9 ± 0.2	0.1 ± 0.0	1.1 ± 0.2	1.9 ± 0.2
GSs/WTs (80/20)	64.3 ± 2.8	2.8 ± 0.3	0.4 ± 0.2	24.0 ± 1.5	4.9 ± 0.5	0.8 ± 0.2	0.1 ± 0.0	1.1 ± 0.2	1.8 ± 0.2
Ca-based-1	70.9 ± 2.9	2.3 ± 0.2	0.6 ± 0.3	16.0 ± 0.8	1.3 ± 0.3	1.5 ± 0.2	2.0 ± 0.1	1.7 ± 0.3	3.6 ± 0.4
Ca-based-2	60.4 ± 2.2	4.1 ± 0.5	1.8 ± 0.4	16.1 ± 0.9	2.8 ± 0.6	2.6 ± 0.3	5.0 ± 0.7	2.5 ± 0.4	4.1 ± 0.5
Ca-based-3	54.5 ± 2.0	5.7 ± 0.8	2.8 ± 0.5	18.0 ± 1.0	2.8 ± 0.6	3.2 ± 0.4	6.2 ± 0.8	2.7 ± 0.5	4.0 ± 0.4
Ca-based-4	58.3 ± 2.2	5.0 ± 0.6	1.6 ± 0.3	18.3 ± 1.0	3.9 ± 0.3	2.9 ± 0.3	5.0 ± 0.7	2.4 ± 0.4	4.0 ± 0.4

The chemical composition of the organic fractions obtained in the catalytic co-pyrolysis experiments is also shown in Table 6. A different composition to that analysed in the bio-oil of the GS pyrolysis run was clearly identified. As in the case of the co-pyrolysis test, a notable reduction in the phenol, ester, and fatty acid contents were achieved, while aromatics, olefins, linear paraffins, and cyclic hydrocarbons were significantly increased. Therefore, this behaviour could be again related to the prevalence of aromatisation and hydrodeoxygenation reactions owing to the significant amount of H_2 produced by both the thermal cracking of WTs and the water gas shift reaction assisted by CaO. These results were in line with the low oxygen content and the marginal amount of water found in these liquids. With regard to the effect of the catalyst on process performance, a slightly different organic fraction composition was evidenced depending on the Ca-based material. There was generally observed to be a reduction in the cyclic-hydrocarbon composition compared to the non-catalytic process. This fact, together with a remarkable increase in ketones, pointed to the cascade of reactions associated with the hydrodeoxygenation of ketones in order to favour cyclic-hydrocarbon production not being fully accomplished. Accordingly, another important deoxygenation route during the catalytic co-pyrolysis experiments seemed to be decarboxylation. This behaviour was in line with previously published works showing that basic metal oxides, including CaO, could act as promoters of this reaction [62]. Interestingly, these ketone compounds could also be considered valuable components of biofuels since they are quite stable during handling and storage. We would like to highlight the results found with the Ca-based-1 catalyst in terms of aromatic content, as it was the only catalyst that led to a higher aromatic content than the non-catalytic co-pyrolysis run. For this catalyst, the aromatisation of olefins and hydrodeoxygenation of phenols may play a key role in the production of aromatics, particularly due to the intensification of H_2 production and the promotion of dehydration and cracking reactions in this catalyst. Given that the total basicity of the Ca-based-1 catalyst determined by the CO_2-TPD data was significantly lower than that determined for the other Ca-based catalysts, and the fact that this catalyst showed a pure CaO crystalline phase with marginal amount of impurities, it could be assumed that these properties of these catalysts could be key parameters facilitating the prevalence of aromatisation and hydrodeoxygenation routes over de-acidification and deoxygenation of the organic fraction through both ketonisation and esterification reactions. Accordingly, the higher content in ketones and esters obtained for the remaining Ca-based catalysts confirmed these assumptions. The main compounds identified after the catalytic process are shown in the supplementary data (Table S7).

At this point, it should be mentioned that although no regeneration cycles were performed in this study, a moderate deactivation of the catalyst by partial carbonation was expected according to previous tests at the pilot scale considering this type of catalysts [14]. This fact would make necessary the incorporation of a purge and input of fresh catalyst in order to maintain bio-oil quality. Fortunately, this purge would only slightly decrease the energy integration of the process, and the economic impact would also be moderate since this is inexpensive and widely available solid.

2.6. Char Characterisation

The solid fraction was also characterised in order to evaluate its potential as a solid fuel. Char produced by GS pyrolysis has a lower heating value (LHV) of 26.9 MJ/kg, whereas the char produced by WT pyrolysis showed an LHV of 30.9 MJ/kg. These values could be expected owing to the high carbon content in WT char fraction, and they were also in line with previous results [30]. At this point, it should be pointed out that the LHVs of the co-pyrolysis and catalytic co-pyrolysis chars were those theoretically expected, approximately 28 MJ/kg. These results confirmed the high potential of this product to be used for further energy purposes and to be used to support the heat requirements of the integrated process. Moreover, and in line with previous results in the fixed-bed reactor, a remarkable decrease in the sulphur content (0.5 wt%) was also achieved for all catalysts in comparison with the theoretically expected value (2.0 wt%). This could be expected since CaO is considered an ideal active solid for H_2S adsorption, forming CaS, and H_2O as products [63]. Thus, char desulphuration enhanced by the addition of Ca-based catalysts was also confirmed at this scale, as previously observed in the fixed bed reactor [30].

3. Materials and Methods

3.1. Biomass, Waste Tyres, and Catalysts

In this study, the biomass used was GSs (Vitis vinifera), obtained as a waste product of the wine industry in the north-east of Spain. The fresh biomass was dried to guarantee moisture levels below 2 wt% before being used directly. The WTs were received in the granulated form with particle sizes between 2 and 4 mm. WTs were supplied without the steel thread and textile netting. Table 7 summarises the main properties (proximate and ultimate analysis and heating value) of both feedstocks. The determination of these properties was carried out in accordance with standard methods [30]. As can be seen in Table 7, significant differences were apparent between the two raw materials. The high oxygen content (33.7 wt%) of GSs should be pointed out, which implies a low LHV (22.2 MJ/Kg), whereas WTs are characterised by high carbon content and low oxygen composition. It is worthy of mention that the heating values attributed to WTs are comparable to or even higher than the heating values characteristic of fossil fuels (% C: 87.9 wt%, % H: 3.3 wt%, LHV: 37 MJ/kg).

Table 7. Main properties of grape seeds and waste tyres.

	GSs [1]	WTs [2]
Ash (wt%)	4.3	3.8
Volatile matter (wt%)	65.1	63.6
Fixed Carbon (wt%)	24.3	31.8
Ultimate analysis (wt%)		
C	53.9	87.9
H	6.6	7.4
N	2.2	0.3
S	0.1	1.1
O	37.2	3.3
LHV (MJ/Kg)	20.5	37.0

[1]: Stabilised for 48 h at room temperature; [2]: As received.

In this study, four different Ca-based catalysts obtained from the calcination of low-cost limestone (Ca-based-1, Ca-based-2, Ca-based-3) and dolomite (Ca-based-4) were selected. Private companies situated in different regions of Europe supplied these solids. The Ca-based-1 catalyst was extracted from a German mine; the Ca-based-2 catalyst was excavated in the north of Spain; and, finally, the Ca-based-3 and Ca-based-4 catalyst samples were mined in the north-east of Spain. All solids were obtained after calcination at 900 °C. Particle size distribution for all solids was in the range of 300–600 μm.

X-ray diffraction (XRD), N_2-physisorption, mercury porosimetry, temperature-programmed desorption of ammonia (NH_3-TPD), temperature-programmed desorption of carbon dioxide (CO_2-TPD), and inductively coupled plasma-optical emission spectroscopy (ICP-OES) were also carried out to complete the characterisation of the catalysts. Experimental details related to XRD N_2-physisorption and mercury porosimetry techniques can be found elsewhere [64,65], and NH_3-TPD and CO_2-TPD determination have also been described by other authors [13].

3.2. Pilot Plant

Experiments were conducted in a pilot plant equipped with an auger reactor operating at atmospheric pressure. N_2 was used as the inert carrier gas. A detailed description of the pilot plant operation can be found in the supplementary information. The process flow diagram of this pilot plant is shown in Figure 4. This Auger reactor is able to process up to approximately 20 kg/h of GSs or WTs. The feeding system comprises two independent stirred hoppers to prevent the formation of voids, which can each be filled with approximately 25 kg of GSs or WTs. One hopper is used to feed the GS and WT mixture, and the other is used to feed the catalysts. Additionally, it should also be pointed out that catalysts were always diluted with an inert (sand), maintaining a (sand + catalyst) to feedstock ratio of 3:1. This stock of solids would be the minimum amount of heat carrier required for a self-sustainable process from an energy perspective [15]. In an integrated process, char combustion would be performed in a secondary reactor, providing the energy for the next pyrolysis step through the cycling of the heat carrier (sand + catalyst).

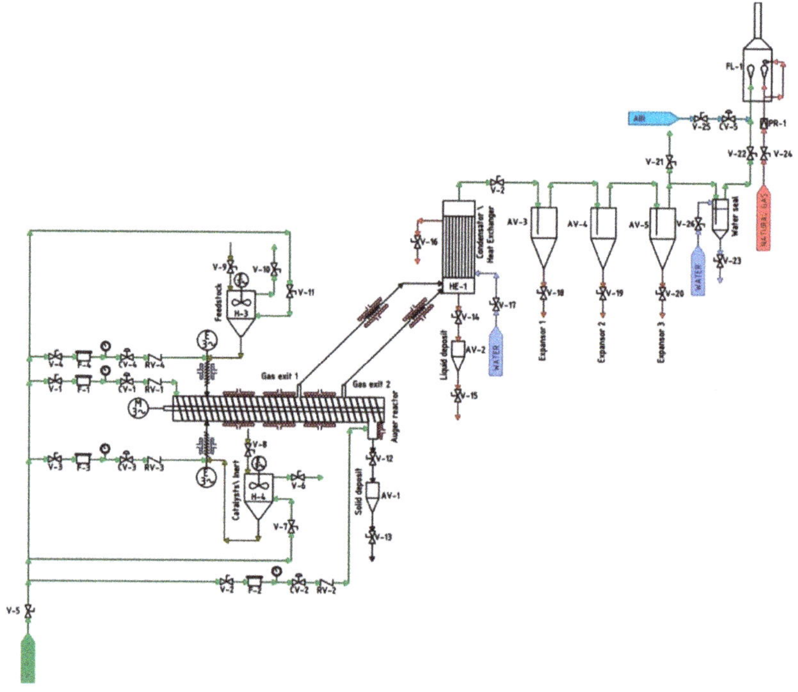

Figure 4. Process flow diagram of the pilot plant used for the catalytic co-pyrolysis experiments.

The reactor is surrounded by three independent electrical heating elements that provide the energy required to carry out the pyrolysis process. The converted product continuously leaves the reactor by gravity at the end of the reactor. The pyrolytic gas produced in the reactor is cooled in a shell and tubes condenser. Before flaring the gas, several samples were analysed by gas chromatography. The pilot plant is also provided with a control and acquisition system to monitor pressure by four pressure transducers and temperature by 10 thermocouples in different locations.

3.3. Catalysts Characterisation

A complete characterisation of the catalysts was carried out by means of XRD, ICP-OES, N_2 adsorption, and Hg porosimetry. The acidic characteristics of the Ca-based catalysts were studied by NH_3-TPD, and the basic characteristics of the catalysts were studied with TPD-CO_2. XRD patterns were measured with a Bruker D8 Advance series II diffractometer (Bruker, Delft, The Netherlands) using monochromatic Cu–Ka radiation (k = 0.1541 nm). Data were collected in the 2 h range from 5° to 80° using a scanning rate of 1°/min. The elemental composition of the different Ca-based catalysts (Al, Ca, Fe, Mg, K, Na, Si, Ti, P, Mn, and S) was determined by ICP-OES in a Perkin Elmer Optima 4300 system (Waltham, MA, Estados Unidos). N_2 physisorption was performed by a Quantachrome Autosorb 1 gas adsorption analyser (Graz, Austria). Prior to the adsorption measurements, the samples were outgassed in situ under a vacuum (4 mbar) at 250 °C for 4 h. The pore structure of the samples was determined by Hg porosimetry in a Micromeritics AutoPore V instrument (Norcross, GA, USA), according to the ISO 15901 standard. TPD techniques were measured by a Micromeritics Pulse Chemisorb 2700 instrument (Norcross, GA, USA) equipped with a thermal conductivity detector (TCD). Detailed information on the process conditions can be found in other research work [13].

3.4. Product Characterisation

The different products obtained after the process (liquid, solid, and gas) were characterised. Liquid and char yields were calculated by weight, and gas yield was calculated by balance. A heterogeneous liquid fraction was obtained, composed of two totally differentiable phases. The samples were centrifuged at 2000 rpm for 10 m, and the two liquid layers (organic/top and aqueous/bottom) were subsequently collected by a decantation process. The organic phase was then analysed in triplicate to determine different physicochemical properties, in accordance with standard methods. For the physical-chemical characterisation of the organic fraction, final composition, calorific value, water content, and pH were determined according to standard procedures [30], and density was determined by gravimetry. The GC-MS technique was also performed to analyse the chemical composition of the organic phase. Experimental details can be found in previous works [30]. The solid fraction was divided into char, CO_2, and coke yields for catalytic experiments. Yields of CO_2, and coke were directly calculated by the % of mass loss calculated from TGA experiments on each catalyst. The solid product (char) was also characterised by determining its calorific value and elemental analysis [30]. The non-condensable gas fraction was determined by GC using a TCD detector coupled to the Hewlett Packard II series [30,66]. Several thermogravimetric analyses were also performed in order to analyse catalyst carbonation.

4. Conclusions

The catalytic co-pyrolysis of both grape seeds and waste tyres using Ca-based catalysts was successfully performed in an Auger reactor at the pilot scale. The results demonstrated the great potential of this process for the production of an improved bio-oil. In general, a deoxygenated bio-oil with improved physical properties, such as viscosity and density, can be obtained using any low-cost Ca-based catalysts. More particularly, upgraded bio-oils can be practically dehydrated (H_2O content = 0.62 wt%) and fully deoxygenated (deoxygenation rate of ~95%), reaching heating values higher than 40 MJ/kg using a Ca-based catalyst with poor total basicity and a pure CaO crystalline phase without impurities. These properties seem to be key parameters that facilitate the prevalence of aromatisation and hydrodeoxygenation routes over the de-acidification and deoxygenation of the organic fraction through both ketonisation and esterification reactions, leading to a highly aromatic bio-oil. Moreover, due to the CO_2-capture and desulphuration effects inherent to these catalysts, a more environmentally friendly gas product was produced, generating H_2 and CH_4 as main components and, thus, increasing the range of potential applications for this fraction. Finally, the solid fraction was also shown to be a clean solid fuel to supply the energy for the proposed catalytic co-pyrolysis process.

Supplementary Materials: The following are available online at http://www.mdpi.com/2073-4344/9/12/992/s1, S.1 Pilot plant operation, Table S1: Yields after Catalytic experiments of GS and WT using Ca-based-1 catalyst, Table S2: Liquid properties after the catalytic experiments of GS and WT using Ca-based-1 catalyst, Table S3: Ultimate composition of the aqueous fraction after catalytic co-pyrolysis experiments, Table S4: Identified compounds (sorted by increasing retention time) in the liquid by GC/MS obtained after grape seeds pyrolysis, Table S5: Identified compounds (sorted by increasing retention time) in the liquid by GC/MS obtained after PS pyrolysis, Table S6: Identified compounds (sorted by increasing retention time) in the liquid by GC/MS obtained after co-pyrolysis of GS and WT experiments, Table S7: Identified compounds (sorted by increasing retention time) in the liquid by GC/MS obtained after Catalytic co-pyrolysis of GS and WT experiments, Table S8: Families and compounds identified after GC/MS.

Author Contributions: Conceptualization, T.G. and R.M.; methodology, O.S.-P., A.V. and J.M.L.; validation, O.S.-P. and A.V.; formal analysis O.S.-P. and A.V.; investigation, A.V., T.G. and R.M.; resources, T.G. and R.M.; data curation, O.S.-P. and A.V.; writing—original draft preparation, O.S.-P. and A.V.; writing—review and editing, T.G. and M.S.C.; supervision, T.G. and R.M.; project administration, T.G. and M.S.C.; funding acquisition, T.G. and R.M.

Funding: This research was funded by MINECO and FEDER for their financial support (Project ENE2015-68320-R) and the Regional Government of Aragon (DGA) under the research groups call.

Conflicts of Interest: The authors declare no conflict of interest.

References

1. Zheng, Y.; Tao, L.; Huang, Y.; Liu, C.; Wang, Z.; Zheng, Z. Improving aromatic hydrocarbon content from catalytic pyrolysis upgrading of biomass on a CaO/HZSM-5 dual-catalyst. *J. Anal. Appl. Pyrolysis* **2019**, *140*, 355–366. [CrossRef]
2. Imran, A.; Bramer, E.A.; Seshan, K.; Brem, G. An overview of catalysts in biomass pyrolysis for production of biofuels. *Biofuel Res. J.* **2018**, *5*, 872–885. [CrossRef]
3. Pires, A.P.P.; Arauzo, J.; Fonts, I.; Domine, M.E.; Arroyo, A.F.; Garcia-Perez, M.E.; Montoya, J.; Chejne, F.; Pfromm, P.; Garcia-Perez, M. Challenges and Opportunities for Bio-oil Refining: A Review. *Energy Fuels* **2019**. [CrossRef]
4. Oasmaa, A.; Van De Beld, B.; Saari, P.; Elliott, D.C.; Solantausta, Y. Norms, standards, and legislation for fast pyrolysis bio-oils from lignocellulosic biomass. *Energy Fuels* **2015**, *29*, 2471–2484. [CrossRef]
5. Stefanidis, S.D.; Kalogiannis, K.G.; Lappas, A.A. Co-processing bio-oil in the refinery for drop-in biofuels via fluid catalytic cracking. *Wiley Interdiscip. Rev.* **2018**, *7*. [CrossRef]
6. Thegarid, N.; Fogassy, G.; Schuurman, Y.; Mirodatos, C.; Stefanidis, S.; Iliopoulou, E.F.; Kalogiannis, K.; Lappas, A.A. Second-generation biofuels by co-processing catalytic pyrolysis oil in FCC units. *Appl. Catal. B Environ.* **2014**, *145*, 161–166. [CrossRef]
7. Baloch, H.A.; Nizamuddin, S.; Siddiqui, M.T.H.; Riaz, S.; Jatoi, A.S.; Dumbre, D.K.; Mubarak, N.M.; Srinivasan, M.P.; Griffin, G.J. Recent advances in production and upgrading of bio-oil from biomass: A critical overview. *J. Environ. Chem. Eng.* **2018**, *6*, 5101–5118. [CrossRef]
8. Yildiz, G.; Ronsse, F.; Duren, R.V.; Prins, W. Challenges in the design and operation of processes for catalytic fast pyrolysis of woody biomass. *Renew. Sustain. Energy Rev.* **2016**, *57*, 1596–1610. [CrossRef]
9. Dickerson, T.; Soria, J. Catalytic Fast Pyrolysis: A Review. *Energies* **2013**, *6*, 514–538. [CrossRef]
10. Rahman, M.M.; Liu, R.; Cai, J. Catalytic fast pyrolysis of biomass over zeolites for high quality bio-oil—A review. *Fuel Process. Technol.* **2018**, *180*, 32–46. [CrossRef]
11. Paasikallio, V.; Lindfors, C.; Kuoppala, E.; Solantausta, Y.; Oasmaa, A.; Lehto, J.; Lehtonen, J. Product quality and catalyst deactivation in a four day catalytic fast pyrolysis production run. *Green Chem.* **2014**, *16*, 3549–3559. [CrossRef]
12. Paasikallio, V.; Lindfors, C.; Lehto, J.; Oasmaa, A.; Reinikainen, M. Short vapour residence time catalytic pyrolysis of spruce sawdust in a bubbling fluidized-bed reactor with HZSM-5 catalysts. *Top. Catal.* **2013**, *56*, 800–812. [CrossRef]
13. Stefanidis, S.D.; Karakoulia, S.A.; Kalogiannis, K.G.; Iliopoulou, E.F.; Delimitis, A.; Yiannoulakis, H.; Zampetakis, T.; Lappas, A.A.; Triantafyllidis, K.S. Natural magnesium oxide (MgO) catalysts: A cost-effective sustainable alternative to acid zeolites for the in situ upgrading of biomass fast pyrolysis oil. *Appl. Catal. B Environ.* **2016**, *196*, 155–173. [CrossRef]
14. Veses, A.; Aznar, M.; Callén, M.S.; Murillo, R.; García, T. An integrated process for the production of lignocellulosic biomass pyrolysis oils using calcined limestone as a heat carrier with catalytic properties. *Fuel* **2016**, *181*, 430–437. [CrossRef]
15. Veses, A.; Aznar, M.; Martínez, I.; Martínez, J.D.; López, J.M.; Navarro, M.V.; Callén, M.S.; Murillo, R.; García, T. Catalytic pyrolysis of wood biomass in an auger reactor using calcium-based catalysts. *Bioresour. Technol.* **2014**, *162*, 250–258. [CrossRef] [PubMed]
16. Kalogiannis, K.G.; Stefanidis, S.D.; Karakoulia, S.A.; Triantafyllidis, K.S.; Yiannoulakis, H.; Michailof, C.; Lappas, A.A. First pilot scale study of basic vs. acidic catalysts in biomass pyrolysis: Deoxygenation mechanisms and catalyst deactivation. *Appl. Catal. B Environ.* **2018**, *238*, 346–357. [CrossRef]
17. Wong, S.L.; Ngadi, N.; Abdullah, T.A.T.; Inuwa, I.M. Current state and future prospects of plastic waste as source of fuel: A review. *Renew. Sustain. Energy Rev.* **2015**, *50*, 1167–1180. [CrossRef]
18. Abnisa, F.; Daud, W.M.A.W. A review on co-pyrolysis of biomass: An optional technique to obtain a high-grade pyrolysis oil. *Energy Convers. Manag.* **2014**, *87*, 71–85. [CrossRef]
19. Uzoejinwa, B.B.; He, X.; Wang, S.; El-Fatah Abomohra, A.; Hu, Y.; Wang, Q. Co-pyrolysis of biomass and waste plastics as a thermochemical conversion technology for high-grade biofuel production: Recent progress and future directions elsewhere worldwide. *Energy Convers. Manag.* **2018**, *163*, 468–492. [CrossRef]
20. Alvarez, J.; Amutio, M.; Lopez, G.; Santamaria, L.; Bilbao, J.; Olazar, M. Improving bio-oil properties through the fast co-pyrolysis of lignocellulosic biomass and waste tyres. *Waste Manag.* **2019**, *85*, 385–395. [CrossRef]

21. Wang, J.; Zhong, Z.; Ding, K.; Li, M.; Hao, N.; Meng, X.; Ruan, R.; Ragauskas, A.J. Catalytic fast co-pyrolysis of bamboo sawdust and waste tire using a tandem reactor with cascade bubbling fluidized bed and fixed bed system. *Energy Convers. Manag.* **2019**, *180*, 60–71. [CrossRef]

22. Shah, S.A.Y.; Zeeshan, M.; Farooq, M.Z.; Ahmed, N.; Iqbal, N. Co-pyrolysis of cotton stalk and waste tire with a focus on liquid yield quantity and quality. *Renew. Energy* **2019**, *130*, 238–244. [CrossRef]

23. Van Nguyen, Q.; Choi, Y.S.; Choi, S.K.; Jeong, Y.W.; Kwon, Y.S. Improvement of bio-crude oil properties via co-pyrolysis of pine sawdust and waste polystyrene foam. *J. Environ. Manag.* **2019**, *237*, 24–29. [CrossRef] [PubMed]

24. Brebu, M.; Yanik, J.; Uysal, T.; Vasile, C. Thermal and catalytic degradation of grape seeds/polyethylene waste mixture. *Cellul. Chem. Technol.* **2014**, *48*, 665–674.

25. Izzatie, N.I.; Basha, M.H.; Uemura, Y.; Hashim, M.S.M.; Afendi, M.; Mazlan, M.A.F. Co-pyrolysis of rubberwood sawdust (RWS) and polypropylene (PP) in a fixed bed pyrolyzer. *J. Mech. Eng. Sci.* **2019**, *13*, 4636–4647. [CrossRef]

26. Martínez, J.D.; Veses, A.; Mastral, A.M.; Murillo, R.; Navarro, M.V.; Puy, N.; Artigues, A.; Bartrolí, J.; García, T. Co-pyrolysis of biomass with waste tyres: Upgrading of liquid bio-fuel. *Fuel Process. Technol.* **2014**, *119*, 263–271. [CrossRef]

27. Hassan, H.; Lim, J.K.; Hameed, B.H. Recent progress on biomass co-pyrolysis conversion into high-quality bio-oil. *Bioresour. Technol.* **2016**, *221*, 645–655. [CrossRef]

28. Zhang, X.; Lei, H.; Chen, S.; Wu, J. Catalytic co-pyrolysis of lignocellulosic biomass with polymers: A critical review. *Green Chem.* **2016**, *18*, 4145–4169. [CrossRef]

29. Zhang, L.; Bao, Z.; Xia, S.; Lu, Q.; Walters, K.B. Catalytic pyrolysis of biomass and polymer wastes. *Catalysts* **2018**, *8*, 659. [CrossRef]

30. Sanahuja-Parejo, O.; Veses, A.; Navarro, M.V.; López, J.M.; Murillo, R.; Callén, M.S.; García, T. Catalytic co-pyrolysis of grape seeds and waste tyres for the production of drop-in biofuels. *Energy Convers. Manag.* **2018**, *171*, 1202–1212. [CrossRef]

31. Iftikhar, H.; Zeeshan, M.; Iqbal, S.; Muneer, B.; Razzaq, M. Co-pyrolysis of sugarcane bagasse and polystyrene with ex-situ catalytic bed of metal oxides/HZSM-5 with focus on liquid yield. *Bioresour. Technol.* **2019**, *289*, 121647. [CrossRef] [PubMed]

32. Wang, J.; Jiang, J.; Zhong, Z.; Wang, K.; Wang, X.; Zhang, B.; Ruan, R.; Li, M.; Ragauskas, A.J. Catalytic fast co-pyrolysis of bamboo sawdust and waste plastics for enhanced aromatic hydrocarbons production using synthesized CeO2/γ-Al2O3 and HZSM-5. *Energy Convers. Manag.* **2019**, *196*, 759–767. [CrossRef]

33. Mishra, R.K.; Iyer, J.S.; Mohanty, K. Conversion of waste biomass and waste nitrile gloves into renewable fuel. *Waste Manag.* **2019**, *89*, 397–407. [CrossRef] [PubMed]

34. Ding, K.; Zhong, Z.; Wang, J.; Zhang, B.; Fan, L.; Liu, S.; Wang, Y.; Liu, Y.; Zhong, D.; Chen, P.; et al. Improving hydrocarbon yield from catalytic fast co-pyrolysis of hemicellulose and plastic in the dual-catalyst bed of CaO and HZSM-5. *Bioresour. Technol.* **2018**, *261*, 86–92. [CrossRef]

35. Zhang, B.; Zhong, Z.; Chen, P.; Ruan, R. Microwave-assisted catalytic fast co-pyrolysis of Ageratina adenophora and kerogen with CaO and ZSM-5. *J. Anal. Appl. Pyrolysis* **2017**, *127*, 246–257. [CrossRef]

36. Gulab, H.; Hussain, K.; Malik, S.; Hussain, Z.; Shah, Z. Catalytic co-pyrolysis of Eichhornia Crassipes biomass and polyethylene using waste Fe and CaCO$_3$ catalysts. *Int. J. Energy Res.* **2016**, *40*, 940–951. [CrossRef]

37. Liu, S.; Xie, Q.; Zhang, B.; Cheng, Y.; Liu, Y.; Chen, P.; Ruan, R. Fast microwave-assisted catalytic co-pyrolysis of corn stover and scum for bio-oil production with CaO and HZSM-5 as the catalyst. *Bioresour. Technol.* **2016**, *204*, 164–170. [CrossRef]

38. Treedet, W.; Suntivarakorn, R. Design and operation of a low cost bio-oil fast pyrolysis from sugarcane bagasse on circulating fluidized bed reactor in a pilot plant. *Fuel Process. Technol.* **2018**, *179*, 17–31. [CrossRef]

39. Campuzano, F.; Brown, R.C.; Martínez, J.D. Auger reactors for pyrolysis of biomass and wastes. *Renew. Sustain. Energy Rev.* **2019**, *102*, 372–409. [CrossRef]

40. Martínez, J.D.; Murillo, R.; García, T.; Veses, A. Demonstration of the waste tire pyrolysis process on pilot scale in a continuous auger reactor. *J. Hazard. Mater.* **2013**, *261*, 637–645. [CrossRef]

41. Veses, A.; Aznar, M.; López, J.M.; Callén, M.S.; Murillo, R.; García, T. Production of upgraded bio-oils by biomass catalytic pyrolysis in an auger reactor using low cost materials. *Fuel* **2015**, *141*, 17–22. [CrossRef]

42. Brassard, P.; Godbout, S.; Raghavan, V. Pyrolysis in auger reactors for biochar and bio-oil production: A review. *Biosyst. Eng.* **2017**, *161*, 80–92. [CrossRef]

43. Lan, Y.; Niu, D.; Liu, Q.; Yang, J.; Yang, Q.; Xu, J. Briquetting burnt dolomite powder for recycling in steel plants. *Nat. Environ. Pollut. Technol.* **2014**, *13*, 649–652.

44. Chinthakuntla, D.; Kumar, M.K.; Chakra, C.S.; Rao, K.; Dayakar, T. Calcium Oxide Nano Particles Synthesized From Chicken Egg Shells by Physical Method. In Proceedings of the International Conference on Emerging Technologies in Mechanical Sciences, Hyderabad, India, 26–27 December 2014.

45. Linggawati, A. Preparation and Characterization of Calcium Oxide Heterogeneous Catalyst Derived from Anadara Granosa Shell for Biodiesel Synthesis. *KnE Eng.* **2016**, *1*. [CrossRef]

46. Lee, S.L.; Wong, Y.C.; Tan, Y.P.; Yew, S.Y. Transesterification of palm oil to biodiesel by using waste obtuse horn shell-derived CaO catalyst. *Energy Convers. Manag.* **2015**, *93*, 282–288. [CrossRef]

47. Correia, L.M.; de Sousa Campelo, N.; Novaes, D.S.; Cavalcante, C.L.; Cecilia, J.A.; Rodríguez-Castellón, E.; Vieira, R.S. Characterization and application of dolomite as catalytic precursor for canola and sunflower oils for biodiesel production. *Chem. Eng. J.* **2015**, *269*, 35–43. [CrossRef]

48. Dabros, T.M.H.; Stummann, M.Z.; Høj, M.; Jensen, P.A.; Grunwaldt, J.D.; Gabrielsen, J.; Mortensen, P.M.; Jensen, A.D. Transportation fuels from biomass fast pyrolysis, catalytic hydrodeoxygenation, and catalytic fast hydropyrolysis. *Prog. Energy Combust. Sci.* **2018**, *68*, 268–309. [CrossRef]

49. Rahman, M.; Chai, M.; Sarker, M.; Nishu; Liu, R. Catalytic pyrolysis of pinewood over ZSM-5 and CaO for aromatic hydrocarbon: Analytical Py-GC/MS study. *J. Energy Inst.* **2019**. [CrossRef]

50. Chen, X.; Li, S.; Liu, Z.; Chen, Y.; Yang, H.; Wang, X.; Che, Q.; Chen, W.; Chen, H. Pyrolysis characteristics of lignocellulosic biomass components in the presence of CaO. *Bioresour. Technol.* **2019**, *287*. [CrossRef]

51. Chen, X.; Chen, Y.; Yang, H.; Wang, X.; Che, Q.; Chen, W.; Chen, H. Catalytic fast pyrolysis of biomass: Selective deoxygenation to balance the quality and yield of bio-oil. *Bioresour. Technol.* **2019**. [CrossRef]

52. Hossain, A.K.; Davies, P.A. Pyrolysis liquids and gases as alternative fuels in internal combustion engines—A review. *Renew. Sustain. Energy Rev.* **2013**, *21*, 165–189. [CrossRef]

53. Biradar, C.H.; Subramanian, K.A.; Dastidar, M.G. Production and fuel quality upgradation of pyrolytic bio-oil from Jatropha Curcas de-oiled seed cake. *Fuel* **2014**, *119*, 81–89. [CrossRef]

54. de Luna, M.D.G.; Cruz, L.A.D.; Chen, W.H.; Lin, B.J.; Hsieh, T.H. Improving the stability of diesel emulsions with high pyrolysis bio-oil content by alcohol co-surfactants and high shear mixing strategies. *Energy* **2017**, *141*, 1416–1428. [CrossRef]

55. Cao, Q.; Jin, L.; Bao, W.; Lv, Y. Investigations into the characteristics of oils produced from co-pyrolysis of biomass and tire. *Fuel Process. Technol.* **2009**, *90*, 337–342. [CrossRef]

56. Li, J.J.; Zhou, T.D.; Tang, X.D.; Chen, X.D.; Zhang, M.; Zheng, X.P.; Wang, C.S.; Deng, C.L. Viscosity reduction process of heavy oil by catalytic co-pyrolysis with sawdust. *J. Anal. Appl. Pyrolysis* **2019**, *140*, 444–451. [CrossRef]

57. Persson, H.; Yang, W. Catalytic pyrolysis of demineralized lignocellulosic biomass. *Fuel* **2019**, *252*, 200–209. [CrossRef]

58. Thoharudin, T.; Nadjib, M.; Santosa, T.H.A.; Juliansyah; Zuniardi, A.; Shihabudin, R. Properties of co-pyrolysed palm kernel shell and plastic grocery bag with CaO as catalyst. In Proceedings of the 3rd International Conference on Biomass: Accelerating the Technical Development and Commercialization for Sustainable Bio-based Products and Energy, Bogor, Indonesia, 1–2 August 2018.

59. Sukumar, V.; Manieniyan, V.; Senthilkumar, R.; Sivaprakasam, S. Production of bio oil from sweet lime empty fruit bunch by pyrolysis. *Renew. Energy* **2020**, *146*, 309–315. [CrossRef]

60. Ghorbannezhad, P.; Kool, F.; Rudi, H.; Ceylan, S. Sustainable production of value-added products from fast pyrolysis of palm shell residue in tandem micro-reactor and pilot plant. *Renew. Energy* **2020**, *145*, 663–670. [CrossRef]

61. Vichaphund, S.; Wimuktiwan, P.; Sricharoenchaikul, V.; Atong, D. In situ catalytic pyrolysis of Jatropha wastes using ZSM-5 from hydrothermal alkaline fusion of fly ash. *J. Anal. Appl. Pyrolysis* **2019**, *139*, 156–166. [CrossRef]

62. Hussmann, G.P.; AMOCO Corporation. Preparation of dialkyl ketones from aliphatic carboxylic acids. U.S. Patent 4754074A, 28 June 1988.

63. Choi, G.-G.; Oh, S.-J.; Kim, J.-S. Scrap tire pyrolysis using a new type two-stage pyrolyzer: Effects of dolomite and olivine on producing a low-sulfur pyrolysis oil. *Energy* **2016**, *114*, 457–464. [CrossRef]

64. Veses, A.; Puértolas, B.; Callén, M.S.; García, T. Catalytic upgrading of biomass derived pyrolysis vapors over metal-loaded ZSM-5 zeolites: Effect of different metal cations on the bio-oil final properties. *Microporous Mesoporous Mater.* **2015**, *209*, 189–196. [CrossRef]
65. Puértolas, B.; Veses, A.; Callén, M.S.; Mitchell, S.; García, T.; Pérez-Ramírez, J. Porosity-Acidity Interplay in Hierarchical ZSM-5 Zeolites for Pyrolysis Oil Valorization to Aromatics. *ChemSusChem* **2015**, *8*, 3283–3293. [CrossRef] [PubMed]
66. Sanahuja-Parejo, O.; Veses, A.; Navarro, M.V.; López, J.M.; Murillo, R.; Callén, M.S.; García, T. Drop-in biofuels from the co-pyrolysis of grape seeds and polystyrene. *Chem. Eng. J.* **2018**. [CrossRef]

© 2019 by the authors. Licensee MDPI, Basel, Switzerland. This article is an open access article distributed under the terms and conditions of the Creative Commons Attribution (CC BY) license (http://creativecommons.org/licenses/by/4.0/).

 *catalysts*

Article

Synthesis of Diesel and Jet Fuel Range Cycloalkanes with Cyclopentanone and Furfural

Wei Wang [1], Shaoying Sun [1], Fengan Han [2], Guangyi Li [2], Xianzhao Shao [1] and Ning Li [2,3,*]

[1] Shaanxi Key Laboratory of Catalysis, School of Chemistry and Environment Science, Shaanxi University of Technology, No. 1 Dong yi huan Road, Hanzhong 723001, China; wangwei@snut.edu.cn (W.W.); sunshaoying1221@163.com (S.S.); xianzhaoshao@snut.eud.cn (X.S.)

[2] CAS Key Laboratory of Science and Technology on Applied Catalysis, Dalian Institute of Chemical Physics, Chinese Academy of Sciences, No. 457 Zhongshan Road, Dalian 116023, China; hanfengan@dicp.ac.cn (F.H.); lgy2010@dicp.ac.cn (G.L.)

[3] Dalian National Laboratory for Clean Energy, No. 457 Zhongshan Road, Dalian 116023, China

* Correspondence: lining@dicp.ac.cn; Tel.: +86-411-84379738

Received: 6 October 2019; Accepted: 24 October 2019; Published: 25 October 2019

Abstract: Diesel and jet fuel range cycloalkanes were obtained in ~84.8% overall carbon yield with cyclopentanone and furfural, which can be produced from hemicellulose. Firstly, 2,5-bis(furan-2-ylmethyl)-cyclopentanone was prepared by the aldol condensation/hydrogenation reaction of cyclopentanone and furfural under solid base and selective hydrogenation catalyst. Over the optimized catalyst (Pd/C-CaO), 98.5% carbon yield of 2,5-bis(furan-2-ylmethyl)-cyclopentanone was acquired at 423 K. Subsequently, the 2,5-bis(furan-2-ylmethyl)-cyclopentanone was further hydrodeoxygenated over the M/H-ZSM-5(Pd, Pt and Ru) catalyst. Overall, 86.1% carbon yield of diesel and jet fuel range cycloalkanes was gained over the Pd/H-ZSM-5 catalyst under solvent-free conditions. The cycloalkane mixture obtained in this work has a high density (0.82 g mL^{-1}) and a low freezing point (241.7 K). Therefore, it can be mixed into diesel and jet fuel to increase their volumetric heat values or payloads.

Keywords: high density fuel; cyclopentanone; furfural; aldol condensation; hydrodeoxygenation

---

1. Introduction

In recent years, the utilization of renewable biomass resources for the production of fuels [1–6] and chemicals [7–12] has attracted worldwide attention. Lignocellulose is the most abundant and cheapest biomass [13]. Diesel and jet fuel are two very important transport fuels. Pioneered by Dumesic [14,15], Huber [15,16], Corma [17,18] and their collaborators, great efforts were devoted to the synthesis of diesel and jet fuel range hydrocarbons using platform compounds derived from lignocellulose in the past decades [19–26].

Furfural is a bulk chemical which has been obtained on a commercial scale by the hydrolysis and dehydration of hemicellulose [27]. In recent years, series of diesel and jet fuel range chain alkanes were produced by the aldol condensation of furfural and acetone [15,19], methyl isobutyl ketone [28], pentanone [29,30], heptanone [31], levulinic acid [32], and angelica lactone [33], followed by hydrogenation/hydrodeoxygenation or hydrodeoxygenation (HDO). Cyclopentanone can be obtained (at high carbon yields of 62–95.8%) through the aqueous phase selectively hydrogenation and rearrangement of furfural [34–38]. High density fuels were prepared by self aldol condensation of cyclopentanone and hydrodeoxygenation [39–42]. In some studies, diesel and jet fuel range cycloalkanes were produced by the aldol condensation of furfural and cyclopentanone [43–46], followed by HDO. Nevertheless, the aldol condensation products obtained in these reports (i.e., 2-(2-furylmethylidene)-cyclopentanone and/or 2,5-bis(2-furylmethylidene)-cyclopeantanone) are solid at room temperature. Hence, organic solvents must be used to improve mass transfer efficiency in subsequent HDO reactions. This will cause lower efficiency and higher energy consumption. As a solution to this problem, we developed a new two-step strategy (see Scheme 1). In the first step, 2,5-bis(furan-2-ylmethyl)-cyclopentanone was generated via the aldol condensation/hydrogenation of cyclopentanone and furfural over the metal and solid base catalysts. Due to the saturation of C=C bonds by hydrogenation, 2,5-bis(furan-2-ylmethyl)cyclopentanone is a liquid at room temperature. As a result, it may be directly used in the hydrodeoxygenation procedure without using any solvent. This is favorable in real application. In the second step, the 2,5-bis(furan-2-ylmethyl)-cyclopentanone was converted to diesel and jet fuel range cycloalkanes by the HDO over Pd/H-ZSM-5 catalyst.

2. Results and Discussion

2.1. Aldol Condensation

First, the aldol condensation of furfural with cyclopentanone was studied over solid base catalysts. According to the analysis of GC (Figure S1 in the Supplementary Materials) and nuclear magnetic resonance (NMR) spectra (Figure S2), the chemical shifts of the target products in ^{1}H NMR and ^{13}C NMR spectra are in good agreement with those reported previously [44]. 2,5-bis(2-furylmethylidene)cyclopentanone was confirmed as the unique product (in Scheme 1). The sequence for 2,5-bis(2-furyl methylidene)-cyclopentanone carbon yields over the solid base catalysts was: CaO > LiAl-HT > MgAl-HT > MgO ≈ CeO$_2$ (see Figure 1). Over the CaO catalyst, 95.4% carbon yield of 2,5-bis(2-furylmethylidene)-cyclopentanone was obtained after reacting for 10 h at 423 K. The yield of 2,5-Bis (2-furylmethylidene)-cyclopentanone on Mg-Al hydrotalcite was 85.7%, while that of 2,5-Bis (2-furylmethylidene)-cyclopentanone on MgO and CeO$_2$ was very low.

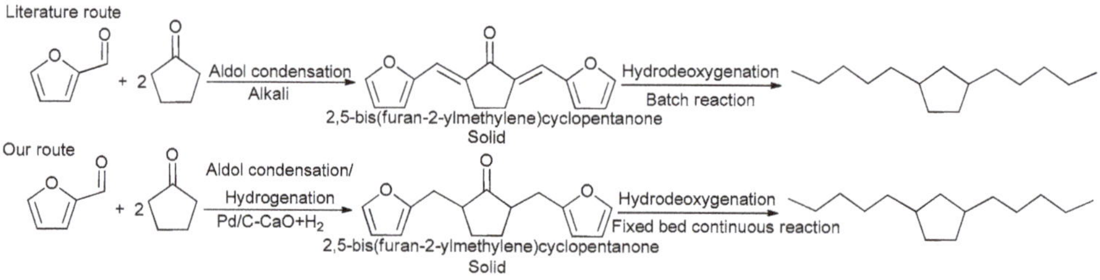

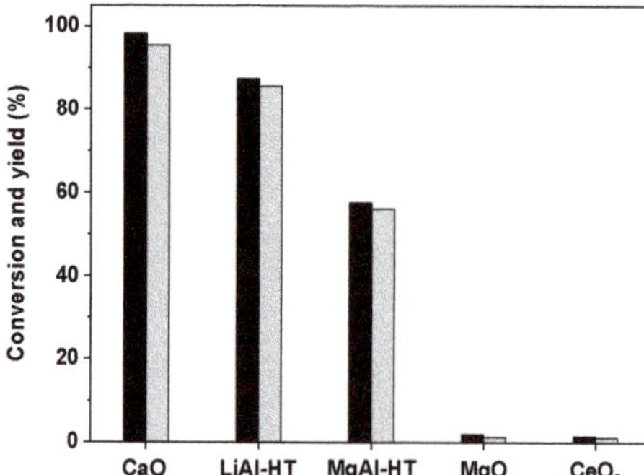

Scheme 1. Synthetic strategies for the production of C$_{15}$ cycloalkanes with furfural and cyclopentanone [45].

Figure 1. Cyclopentanone conversion (black columns) and carbon yield (gray columns) of 2,5-bis(2-furylmethylidene)-cyclopentanone on various solid base catalysts. Experimental conditions: 10 h, 423 K, and 0.84 g cyclopentanone (10 mmol), 1.92 g furfural (20 mmol), and 0.05 g catalyst were used in the tests.

According to the CO_2-TPD results (see Figure 2 and Table 1), base site amounts of CaO, LiAl-HT, MgAl-HT, MgO and CeO_2 catalysts were 0.16, 0.15, 0.12, 0.04 and 0.02 mmol g^{-1}, respectively. The base site amounts of the catalysts were in good agreement with the activity of aldol condensation of cyclopentanone with furfural. The high activity of CaO can be illustrated by its higher alkali strength and higher amount of base sites. Taking into consideration the higher activity, low cost, and good availability of CaO, we consider it as a potential catalyst in future industrial applications.

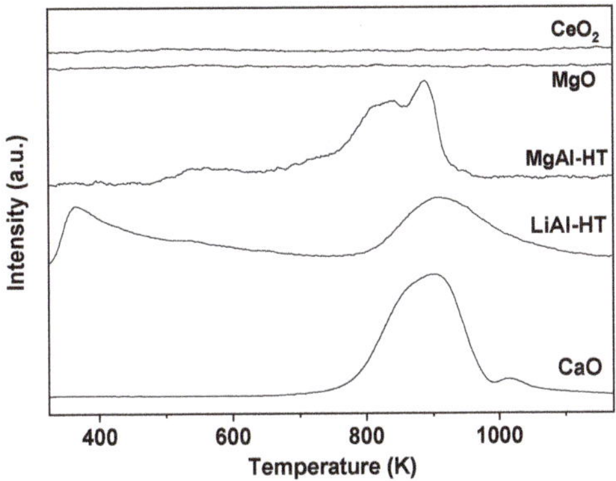

Figure 2. CO_2-TPD charts of the various solid base catalysts (the specific information of the CO_2-TPD experiment is shown in the Supplementary Materials).

Table 1. The number of base sites over various solid base catalysts.

Catalysis	Base Sites Amount (mmol g^{-1})
CaO	0.16
LiAl-HT	0.15
MgAl-HT	0.12
MgO	0.04
CeO_2	0.02

The influences of catalyst dosage on cyclopentanone conversion and 2,5-bis(2-furylmethylidene) cyclopentanone carbon yield over CaO catalyst were studied as well (see Figure S3a). The results show that the conversion of cyclopentanone raised with the increase of catalyst dosage from 0.01 to 0.07 g. In the meantime, yields of 2, 5-bis(2-furylmethylidene) cyclopentanone increased with the catalyst dosage first, and then leveled off. When the amount of catalyst was 0.05 g, the yield of 2,5-bis(2-furylmethylidene) cyclopentanone was the highest. Therefore, the amount of catalyst in the following part was 0.05 g.

The effects of the reaction temperature on the conversion of cyclopentanone and yield of 2,5-bis(2-furylmethylidene) cyclopentanone were studied. The results are shown in Figure S3b. The conversion of cyclopentanone increased first, and then completely converted as the reaction temperature grew from 373 K to 443 K. At the same time, yield of 2,5-bis(2-furylmethylidene) cyclopentanone grew at first as the reaction temperature ascended, and then remained constant. Consequently, the reaction temperature of 423 K was chosen.

Under the optimum reaction conditions (0.05 g CaO catalyst, 10 h, and 423 K), 98.2% conversion of cyclopentanone and 95.4% carbon yield of 2,5-bis(2-furylmethylidene)-cyclopentanone were achieved.

2.2. One Pot Aldol Condensation/Hydrogenation

Subsequently, we developed the one-pot synthesis of 2,5-bis(furan-2-ylmethyl)-cyclopentanone by the aldol condensation/hydrogenation reaction of furfural, cyclopentanone, and hydrogen under the co-catalysis of Pd/C and CaO. Based on GC, NMR and MS spectra analysis (Figures S4–S6),

2,5-bis(furan-2-ylmethyl)-cyclopentanone was identified as the main product. At room temperature, the 2,5-bis(furan-2-ylmethyl)-cyclopentanone exists in liquid state. Consequently, it can be directly used in solvent-free HDO process. According to Scheme 2, 2,5-bis(furan-2-ylmethyl)-cyclopentanone was generated by the hydrogenation of 2,5-bis(2-furylmethylidene)-cyclopentanone from the aldol condensation of furfural and cyclopentanone. To illustrate the advantages of one-pot condensation hydrogenation, we compared the solvent-free, methanol-solvent hydrogenation of 2,5-bis(2-furylmethylidene)-cyclopentanone and one-pot condensation/hydrogenation of furfural and cyclopentanone by Pd/C + CaO. The experimental results are shown in Figure 3. Under the condition of solvent-free Pd/C hydrogenation, conversion of 2,5-bis (2-furylmethylidene)-cyclopentanone and carbon yields of 2,5-bis(furan-2-ylmethyl)-cyclopentanone were 1.1% and 0.8%, respectively.

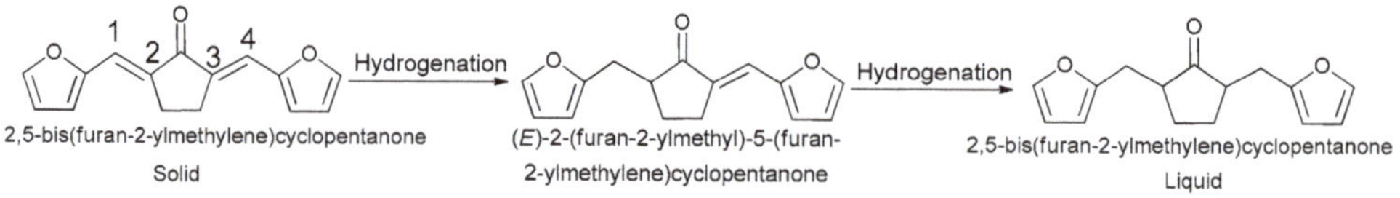

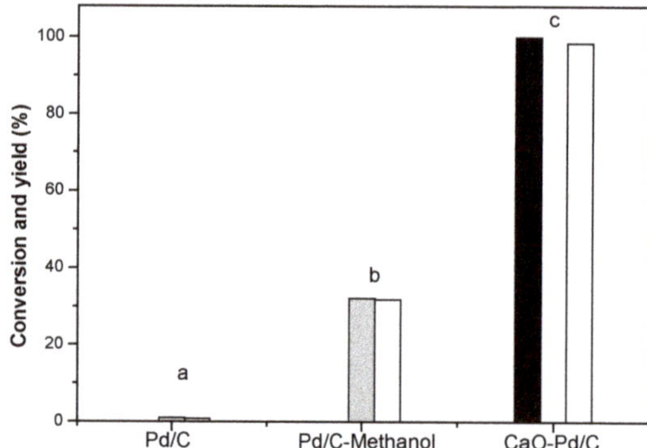

Scheme 2. Synthetic strategies for 2,5-bis(furan-2-ylmethyl)-cyclopentanone from 2,5-bis(2-furylmethylde ne)-cyclopentanone hydrogenation [47].

Figure 3. Cyclopentanone conversions (black columns), 2,5-bis (2-furylmethylidene)-cyclopentanone conversion (gray columns) and carbon yields of 2,5-bis(furan-2-ylmethyl)-cyclopentanone (white columns) over Pd/C catalysts. Experimental conditions: 10 h, 423 K, and 4.0 MPa H_2; (**a**) 2.40 g 2,5-bis(2-furylmethylidene)-cyclopentanone, 0.05 g Pd/C; (**b**) 2.40 g 2,5-bis (2-furylmethylidene)-cyclopentanone, 0.05 g Pd/C and 10.0 mL methanol; and (**c**) 0.84 g cyclopentanone (10 mmol), 1.92 g furfural (20 mmol); 0.05 g CaO and 0.05 g metal catalyst were used in the tests.

When methanol solvent was added to the hydrogenation process, the hydrogenation activity was significantly improved. Conversion of 2,5-bis (2-furylmethylidene)-cyclopentanone and carbon yields of 2,5-bis(furan-2-ylmethyl)-cyclopentanone were 32.1% and 31.8%, respectively. Because methanol can dissolve 2,5-bis (2-furylmethylidene)-cyclopentanone and increase the contact area with hydrogenation active center, the hydrogenation reaction activity of 2,5-bis (2-furylmethylidene)-cyclopentanone was accelerated. Under one-pot condensation/hydrogenation of furfural and cyclopentanone by Pd/C+CaO, cyclopentanone was completely converted and high carbon yield (98.5%) of 2,5-bis(furan-2-ylmethyl)-cyclopentanone was achieved. During one-pot condensation/hydrogenation of furfural and cyclopentanone, furfural and cyclopentanone were not only raw materials, but also as solvents. 2,5-Bis (2-furylmethylidene)-cyclopentanone produced by furfural and cyclopentanone can be dissolved well, which increased the contact area with hydrogenation active center. The reaction rate was accelerated.

Moreover, we studied the activities of CaO + other metal catalysts (PdC, Pt/C, Ru/C, Raney Co and Raney Ni). Based on the results shown in Figure 4, the CaO + Pd/C displayed the best performance among the investigated catalysts. Over it, 100% conversion of cyclopentanone and 98.5% carbon yield of 2,5-bis(furan-2-ylmethyl)-cyclopentanone were obtained, when the reaction was carried out for 10 h at 423 K. Pd/C can selectively hydrogenate carbon–carbon double bonds ($C_1=C_2$ and $C_3=C_4$) of 2,5-bis(2-furylmethylidene)-cyclopentanone, but Pt/C, Ru/C, Raney Co and Raney Ni cannot selectively hydrogenate its carbon-carbon double bonds ($C_1=C_2$ and $C_3=C_4$). The product of carbon-carbon double bonds ($C_1=C_2$ and $C_3=C_4$) of 2,5-bis(2-furylmethylidene)-cyclopentanone hydrogenated is liquid. These products as solvents promoted selective hydrogenation of 2,5-bis(2-furylmethylidene)-cyclopentanone. This can be explained by the high activity of palladium catalyst for high yield of one-pot synthesis of 2,5-bis(furan-2-ylmethyl)-cyclopentanone.

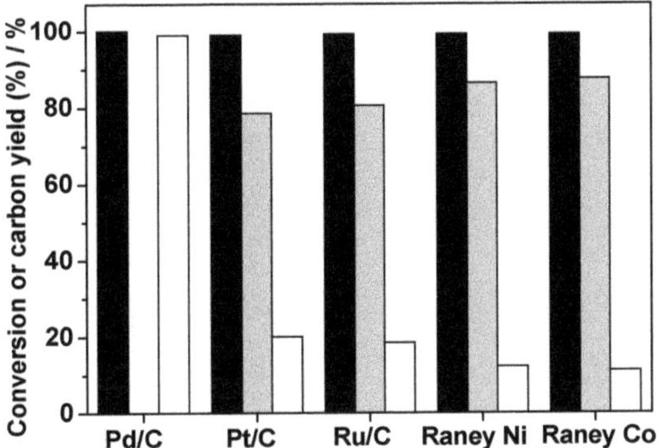

Figure 4. Cyclopentanone conversions (black columns) and carbon yields of 2,5-bis(2-furylmethylidene)-cyclopentanone (gray columns), and 2,5-bis(furan-2-ylmethyl)-cyclopentanone (white columns) over CaO and metal catalysts. Experimental conditions: 10 h, 423 K, and 4.0 MPa H_2; 0.84 g cyclopentanone (10 mmol), 1.92 g furfural (20 mmol), 0.05 g CaO and 0.05 g metal catalyst were used in the tests.

The influences of catalyst dosage and reaction time on the 2,5-bis(furan-2-ylmethyl)-cyclopentanone carbon yield over the Pd/C-CaO catalyst were investigated (see Figure 5).

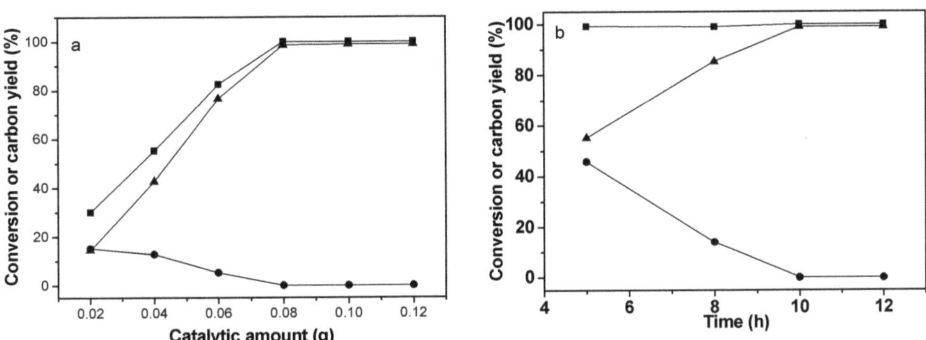

Figure 5. Cyclopentanone conversion (■) and carbon yields of 2,5-bis(furan-2-ylmethyl)-cyclopentanone (▲) and 2,5-bis(2-furylmethylidene)-cyclopentanone (●) over the Pd/C-CaO catalyst. Experimental conditions: 423 K, 10 h, and 4.0 MPa H_2; 0.84 g cyclopentanone (10 mmol), 1.92 g furfural (20 mmol), 0.08 g Pd/C-CaO catalyst were used in the tests. (**a**) The effect of catalyst dosage. (**b**) The effect of time.

During the change of Pd/C-CaO catalyst from 0.02 g to 0.12 g, the yield of 2,5-bis(furan-2-ylmethyl)-cyclopentanone increased and the yield of 2,5-bis(2-furylmet-hylidene)-cyclopentanone decreased. While the Pd/C-CaO catalyst was 0.08 g, the yield of 2,5-bis(furan-2-ylmethyl)-cyclopentanone reached 98.5%, while cyclopentanone and 2,5-bis(2-furyl methylidene)-cyclopentanone were completely converted. When the reaction time was 10 h, yield of 2,5-bis(furan-2-ylmethyl)-cyclopentanone was highest and 2,5-bis(2-furyl methylidene)-cyclopentanone was completely converted. Under the optimized conditions (423 K, 10 h, and 0.08 g Pd/C-CaO), high carbon yield (98.5%) of 2,5-bis(furan-2-ylmethyl)-cyclopentanone was achieved over the Pd/C-CaO catalyst.

The re-usability of the Pd/C-CaO catalyst was studied as well. To eliminate the disturbance of the residues, the used catalysts were washed thoroughly with tetrahydrofuran (THF) at room temperature after each usage, and then dried for 1 h at 353 K. Based on the results in Figure 6, the Pd/C-CaO catalyst was stable under the investigated reaction conditions.

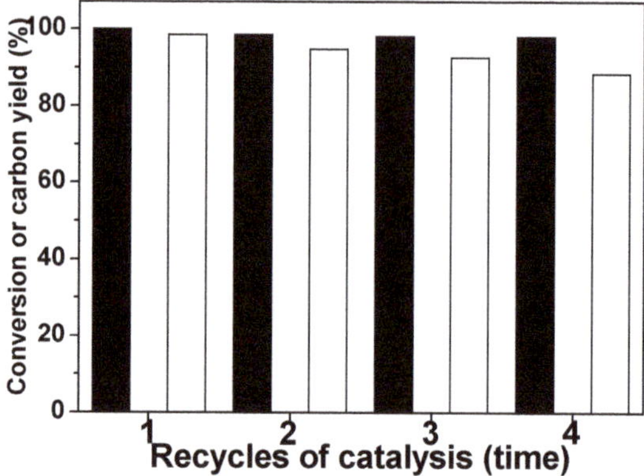

Figure 6. Cyclopentanone conversion (black columns) and carbon yields of 2,5-bis(furan-2-ylmethyl)-cyclopentanone (white columns) and 2,5-bis(2-furylmethylidene)-cyclopentanone (gray columns) over the Pd/C-CaO catalyst. Experimental conditions: 423 K, 10 h, and 4.0 MPa H_2; 0.84 g cyclopentanone (10 mmol), 1.92 g furfural (20 mmol), 0.08 g Pd/C-CaO were used for the tests.

The 2,5-bis(furan-2-ylmethyl)-cyclopentanone carbon yield varied slightly even after four usages (after the third operation, the slight decrease of activity can be attributed to the loss of catalyst in the process of filtration and drying). Considering the good stability and high activity of the Pd/C-CaO catalyst, we consider Pd/C-CaO catalyst as a potential catalyst for industrial application in the future.

2.3. Hydrodeoxygenation (HDO)

The solvent-free HDO of 2,5-bis(furan-2-methyl)-cyclopentanone was investigated over the M/H-ZSM-5 (M = Pt, Pd or Ru) catalysts. Based on GC-MS analysis (Figures S7 and S8), the 2,5-bis(furan-2-ylmethyl)cyclopentanone was totally converted at 573 K, high carbon yields (>66%) of cycloalkanes was obtained over the M/H- ZSM-5 catalysts (see Figure 7). By comparison, the carbon yields (or selectivity) of diesel and jet fuel range cycloalkanes (86.1% and 83.5%) over the 5 wt.% Pd/H-ZSM-5 and 5 wt.% Pt/H-ZSM-5 catalysts were evidently higher than those on the 5 wt.% Ru/H-ZSM-5 catalyst (66.8%). Furthermore, it was also noticed that the carbon yields of C1–C4 light alkanes over the 5 wt.% Ru/H-ZSM-5 was evidently higher than those over the 5 wt.% Pd/H-ZSM-5 and 5 wt.% Pt/H-ZSM-5 catalysts. The lower selectivity of diesel and jet fuel range alkanes over the Ru/H-ZSM-5 catalyst can be explained by the high hydrogenolysis (or methanation) activity of Ru.

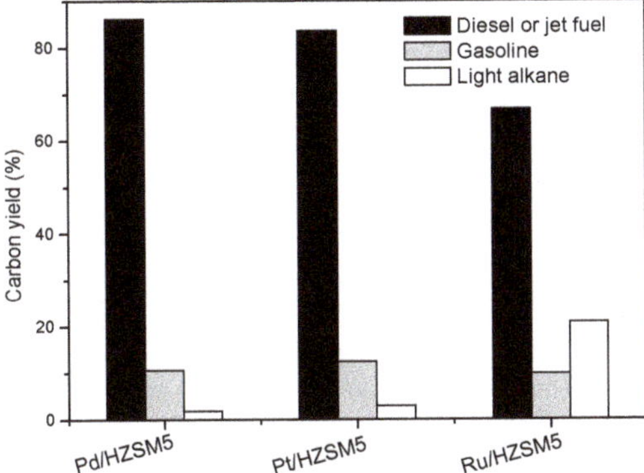

Figure 7. Carbon yields of C_9–C_{15} diesel alkanes (black columns), C_5–C_8 gasoline alkanes (grey columns), and C_1–C_4 alkanes (white columns) over M/H-ZSM-5 (M = Pt, Pd or Ru) catalysts. Experimental conditions: 1.8 g catalyst, 6.0 MPa H_2, 573 K, 0.04 mL min^{-1} liquid feedstock flow rate, and 120 mL min^{-1} H_2 flow rate.

The stability of Pd/H-ZSM-5 catalyst in the HDO process was studied as well. Based on the results shown in Figure 8, during 24 h of continuous test, the Pd/H-ZSM-5 catalyst was steady under the investigated conditions. According to our determination, the density and the freezing point of the branched cycloalkane mixtures gained in this work were 0.82 g mL^{-1} and 241.7 K, respectively.

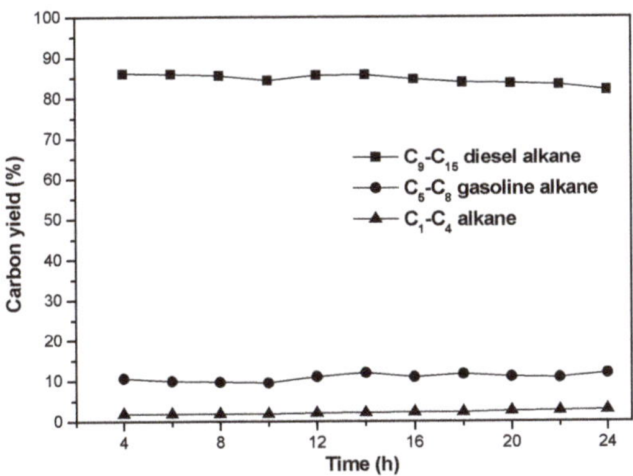

Figure 8. Carbon yields of different hydrocarbons over the Pd/H-ZSM-5 catalyst. Experimental conditions: 1.80 g catalyst, 6.0 MPa H_2, 573 K, 2,5-bis(furan-2-ylmethyl)-cyclopentanone flow rate of 0.04 mL min^{-1}, and hydrogen flow rate of 120 mL min^{-1}.

3. Materials and Methods

3.1. Catalyst Preparation

The CaO, MgO and CeO_2 catalysts were purchased from Aladdin Chemical Reagent Co., Ltd. (Shanghai, China). The MgAl hydrotalcite (MgAl-HT) catalyst was homemade according to the method in [15]. To do this, an aqueous solution of magnesium nitrate and aluminum nitrate was added into another aqueous solution of sodium hydroxide and sodium carbonate. This process was conducted at 343 K under vigorous mechanical agitation. After aging at this temperature for 24 h, the as-obtained solid was filtered, repeatedly washed with water until the pH value of the filtrate changed to 7 and dried at 353 K. The dried precursor was calcined for 8 h at 723 K in nitrogen atmosphere. The LiAl hydrotalcite (LiAl-HT) catalyst was synthesized by the method introduced in our previous work.

Typically, an aqueous solution of Al(NO$_3$)$_3$·9H$_2$O (125 mL 0.4 mol L^{-1}) was slowly added into 300 mL solution of lithium hydroxide (1.5 mol L^{-1}) and sodium carbonate (0.08 mol L^{-1}). The dried precursor was activated by nitrogen flow for 8 h at 723 K.

The Raney Co and Raney nickel catalysts were obtained from Dalian Tongyong Chemical Co., Ltd. The noble metal catalysts loaded on activated carbon (denoted as M/C, M = Pt, Pd or Ru) were obtained from Aladdin Chemical Reagent Co., Ltd. The metal contents were 5 wt.% in the M/C catalysts on the basis of the information from the supplier. The H-ZSM-5 loaded Pt, Pd, and Ru catalysts were homemade by the incipient wetness impregnation of H-ZSM-5 (SiO$_2$/Al$_2$O$_3$ = 25, provided by Nankai University) and the aqueous solutions of H$_2$PtCl$_6$, PdCl$_2$ and RuCl$_3$, respectively. For comparison, the metal contents of the M/H-ZSM-5 (M = Pt, Pd, and Ru) catalysts were fixed as 5 wt.%.

Pd/C-CaO catalyst was prepared by grinding method. Calcium oxide and palladium–carbon were mixed at a mass ratio of 1:1, and then grinded in a mortar until the mixture was uniform.

3.2. Aldol Condensation

The synthesis of 2,5-bis(2-furylmethylidene)-cyclopentanone was realized directly by the aldol condensation of furfural and cyclopentanone in a stainless-steel kettle reactor. In each test, 0.84 g cyclopentanone, 1.92 g furfural and 0.05 g solid base were utilized. The stainless-steel kettle reactor was flushed with nitrogen three times before the test. The mixture of reactant and catalyst was stirred for 10 h at 423 K. Afterwards, the stainless-steel kettle reactor was cooled using ice water. The reaction product was dissolved in 40 mL tetrahydrofuran containing internal standard of cyclohexanone. The liquid products were analyzed by an Agilent 7890A gas chromatograph (GC, Shanghai, China).

3.3. One Pot Aldol Condensation

The direct synthesis of 2,5-bis(furan-2-ylmethyl)-cyclopentanone was conducted by the one-pot cascade reaction of cyclopentanone, furfural, and hydrogen using a 20 mL stainless-steel kettle reactor (Internal diameter: 2.0 cm, high: 8.0 cm) with hydrogen pressure gauge. In each test, 0.84 g cyclopentanone, 1.92 g furfural, 0.05 g solid base and 0.05 g Raney metal (or M/C) catalyst were utilized. The reactor was flushed by hydrogen three times before the test. Subsequently, hydrogen was added into the reactor to make the system pressure up to 4.0 MPa. The mixture of reactant and catalyst was stirred for 10 h at 423 K, and then quenched to room temperature. The liquid product was isolated with catalyst by centrifuge and analyzed by the Agilent 7890A gas chromatograph.

3.4. Hydrodeoxygenation

The hydrodeoxygenation of 2,5-bis(furan-2-methyl)-cyclopentanone was implemented at 573 K under solvent-free condition. In each test, 1.8 g Pd-HZSM-5 catalyst were used. This catalyst was reduced in-situ in a stainless-steel tube reactor (ϕ6.0 mm × 1.5 mm, length: 350 mm) for 2 h at 773 K by hydrogen flow (120 ml min^{-1}). Then, the reactor temperature dropped to 573 K and stabilized at this value for 0.5 h. 2,5-Bis (furan-2-methyl)-cyclopentanone was transported to the reactor by high pressure pump (0.04 mL min^{-1}) with hydrogen (120 mL min^{-1}). Mixture products were collected in the gas–liquid separation tank and back pressure regulator (maintaining system pressure at 6.0 MPa). Gas phase samples were analyzed on-line by Agilent 6890N GC. The liquid samples were collected regularly from the bottom of the separation tank and diluted and analyzed by Agilent 7890A gas chromatograph.

4. Conclusions

Diesel and jet fuel range cycloalkanes were synthesized from furfural and cyclopentanone by the one-pot aldol condensation/hydrogenation, and hydrodeoxygenation (HDO) in solvent-free conditions. Among the investigated catalysts, Pd/C-CaO displayed the highest activity for aldol condensation/hydrogenation reaction. Over it, high carbon yield (98.5%) of 2,5-bis(furan-2-ylmethyl)-cyclopentanone was acquired under mild conditions (4.0 MPa H$_2$, and 423 K).

2,5-bis(furan-2-ylmethyl)-cyclopentanone exists in liquid state at room temperature. As a result, it can be directly used for the solvent-free hydrodeoxygenation process. This is advantageous in industrial application. Pd/H-ZSM-5 was discovered to be a highly active and stable catalyst for the HDO of 2,5-bis(furan-2-ylmethyl)-cyclopentanone. Over it, 86.1% carbon yield of diesel and jet fuel range cycloalkanes was reached. No significant inactivation was noticed during the 24 h time on stream. According to our measurements, the cycloalkane mixture as obtained has a freezing point of 241.7 K and a density of 0.82 g mL^{-1}. As a potential application, it can be mixed with diesel and jet fuel to enhance their volume calorific values or payloads.

Supplementary Materials: The following are available online at http://www.mdpi.com/2073-4344/9/11/886/s1, Figure S1: GC chromatogram of the aldol condensation product of cyclopentanone and furfural, Figure S2: ^{1}H and ^{13}C NMR spectra of 2,5-bis (2-furylmethylidene) cyclopentanone, Figure S3: The influences of catalyst dosage and temperature on cyclopentanone conversion, and 2,5-bis(2-furylmethylidene)cyclopentanone conversion carbon yield, Figure S4: GC chromatogram of the 2,5-bis(furan-2-ylmethyl)cyclopentanone from the one-pot aldol condensation/hydrogenation reaction, Figure S5: ^{1}H and ^{13}C NMR spectra of 2,5-bis(furan-2-ylmethyl) cyclopentanone, Figure S6: Mass spectrogram of the 2,5-bis(furan-2-ylmethyl)cyclopentanone, Figure S7: GC chromatogram of the products from the solvent-free HDO of 2,5-bis(furan-2-ylmethyl)-cyclopentanone, Figure S8: Mass spectrogram of the 1,3-dipentylcyclopentane.

Author Contributions: Data curation, W.W. and S.S.; Formal analysis, F.H.; Investigation, G.L. and X.S.; Writing—review & editing, N.L.

Funding: This research was funded by the National Natural Science Foundation of China (Nos. 21776273, 21721004, and 21690082), DNL Cooperation Fund, CAS (DNL180301), the Strategic Priority Research Program of the Chinese Academy of Sciences (XDB17020100), the National Key Projects for Fundamental Research and Development of China (2016YFA0202801), and Dalian Science Foundation for Distinguished Young Scholars (No. 2015R005). Dr. Wei Wang appreciates Natural Science Basic Research Plan in Shaanxi Province of China (2018JM2048), the Open Project of Shaanxi Key Laboratory of Catalysis (No. SLGPT2019KF01-24), and the Funds of Research Programs of Shaanxi University of Technology (No. SLGQD13(2)-1) for financial support.

Acknowledgments: This work was supported by Natural Science Basic Research Plan in Shaanxi Province of China (2018JM2048) and the Open Project of Shaanxi Key Laboratory of Catalysis (No. SLGPT2019KF01-24).

Conflicts of Interest: The authors declare no conflict of interest.

References

1. Huber, G.W.; Iborra, S.; Corma, A. Synthesis of Transportation Fuels from Biomass: Chemistry, Catalysts, and Engineering. *Chem. Rev.* **2006**, *106*, 4044–4098. [CrossRef]

2. Harvey, B.G.; Wright, M.E.; Quintana, R.L. High-Density Renewable Fuels Based on the Selective Dimerization of Pinenes. *Energy Fuels* **2010**, *24*, 267–273. [CrossRef]

3. Alonso, D.M.; Wettstein, S.G.; Dumesic, J.A. Bimetallic catalysts for upgrading of biomass to fuels and chemicals. *Chem. Soc. Rev.* **2012**, *41*, 8075–8098. [CrossRef]

4. Zhao, C.; Kou, Y.; Lemonidou, A.A.; Li, X.; Lercher, J.A. Highly Selective Catalytic Conversion of Phenolic Bio-Oil to Alkanes. *Angew. Chem.* **2009**, *121*, 4047–4050. [CrossRef]

5. Bond, J.Q.; Upadhye, A.A.; Olcay, H.; Tompsett, G.A.; Jae, J.; Xing, R.; Alonso, D.M.; Wang, D.; Zhang, T.; Kumar, R.; et al. Production of renewable jet fuel range alkanes and commodity chemicals from integrated catalytic processing of biomass. *Energy Environ. Sci.* **2014**, *7*, 1500–1523. [CrossRef]

6. Shylesh, S.; Gokhale, A.A.; Ho, C.R.; Bell, A.T. Novel Strategies for the Production of Fuels, Lubricants, and Chemicals from Biomass. *Accounts Chem. Res.* **2017**, *50*, 2589–2597. [CrossRef]

7. Li, C.; Zhao, X.; Wang, A.; Huber, G.W.; Zhang, T. Catalytic Transformation of Lignin for the Production of Chemicals and Fuels. *Chem. Rev.* **2015**, *115*, 11559–11624. [CrossRef]

8. Zakzeski, J.; Bruijnincx, P.C.A.; Jongerius, A.L.; Weckhuysen, B.M. The Catalytic Valorization of Lignin for the Production of Renewable Chemicals. *Chem. Rev.* **2010**, *110*, 3552–3599. [CrossRef]

9. Besson, M.; Gallezot, P.; Pinel, C. Conversion of biomass into chemicals over metal catalysts. *Chem. Rev.* **2014**, *114*, 1827–1870. [CrossRef]

10. Wang, X.; Rinaldi, R. A Route for Lignin and Bio-Oil Conversion: Dehydroxylation of Phenols into Arenes by Catalytic Tandem Reactions. *Angew. Chem.* **2013**, *125*, 11713–11717. [CrossRef]

11. Zan, Y.; Miao, G.; Wang, H.; Kong, L.; Ding, Y.; Sun, Y. Revealing the roles of components in glucose selective hydrogenation into 1,2-propanediol and ethylene glycol over Ni-MnO-ZnO catalysts. *J. Energy Chem.* **2019**, *38*, 15–19. [CrossRef]

12. Fang, S.; Cui, Z.; Zhu, Y.; Wang, C.; Bai, J.; Zhang, X.; Xu, Y.; Liu, Q.; Chen, L.; Zhang, Q.; et al. In situ synthesis of biomass-derived Ni/C catalyst by self-reduction for the hydrogenation of levulinic acid to γ-valerolactone. *J. Energy Chem.* **2019**, *37*, 204–214. [CrossRef]

13. Liu, C.; Wang, H.; Karim, A.M.; Sun, J.; Wang, Y. Catalytic fast pyrolysis of lignocellulosic biomass. *Chem. Soc. Rev.* **2014**, *43*, 7594–7623. [CrossRef]

14. Kunkes, E.L.; Simonetti, D.A.; West, R.M.; Serrano-Ruiz, J.C.; Gartner, C.A.; Dumesic, J.A.; Baldauf, S.L. Catalytic Conversion of Biomass to Monofunctional Hydrocarbons and Targeted Liquid-Fuel Classes. *Science* **2008**, *322*, 417–421. [CrossRef]

15. Huber, G.W. Production of Liquid Alkanes by Aqueous-Phase Processing of Biomass-Derived Carbohydrates. *Science* **2005**, *308*, 1446–1450. [CrossRef]

16. Xing, R.; Subrahmanyam, A.V.; Olcay, H.; Qi, W.; Van Walsum, G.P.; Pendse, H.; Huber, G.W. Production of jet and diesel fuel range alkanes from waste hemicellulose-derived aqueous solutions. *Green Chem.* **2010**, *12*, 1933. [CrossRef]

17. Corma, A.; De La Torre, O.; Renz, M.; Villandier, N. Production of High-Quality Diesel from Biomass Waste Products. *Angew. Chem.* **2011**, *50*, 2375–2378. [CrossRef]

18. Corma, A.; Arias, K.S.; Climent, M.J.; Iborra, S. Synthesis of high quality alkyl naphthenic kerosene by reacting an oil refinery with a biomass refinery stream. *Energy Environ. Sci.* **2015**, *8*, 317–331.

19. Xia, Q.-N.; Cuan, Q.; Liu, X.-H.; Gong, X.-Q.; Lu, G.-Z.; Wang, Y.-Q. Pd/NbOPO 4 Multifunctional Catalyst for the Direct Production of Liquid Alkanes from Aldol Adducts of Furans. *Angew. Chem.* **2014**, *53*, 9755–9760. [CrossRef]

20. Sacia, E.R.; Balakrishnan, M.; Deaner, M.H.; Goulas, K.A.; Toste, F.D.; Bell, A.T. Highly Selective Condensation of Biomass-Derived Methyl Ketones as a Source of Aviation Fuel. *ChemSusChem* **2015**, *8*, 1726–1736. [CrossRef]

21. Nie, G.; Zhang, X.; Han, P.; Xie, J.; Pan, L.; Wang, L.; Zou, J.-J. Lignin-derived multi-cyclic high density biofuel by alkylation and hydrogenated intramolecular cyclization. *Chem. Eng. Sci.* **2017**, *158*, 64–69. [CrossRef]

22. Tang, H.; Chen, F.; Li, G.; Yang, X.; Hu, Y.; Wang, A.; Cong, Y.; Wang, X.; Zhang, T.; Li, N. Synthesis of jet fuel additive with cyclopentanone. *J. Energy Chem.* **2019**, *29*, 23–30. [CrossRef]

23. Nie, G.; Zhang, X.; Pan, L.; Wang, M.; Zou, J.-J. One-pot production of branched decalins as high-density jet fuel from monocyclic alkanes and alcohols. *Chem. Eng. Sci.* **2018**, *180*, 64–69. [CrossRef]

24. Zhao, C.; Camaioni, D.M.; Lercher, J.A. Selective catalytic hydroalkylation and deoxygenation of substituted phenols to bicycloalkanes. *J. Catal.* **2012**, *288*, 92–103. [CrossRef]

25. Li, H.; Riisager, A.; Saravanamurugan, S.; Pandey, A.; Sangwan, R.S.; Yang, S.; Luque, R. Carbon-Increasing Catalytic Strategies for Upgrading Biomass into Energy-Intensive Fuels and Chemicals. *ACS Catal.* **2017**, *8*, 148–187. [CrossRef]

26. Jing, Y.; Xia, Q.; Xie, J.; Liu, X.; Guo, Y.; Zou, J.-J.; Wang, Y. Correction to Robinson Annulation-Directed Synthesis of Jet-Fuel-Ranged Alkylcyclohexanes from Biomass-Derived Chemicals. *ACS Catal.* **2018**, *8*, 3280–3285. [CrossRef]

27. Mariscal, R.; Maireles-Torres, P.; Ojeda, M.; Sádaba, I.; Granados, M.L. Furfural: a renewable and versatile platform molecule for the synthesis of chemicals and fuels. *Energy Environ. Sci.* **2016**, *9*, 1144–1189. [CrossRef]

28. Yang, J.; Li, N.; Li, G.; Wang, W.; Wang, A.; Wang, X.; Cong, Y.; Zhang, T. Solvent-Free Synthesis of C 10 and C 11 Branched Alkanes from Furfural and Methyl Isobutyl Ketone. *ChemSusChem* **2013**, *6*, 1149–1152. [CrossRef]

29. Chen, F.; Li, N.; Li, S.; Yang, J.; Liu, F.; Wang, W.; Wang, A.; Cong, Y.; Wang, X.; Zhang, T. Solvent-free synthesis of C9 and C10 branched alkanes with furfural and 3-pentanone from lignocellulose. *Catal. Commun.* **2015**, *59*, 229–232. [CrossRef]

30. Yang, J.; Li, N.; Li, S.; Wang, W.; Li, L.; Wang, A.; Wang, X.; Cong, Y.; Zhang, T. Synthesis of diesel and jet fuel range alkanes with furfural and ketones from lignocellulose under solvent free conditions. *Green Chem.* **2014**, *16*, 4879–4884. [CrossRef]

31. Jing, Y.; Xin, Y.; Guo, Y.; Liu, X.; Wang, Y. Highly efficient Nb$_2$O$_5$ catalyst for aldol condensation of biomass-derived carbonyl molecules to fuel precursors. *Chin. J. Catal.* **2019**, *40*, 1168–1177. [CrossRef]

32. Li, C.; Ding, D.; Xia, Q.; Liu, X.; Wang, Y. Conversion of raw lignocellulosic biomass into branched long-chain alkanes through three tandem steps. *ChemSusChem* **2016**, *9*, 1712–1718. [CrossRef]

33. Xu, J.; Li, N.; Yang, X.; Li, G.; Wang, A.; Cong, Y.; Wang, X.; Zhang, T. Synthesis of Diesel and Jet Fuel Range Alkanes with Furfural and Angelica Lactone. *ACS Catal.* **2017**, *7*, 5880–5886. [CrossRef]

34. Hronec, M.; Fulajtarová, K. Selective transformation of furfural to cyclopentanone. *Catal. Commun.* **2012**, *24*, 100–104. [CrossRef]

35. Hronec, M.; Fulajtárová, K.; Vávra, I.; Soták, T.; Dobročka, E.; Mičušík, M. Carbon supported Pd–Cu catalysts for highly selective rearrangement of furfural to cyclopentanone. *Appl. Catal. B Environ.* **2016**, *181*, 210–219. [CrossRef]

36. Yang, Y.; Du, Z.; Huang, Y.; Lu, F.; Wang, F.; Gao, J.; Xu, J. Conversion of furfural into cyclopentanone over Ni–Cu bimetallic catalysts. *Green Chem.* **2013**, *15*, 1932–1940. [CrossRef]

37. Li, Y.; Guo, X.; Liu, D.; Mu, X.; Chen, X.; Shi, Y. Selective Conversion of Furfural to Cyclopentanone or Cyclopentanol Using Co-Ni Catalyst in Water. *Catalysts* **2018**, *8*, 193. [CrossRef]

38. Wang, Y.; Zhao, D.; Rodríguez-Padrón, D.; Len, C. Recent Advances in Catalytic Hydrogenation of Furfural. *Catalysts* **2019**, *9*, 796. [CrossRef]

39. Sheng, X.; Xu, Q.; Wang, X.; Li, N.; Jia, H.; Shi, H.; Niu, M.; Zhang, J.; Ping, Q. Waste Seashells as a Highly Active Catalyst for Cyclopentanone Self-Aldol Condensation. *Catalysts* **2019**, *9*, 661. [CrossRef]

40. Yang, J.; Li, N.; Li, G.; Wang, W.; Wang, A.; Wang, X.; Cong, Y.; Zhang, T. Synthesis of renewable high-density fuels using cyclopentanone derived from lignocellulose. *Chem. Commun.* **2014**, *50*, 2572–2574. [CrossRef]

41. Sheng, X.; Li, G.; Wang, W.; Cong, Y.; Huber, G.W.; Zhang, T. Dual-bed Catalyst System for the Direct Synthesis of High Density Aviation Fuel with Cyclopentanone from Lignocellulose. *AIChE J.* **2016**, *62*, 2754–2761. [CrossRef]

42. Wang, W.; Li, N.; Li, G.; Li, S.; Wang, W.; Wang, A.; Cong, Y.; Wang, X.; Zhang, T. Synthesis of Renewable High-Density Fuel with Cyclopentanone Derived from Hemicellulose. *ACS Sustain. Chem. Eng.* **2017**, *5*, 1812–1817. [CrossRef]

43. Hronec, M.; Fulajtarová, K.; Liptaj, T.; Stolcova, M.; Prónayová, N.; Soták, T. Cyclopentanone: A raw material for production of C15 and C17 fuel precursors. *Biomass Bioenergy* **2014**, *63*, 291–299. [CrossRef]

44. Wang, W.; Ji, X.; Ge, H.; Li, Z.; Tian, G.; Shao, X.; Zhang, Q. Synthesis of C 15 and C 10 fuel precursors with cyclopentanone and furfural derived from hemicellulose. *RSC Adv.* **2017**, *7*, 16901–16907. [CrossRef]

45. Deng, Q.; Xu, J.; Han, P.; Pan, L.; Wang, L.; Zhang, X.; Zou, J.-J. Efficient synthesis of high-density aviation biofuel via solvent-free aldol condensation of cyclic ketones and furanic aldehydes. *Fuel Process. Technol.* **2016**, *148*, 361–366. [CrossRef]

46. Ao, L.; Zhao, W.; Guan, Y.-S.; Wang, D.-K.; Liu, K.-S.; Guo, T.-T.; Fan, X.; Wei, X.-Y. Efficient synthesis of C15 fuel precursor by heterogeneously catalyzed aldol-condensation of furfural with cyclopentanone. *RSC Adv.* **2019**, *9*, 3661–3668. [CrossRef]

47. Hronec, M.; Fulajtarová, K.; Liptaj, T.; Prónayová, N.; Soták, T. Bio-derived fuel additives from furfural and cyclopentanone. *Fuel Process. Technol.* **2015**, *138*, 564–569. [CrossRef]

© 2019 by the authors. Licensee MDPI, Basel, Switzerland. This article is an open access article distributed under the terms and conditions of the Creative Commons Attribution (CC BY) license (http://creativecommons.org/licenses/by/4.0/).

Article

Gel-Type and Macroporous Cross-Linked Copolymers Functionalized with Acid Groups for the Hydrolysis of Wheat Straw Pretreated with an Ionic Liquid

Giulia Lavarda [1,2], Silvia Morales-delaRosa [2,*], Paolo Centomo [1,*], Jose M. Campos-Martin [2], Marco Zecca [1] and Jose L. G. Fierro [2]

[1] Dipartimento di Scienze Chimiche, Università di Padova, via Marzolo 1, I-35131 Padua, Italy
[2] Sustainable Energy and Chemistry Group. Instituto de Catálisis y Petroleoquímica, CSIC, Marie Curie, 2, Cantoblanco, 28049 Madrid, Spain
* Correspondence: smorales@icp.csic.es (S.M.-d.); paolo.centomo@unipd.it (P.C.)

Received: 12 July 2019; Accepted: 6 August 2019; Published: 8 August 2019

Abstract: Several sulfonated cross-linked copolymers functionalized with hydroxyl and carboxylic groups have been synthesized. The amount of the cross-linking monomer was tailored (from 4% up to 40%) to tune the resulting micro- and nano-morphologies, and two types of catalysts, namely, gel-type and macroreticular catalysts, were obtained. These copolymers were employed in the catalytic hydrolysis of wheat straw pretreated in 1-ethyl-3-methylimidazolium acetate to obtain sugars. Remarkably, the presence of additional oxygenated groups enhances the catalytic performances of the polymers by favoring the adsorption of β-(1,4)-glucans and makes these materials significantly more active than an acidic resin bearing only sulfonic groups (i.e., Amberlyst 70). In addition, the structure of the catalyst (gel-type or macroreticular) appears to be a determining factor in the catalytic process. The gel-type structure provides higher glucose concentrations because the morphology in the swollen state is more favorable in terms of the accessibility of the catalytic centers. The observed catalytic behavior suggests that the substrate diffuses within the swollen polymer matrix and indirectly confirms that the pretreatment based on dissolution/precipitation in ionic liquids yields a substantial enhancement of the conversion of lignocellulosic biomass to glucose in the presence of heterogeneous catalysts.

Keywords: heterogeneous catalysts; ionic liquid; lignocellulosic biomass; acidic resin catalysts

---

1. Introduction

Sustainable energy is increasingly being applied to all social, economic and political fields worldwide. The European Union (EU) has a political objective to decarbonize its economy by 2050 and reduce 80%–95% of its greenhouse gas (GHG) emissions [1,2].

In addition, market requirements are trending towards an ecological compromise, and society is increasingly becoming aware of the need to consume products that have been obtained through processes with minimized environmental impacts. As a result, the use of bioproducts, renewable energy [3] and biorefineries is constantly expanding to various industries, such as the chemical, pharmaceutical, paper and food industries, as well as others.

The total amount of lignocellulosic biomass produced in the EU-28 is calculated to be 419 Mt, among which wheat is the main contributor (155 Mt, 37%). Thus, these residues are highly suitable raw materials for both producing second-generation biofuels and preparing high value-added compounds in the EU [4].

The transformation of lignocellulosic biomass into valuable chemicals requires, almost inevitably, a previous treatment (physical, chemical, biochemical, biological or a combination of them) of the substrate [5]. In this way, the particle size can be reduced, the porosity can be improved, and the crystallinity of the cellulose can be altered. After fractioning its main components, e.g., cellulose, hemicellulose and lignin, biomass can be converted into a wide variety of industrial products, such as biofuels, biomaterials, cellulose pulps, cellulose nanofibers, oligosaccharides and a large number of by-products, in addition to lignin derivatives.

Numerous studies have shown ionic liquids (ILs) to be effective at solubilizing lignocellulosic biomass, allowing for subsequent regeneration by precipitation with anti-solvents. Depending on the anti-solvent used, it is possible to achieve selective precipitation and separate the lignocellulosic biomass into its main components [6–14]. Among the most commonly employed ILs, 1-ethyl-3-methylimidazolium acetate ([EMIM]OAc) holds a privileged position due to its physico-chemical properties.

The acid hydrolysis of the polysaccharide fractions of lignocellulosic biomass, which leads to monosaccharides that can be converted into fine chemicals or platform molecules used for industrial applications, can be performed in the presence of enzymes, homogeneous catalysts or heterogeneous catalysts [15–21]. During the last few years, increasing attention has been focused on the heterogeneous approach, which leads to several advantages in terms of the process design and plant costs [22–25]. Some of us have investigated using resins as catalysts or scaffolds in material chemistry [26–30]. These resins could be more advantageous materials than the most commonly employed materials used for the acid hydrolysis of lignocellulosic biomass (e.g., zeolites, carbon and hybrid materials), especially due to their structural and functional versatility and stability in reaction media. In fact, the morphology and chemical properties of these resins can be easily tuned by modifying the nature and amount of structural, cross-linking and functional monomers used [31–33].

The presence of additional polar groups in combination with sulfonic groups in carbon-based materials clearly enhances the hydrolysis of cellulose [34–37] by favoring the adsorption of β-(1,4)-glucans. This effect was observed also in other supports, such as silica and resins [38–40]. However, the effect on the catalytic performances of the polar group amount and of the synergy between the presence of additional acid groups and the morphology of the polymer matrix has not yet been studied.

In this work, we describe the design and synthesis of several novel cross-linked copolymers, as well as the ability of these copolymers to catalyze the acid hydrolysis of wheat straw pretreated with [EMIM]OAc. For these catalysts, the amount of the cross-linking monomer has been tailored to tune the morphology of the materials from a gel-type matrix to a macroreticular structure. Moreover, the chemical properties of the supports have been controlled by decorating the polymers with carboxylic and hydroxyl groups. The percentage of the functional monomer was adjusted to study the effect of the number of polar groups on the performance of the catalyst.

2. Results and Discussion

2.1. Synthesis of the Catalysts

Two copolymers containing 2-acrylamido-2-methyl-1-propanesulfonic acid (AMPS), 2-hydroxyethyl methacrylate (HEMA) and ethylene dimethacrylate (EDMA) and four copolymers containing AMPS, methacrylic acid (MAA) and EDMA were prepared by free radical polymerization using N,N-Dimethylformamide (DMF) as the solvent.

For both the AHE (AMPS-HEMA-EDMA) and AME (AMPS-MAA-EDMA) polymers (see Sections 3.2 and 3.3 of this work), the ratio of HEMA or MAA to AMPS was tuned from 1:2 to 1:3 to obtain materials with different amounts of polar groups. In the case of the AHE catalysts, the molar amount of the cross-linking monomer (EDMA) was established to obtain gel-type textures. For the AME materials, the molar percentage of EDMA was varied from 4% to 40%, and two kinds of catalysts were obtained, i.e., gel-type structures and rigid macroreticular resins. The notations AHE Y-X and AME Y-X, where Y and X are the nominal molar ratios of AMPS and EDMA, respectively, will be used henceforward to refer to the polymer materials. The expected chemical structures of the AHE and AME materials are reported in Figures 1 and 2.

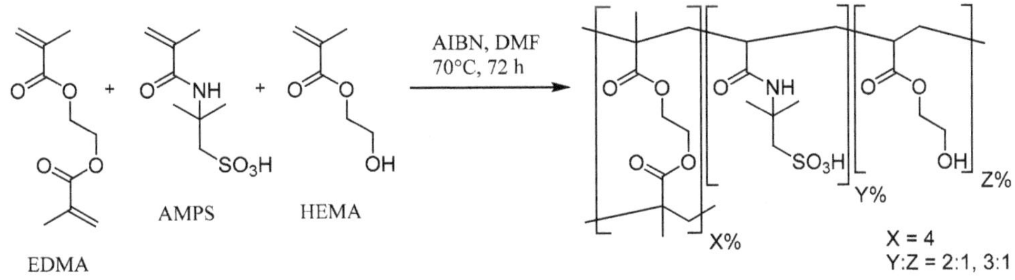

Figure 1. Chemical structure of the AHE copolymers.

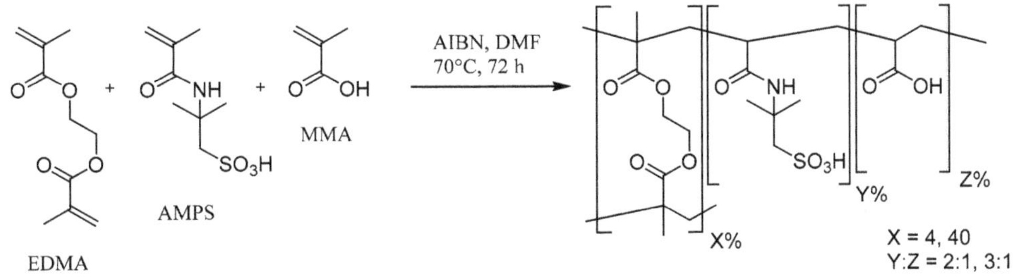

Figure 2. Chemical structure of the AME copolymers.

2.2. Elemental Analysis and Cation Exchange Capacity (CEC) of the Catalysts

The elemental analysis and determination of the cation exchange capacity (CEC) by back-titration confirmed the composition of the synthetized AME and AHE materials (Tables 1 and 2). Slight differences between the expected elemental composition (calculated from the composition of the monomer mixtures) and the experimental results may be due to the incomplete polymerization of the monomers or the presence of moisture in the materials. As a matter of fact, the presence of a substantial amount of water in acrylic resins is well known, particularly in the case of polar monomers, which makes quite reliable the hypothesis of the presence of water in the AHE and AME materials. Regarding the cation exchange capacity of the AME and AHE catalysts, back-titration pointed out that the actual CEC of the materials is somewhat lower than the nominal CEC (calculated from the AMPS molar content), although the results appear in good agreement.

Table 1. The expected molar composition of the AHE materials.

		χ (%)			Weight Ratios (%)				CEC
		AMPS	**HEMA**	**EDMA**	**C**	**H**	**N**	**S**	**(mmol H$^+$/g)**
AHE 64-4	Experimental	-	-	-	45.36	6.56	4.75	10.29	3.12
	Theoretical [1]	64	32	4	45.12	6.06	4.95	11.34	3.48 [2]
AHE 72-4	Experimental	-	-	-	43.94	6.49	5.27	11.53	3.47
	Theoretical [1]	72	24	4	43.87	6.59	5.35	12.26	3.78 [2]

[1] Excludes the presence of contaminants; [2] cation exchange capacity (CEC) was calculated from the AMPS content.

The explanations for these discrepancies are the same as those mentioned in the elemental analysis. In the case of the AME polymers, the possibility that the CEC values were underestimated due to the re-acidification of the more accessible carboxylic groups during the titration must also be considered.

Table 2. The expected molar composition of the AME materials.

		χ (%)			Weight Ratios (%)				CEC (mmol H$^+$/g)
		AMPS	**MAA**	**EDMA**	**C**	**H**	**N**	**S**	
AME 64-4	Experimental	-	-	-	43.4	6.42	5.08	10.75	5.49
	Theoretical [1]	64	32	4	44.01	6.48	5.23	12.21	5.71 [2]
AME 72-4	Experimental	-	-	-	42.53	6.42	5.45	11.56	5.3
	Theoretical [1]	72	24	4	43.23	6.44	5.67	12.98	5.4 [2]
AME 40-40	Experimental	-	-	-	50.82	6.65	3.01	6.3	2.99
	Theoretical [1]	40	20	40	50.8	6.74	3.12	7.15	3.34 [2]
AME 45-40	Experimental	-	-	-	49.97	6.65	3.36	6.66	2.98
	Theoretical [1]	45	15	40	50.19	6.71	3.4	7.78	3.24 [2]

[1] Excludes the presence of contaminants; [2] CEC was calculated from the AMPS content.

2.3. FTIR Spectroscopy of Catalysts

From the FTIR spectra of the AHE materials (Figure 3), the broad band at approximately 3500 cm^{-1} is due to the N-H stretching of the acrylamide moiety (AMPS), the O-H stretching of the hydroxyl group of HEMA and the sulfonic group of AMPS and the water contribution. The group of bands at 2858–2964 cm^{-1} can be attributed to the stretching of C-H bonds. The signal corresponding to the C=O stretching of the ester group of HEMA and EDMA can be observed at 1732 cm^{-1}. The signal at 1644–1647 cm^{-1} can be ascribed to the stretching of the carbonyl groups of the acrylamide moiety (amide I band), whereas the weak signal at 1557 cm^{-1} is due to N-H bending (amide II band). The stretching of the C-N bond of the acrylamide group produces a signal at 1408 cm^{-1}. The strong bands at 1262 cm^{-1} and 1097–1099 cm^{-1} can be attributed to the asymmetric and symmetric stretching of the sulfonic group of AMPS, respectively, whereas the band at 1024–1025 cm^{-1} can be ascribed to the asymmetric O-C-C stretching of HEMA and EDMA. The strong band at 802–803 cm^{-1} is due to the out-of-plane wagging vibration of the N-H bonds of the primary amide group. Finally, it is possible to exclude the presence of a significant amount of unreacted C=C double bonds, due to the lack of signals at 1637 and 1610 cm^{-1}, which are expected for alkenes conjugated to carbonyl groups [41].

In the FTIR spectra of the AME materials (Figure 4), the N-H stretching of the acrylamide moiety and the O-H stretching of the sulfonic group in the AMPS monomer (as well as the water contribution) generate the broad band at approximately 3500 cm^{-1}. The C=O stretching of the ester group in the cross-linking monomer produces a band at 1725–1734 cm^{-1}. The C=O stretching of the acrylamide moiety generates a peak at 1652–1646 cm^{-1} (amide I band), whereas N-H bending is associated with a weak signal at 1557–1555 cm^{-1}. The signal at 1400 cm^{-1} is associated with the stretching of C-N. In addition, the asymmetric and symmetric stretching of the sulfonic group of AMPS produces two intense signals at 1261 cm^{-1} and 1101–1105 cm^{-1}, respectively, whereas the coupled C-C(=O)-O and O-C-C vibrations produce a band at 1035 cm^{-1} and contribute to the strong signal at 1261 cm^{-1}. The strong signal at 800–802 cm^{-1} is related to the out-of-plane wagging of the N-H bond in AMPS. As observed for the AHE resins, the lack of bands related to C=C stretching confirms the absence of appreciable amounts of unreacted monomers.

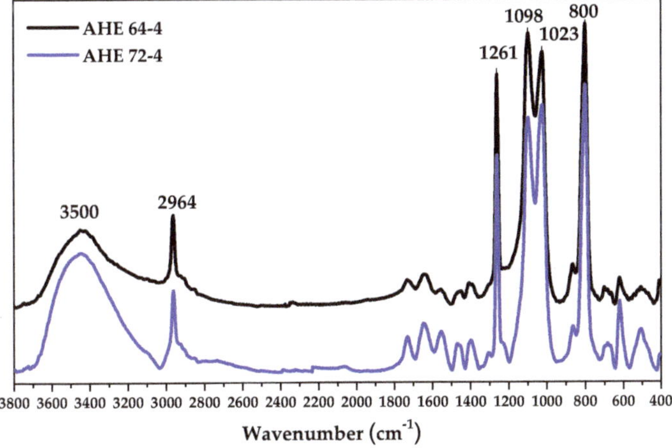

Figure 3. FTIR absorption spectra of the AHE catalysts.

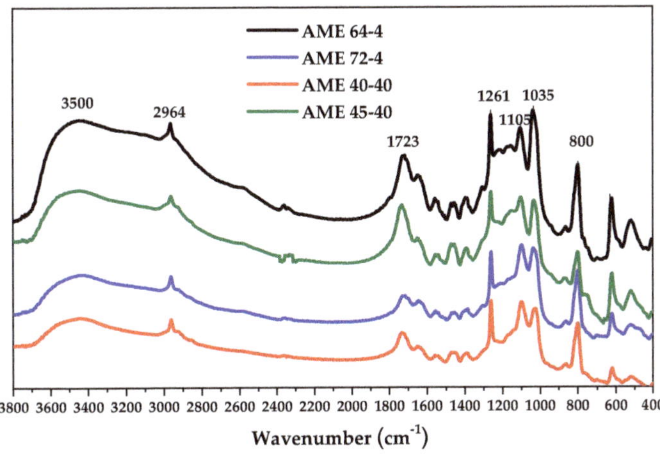

Figure 4. FTIR absorption spectra of the AME catalysts.

2.4. X-Ray Diffraction Analysis of the Wheat Straw Samples

The X-ray diffraction pattern of the untreated wheat straw (Figure 5) shows the characteristic peaks of the raw cellulose crystalline structure, although the signals are slightly weaker and broader with respect to the pure commercial cellulose. This effect can be attributed to the presence of amorphous components, such as hemicellulose and lignin, which do not have ordered structures. The strong peak at 23° corresponds to the diffraction generated by the (200) plane, whereas the wide peak between 15° and 17° represents the combination of the two reflections corresponding to (110) and (1$\bar{1}$0). The combined signal centered at 34° consists of several overlapping signals, among which is the signal generated from the (004) plane [42].

The X-ray diffraction pattern of the IL-treated wheat straw shows important changes with respect to the original sample (Figure 5). The intensity of the diffractogram is very low, indicating a clear loss in the initial crystallinity. Only a low intense and broad peak is detected at approximately 23°, which indicates the incipient formation of cellulose II, a structure that can appear in dissolved/regenerated cellulose.

This behavior has already been reported in previous works published by us and other authors using different samples of lignocellulosic biomass after pretreatment with ILs [17,21,43].

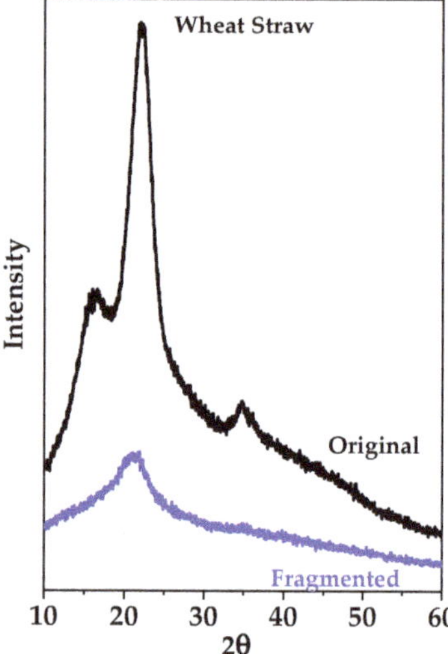

Figure 5. XRD characterization of the original wheat straw and solid fraction (rich in sugars) after pretreatment with [EMIM]OAc.

2.5. SEM Analysis of the Wheat Straw Samples

From the SEM micrographs of the original raw wheat straw samples, which were obtained at different magnifications, a structure consisting of vascular bundles forming hollows along small fragments (0.250 µm) is observed (Figure 6). In contrast, the treated sample shows a dramatic change in the structure and morphology of the material (Figure 7). These structural changes have been previously described for cellulose or lignocellulosic biomass samples treated with [EMIM]OAc and other ILs [21,39]. The surface of the pretreated solid (Figure 7) appears to be more porous than in the untreated wheat straw, making the biomass more susceptible to hydrolysis. The presence of remaining fragments of raw wheat straw (Figure 7) may be due to incomplete dissolution of the biomass, which allowed for the partial retention of the initial structure.

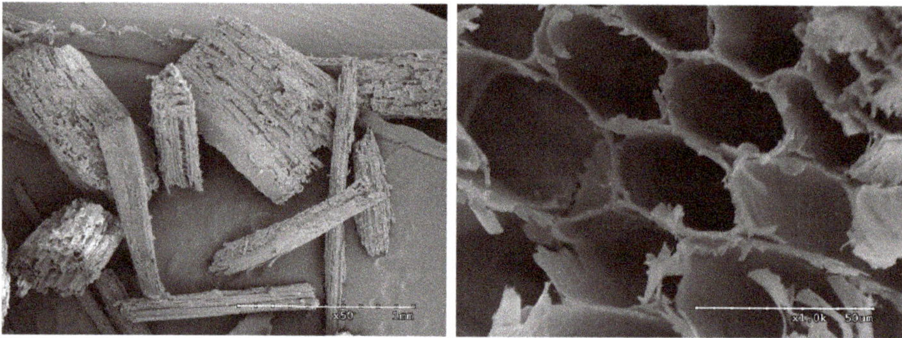

Figure 6. SEM micrographs of the raw wheat straw.

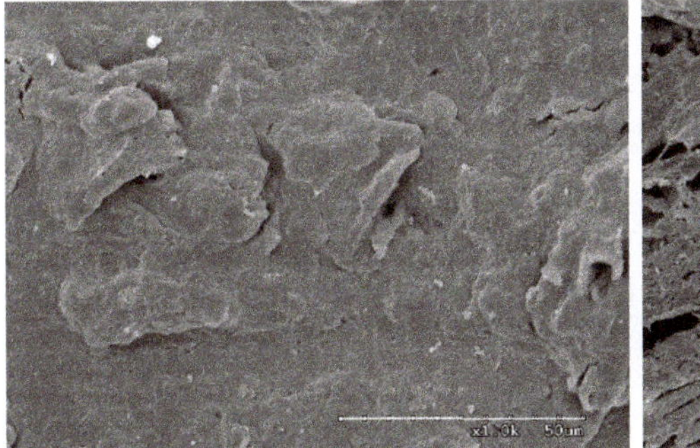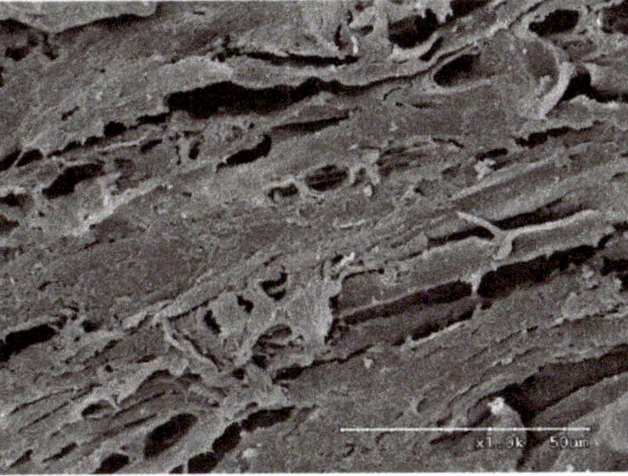

Figure 7. SEM micrographs of the wheat straw after dissolution in [EMIM]OAc (at 105 °C) and precipitation with acetone:water (1:1).

These results agree with the data obtained from X-ray diffraction, as both techniques clearly indicate that the original structure of the wheat straw is altered after pretreatment with [EMIM]OAc and subsequent fractional precipitation in the presence of acetone:water (1:1).

2.6. FTIR Spectroscopy of the Wheat Straw Samples

The changes in the composition of the wheat straw samples after pretreating with [EMIM]OAc were studied by attenuated total reflectance-Fourier transform infrared spectroscopy (ATR-FTIR). The region between 1800 and 850 cm^{-1} of the spectra is quite complex, and many peaks related to the main biomass component can be detected, allowing for changes in the composition of the untreated and treated lignocellulosic biomass to be monitored [44,45].

The FTIR spectrum of the starting material (Figure 8) shows several signals attributed to the main components of the lignocellulosic biomass (cellulose, hemicellulose and lignin). In particular, the band at 1715 cm^{-1} is associated with the C=O stretching of unconjugated carbonyl groups, whereas the signal at 1370 cm^{-1} is associated with the C-H symmetric deformation that is typical in cellulose and hemicellulose. The band at 1224 cm^{-1} can be attributed to the aryl C-O stretching vibration mode of phenolic hydroxyl groups and is thus assigned to lignin; in addition, the signal at 1260 cm^{-1} is attributed to the methoxy substituents of the aromatic rings in lignin [21,46]. The other weaker bands at 1422 and 1460 cm^{-1} are attributed to the asymmetric component of the C-H deformations in lignin and carbohydrates, respectively. The band at 1508 cm^{-1} is associated with the vibrations of the aromatic C-C skeletons of lignin, the signal at 1158 cm^{-1} is attributed to the C-O-C vibration in cellulose and hemicellulose, and the intense wide peak at 1035 cm^{-1} is associated with the C-O stretching vibrations of cellulose and hemicellulose. The absorption band at 899 cm^{-1} is assigned to the asymmetric C-O-C stretching of the β-(1,4)-glycosidic linkage [44,47].

On the other hand, from the spectrum of the IL-treated sample (Figure 8), many of the peaks associated with the presence of lignin disappear or decrease in intensity, and the peaks characteristic of cellulose and hemicellulose are clearly the predominant peaks. The spectra of the treated samples show an intense absorbance at 1640 cm^{-1} due to the presence of absorbed water. In summary, the FTIR analysis indicates a dramatic change in the chemical composition of the samples after treatment in [EMIM]OAc and subsequent regeneration in presence of acetone:water, which eliminates most of the lignin.

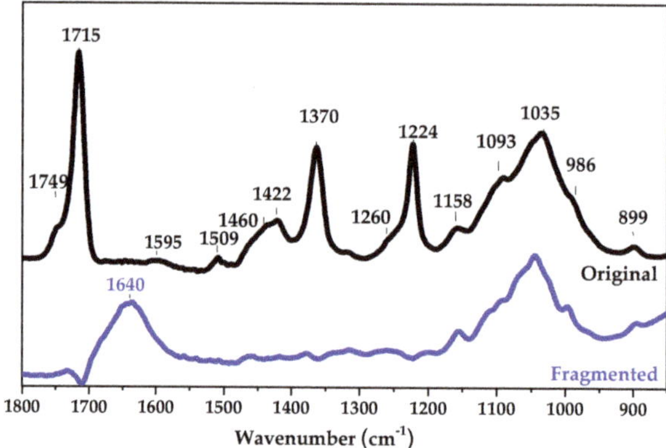

Figure 8. FTIR absorption spectra of the original wheat straw and the solid fraction (rich in sugars) obtained after pretreatment with [EMIM]OAc.

2.7. Catalytic Hydrolysis

To study the effect of the morphology and the presence of additional functional groups on the properties of the catalysts, we compared the reactivity of the AME and AHE catalysts with that of the commercial catalyst Amberlyst 70 in hydrolyzing wheat straw fragmented with [EMIM]OAc. In all the catalytic tests, a suitable amount of the synthesized resin catalyst was employed to keep the concentration of the sulfonic groups constant in the reaction mixture. The changes in the concentration of the main products obtained by hydrolyzing the biomass (glucose, levulinic acid, xylose and furfural) over time are depicted in Figures 9 and 10. Due to the quick decomposition of 5-HMF to levulinic acid in water, the concentration of 5-HMF was generally very low [17]. For this reason, the evolution of the concentration of this product over time is not reported.

Remarkably, the glucose and levulinic acid concentrations are definitely higher in the presence of the AME and AHE copolymers, both gel-type and macroreticular (Figures 9 and 10) than in the presence of the commercial acid catalyst Amberlyst 70. This evidence does not agree with what was expected based on the acid strength of the alkyl and aryl sulfonic groups of the tested catalyst (pKa = 1.9 for propanesulfonic acid; −2.7 for benzenesulfonic acid) [48]. This suggests that the hydroxyl groups (AHE) and carboxylic groups (AME) of the synthetized resins, which are not present in the Amberlyst 70 reference catalyst, act as promoters. Their direct co-catalytic role seems to be ruled out by the relatively low acid strength of the carboxylic and especially of the hydroxyl groups in comparison with the sulfonic groups. For instance, the pK_a of propionic acid, which can be taken as representative of the carboxylic moieties of AME resins, is 4.88 [49] three order of magnitude lower than the pK_a of propanesulfonic acid. In addition, the carboxylic groups of AME resins are lower in amount in comparison with the sulfonic acid groups.

Therefore, the promoting effect of these protic polar groups is likely physico-chemical rather than chemical. On the one hand, their presence could simply affect the hydrophilicity and the swelling behavior in the water of the catalysts, making them more readily accessible to the reactant's molecules. Swelling is of the utmost importance in heterogeneous solid–liquid processes involving cross-linked organic polymers. In this context, a clear difference in the behavior of the AME catalysts can be observed depending on the extent of cross-linking. In particular, the glucose concentration achieved over AME 64-4 and AME 72-4 (gel-type) is higher than that achieved over AME 40-40 and AME 45-40 (macroreticular). Whereas the 4% cross-linked resins are gel-type, the 40% are macroreticular [31,32]. These two classes of polymers differ in many ways; the most important way being their swelling behavior. The gel-type resins are glassy nonporous materials when dry, but, due to the low cross-linking degree, a suitable liquid can permeate the whole polymer framework, making the polymer to completely swell and thus producing a fully accessible gel-phase that extends throughout the whole volume of the resin. In contrast, the macroreticular resins have permanent mesopores, arising from the partial agglomeration of small, highly cross-linked and dense polymer particles. In this case, only a relatively thin layer of the polymer framework, just beneath the pore walls, can swell. Accordingly, only a small fraction of the polymer framework is involved in forming the accessible gel-phase. This implies that the number of accessible acidic groups per unit of catalyst volume is much higher in the case of sulfonated gel-type resins, which are consequently more catalytically active, than in macroreticular resins. This is somewhat counter-intuitive because macroreticular resins have permanent pores and moderate specific surface areas and are generally believed to be more active catalysts than gel-type resins. However, this simplistic view does not consider that macroreticular resins, due to their relatively high cross-linking degree, have more rigid frameworks, which eventually lower their efficiency and the apparent reaction rate. According to the previous discussion, the generally best performance of the gel-type catalysts investigated herein can be likely attributed to their higher swelling degree. For them, it can be observed that the curves of glucose production are practically superimposed for AHE 64-4 and AHE 72-4 and very close to each other up to 120 min for AME 64-4 and AME 72-4. The promoting effect in the gel-type catalysts, if any, seems little dependent on the OH/SO_3H or $COOH/SO_3H$ ratio. As the catalytic performance of the gel-type resins is the most dependent on the swelling behavior, this implies that for them the OH/SO_3H or $COOH/SO_3H$ ratio has a limited effect on the production of glucose.

To get a deeper insight into the origin of the promoting ability of the hydroxyl and carboxylic groups, macroreticular catalysts can be better compared. The results of AME 40-40 and AME 45-40 not only show that they produce a higher amount of glucose than Amberlyst 70, but also that a higher $COOH/SO_3H$ ratio leads to a higher concentration of glucose. This also true for the gel-type AME 64-4 and AME 72-4 at relatively long reaction time. The beneficial role of polar groups has already been observed in previous works with different heterogeneous catalysts functionalized with sulfonic groups [22] and other polar functional groups [34–38,41]. In some of these examples, the supports were rigid inorganic oxides, whereby effects connected to the swelling of the catalyst do not make sense. Moreover, it has already been reported that in resin catalysts, additional functional groups can favor the adsorption of the reagents with no appreciable increase of their swelling degree ("assisted adsorption"), provided they have some kind of affinity towards the molecules of the reagents. In that case, the "assisted adsorption" was based on hydrophobic interaction [50–52]. It can be therefore argued that also in AME 40-40 and AME 45-40 there is an "assisted adsorption" of the β-(1,4)-glucans sustained by the interactions of their molecules with hydroxyl and carboxylic groups. The effect of the assisted adsorption is to bring to closer contact the β-(1,4)-glucans with the active sulfonic groups or to increase their concentrations nearby the active sites. In view of the polar, protic nature of the hydroxyl and carboxylic groups they most likely interact with β-(1,4)-glucans through hydrogen bonding, in which they can act both as donors or acceptors.

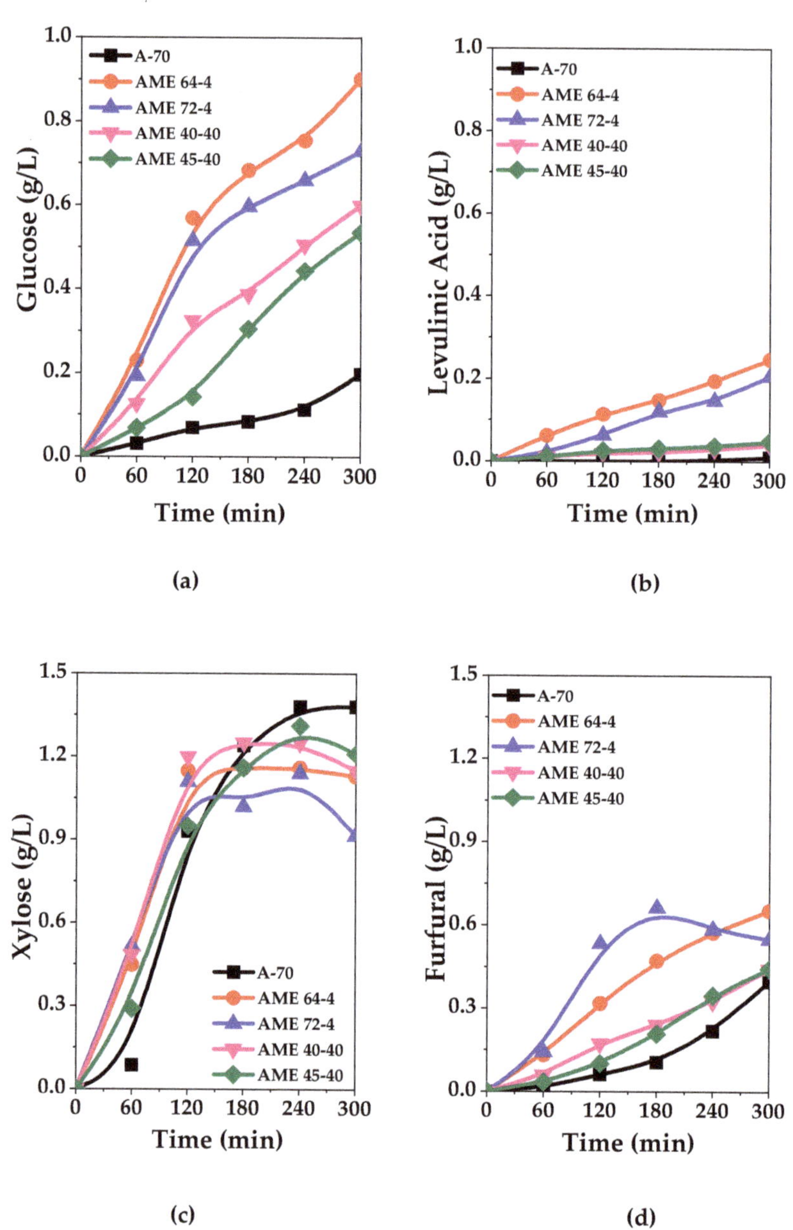

(a)

(b)

(c)

(d)

Figure 9. The concentration of (**a**) glucose, (**b**) levulinic acid, (**c**) xylose and (**d**) furfural over time in the hydrolysis of wheat straw catalyzed by the AME materials at 140 °C.

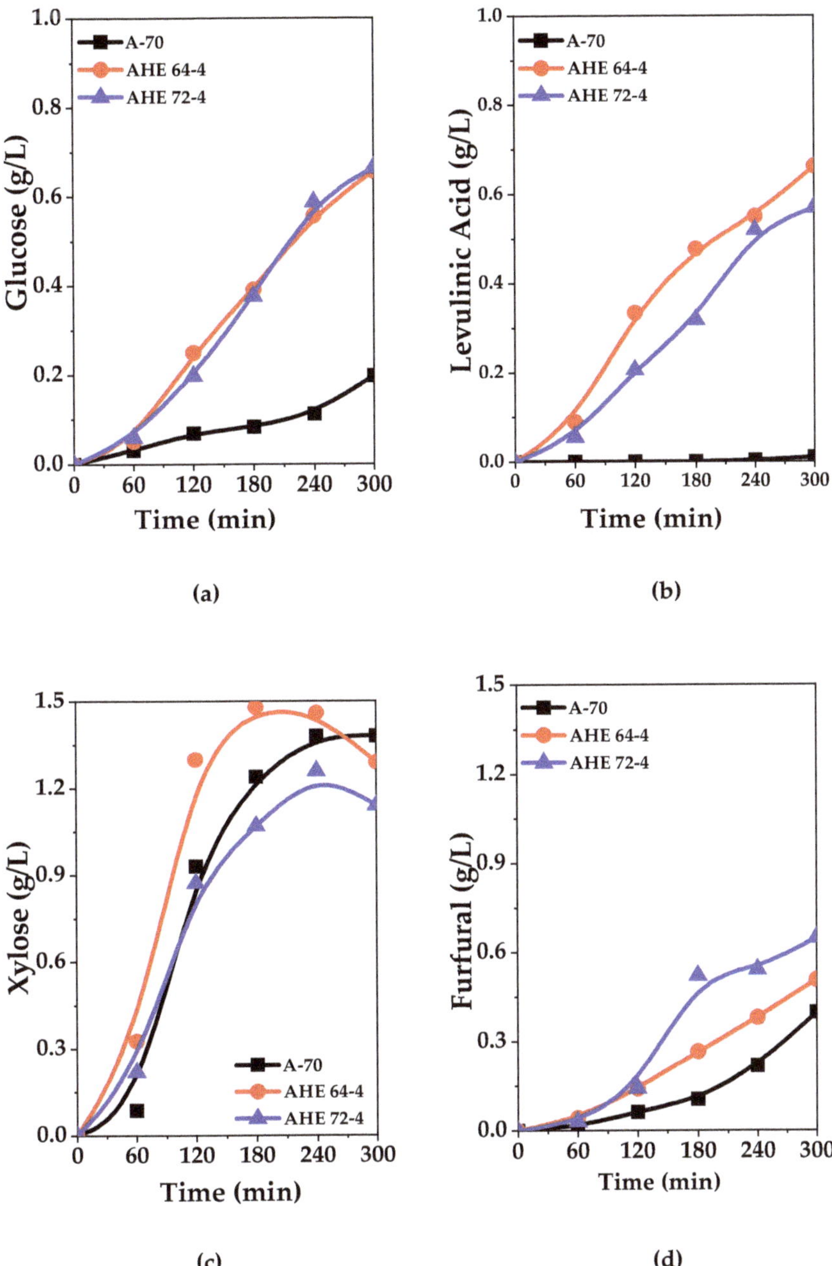

Figure 10. The concentration of (**a**) glucose, (**b**) levulinic acid, (**c**) xylose and (**d**) furfural over time in the hydrolysis of wheat straw catalyzed by the AHE materials at 140 °C.

The concentrations of the products of the hydrolysis of hemicellulose, i.e., xylose and its dehydration product (furfural), are also reported in Figures 9 and 10 as functions of time. The concentration of xylose increases at short reaction times, reaches a maximum and eventually drops at long reaction times. This can be expected for an intermediate product like xylose, which can be transformed into other products (e.g., furfural) in further consecutive reactions. As expected, the concentration of xylose increases faster than that of glucose, because hemicellulose is easier to hydrolyze than cellulose [5,21,53,54]. This also explains why the concentration of xylose, over time, appears practically independent of the catalyst (including A-70). However, again, the products are generally formed faster over the gel-type catalysts than over the macroreticular catalysts.

3. Materials and Methods

3.1. Materials

Ethylene dimethacrylate (EDMA, Aldrich, St. Louis, MI, USA) and methacrylic acid (MAA, Aldrich, St. Louis, MI, USA) were purified upon distillation prior to use. 2-Acrylamido-2-methyl-1-propane-sulfonic acid (AMPS, Aldrich, St. Louis, MI, USA), 2-hydroxyethyl methacrylate (HEMA, tech grade 97%, Aldrich, St. Louis, MI, USA), azo-bis-isobutyronitrile (AIBN, Janssen Chimica, Beerse, Belgium), dimethylformamide (DMF, Analyticals Carlo Erba, Milano, Italy), methanol (MeOH, Aldrich, St. Louis, MI, USA), sulfuric acid 95%–97% (Aldrich, St. Louis, MI, USA), 1-ethyl-3-methylimidazole acetate ([EMIM]OAc, Aldrich, St. Louis, MI, USA), Avicel cellulose (Fluka Analyticals, Munich, Germany), acetone (Scharlau, Barcelona, Spain), D-(+)-glucose, D-(+)-xylose, 5-hydroxymethylfurfural (5-HMF, Alfa Aesar, Ward Hill, MS, USA), levulinic acid (LA, Aldrich, St. Louis, MI, USA), furfural (Aldrich, St. Louis, MI, USA) and Amberlyst-70 (A-70, Dow chemical, Midland, MI, USA) were used as received. The lignocellulosic biomass (wheat straw) was characterized and supplied by the Unit of Biofuels from the Centre of Energy, Environmental and Technological Research (CIEMAT, Madrid, Spain).

3.2. Synthesis of the AHE Copolymers

Two copolymers containing AMPS, HEMA and EDMA were prepared by adapting the synthetic procedure described by Zecca et al. [55]. The polymers were labeled as AHE Y-X, where Y and X are the nominal molar ratios of AMPS and the cross-linker (EDMA). AHE 64-4 and AHE 72-4 are characterized by a 4% nominal EDMA molar fraction and a nominal AMPS:HEMA molar ratio of 2:1 and 3:1, respectively. As an example, the procedure used to prepare AHE 64-4 is reported. DMF (10 mL) was added at room temperature to a mixture of AMPS (3.64 g), HEMA (1.14 g), freshly distilled EDMA (0.22 g) and AIBN (0.05 g). The heterogeneous mixture was magnetically stirred at room temperature until AMPS and AIBN were completely dissolved (approximately 30 min). The reaction mixture was held at 70 °C for 72 h. A yellowish solid was obtained. After cooling the system to room temperature, the solid was manually crushed with a steel spatula and swollen in methanol (80 mL) for 3 h. Then, the material was washed with methanol in a Soxhlet extractor for 72 h. The swollen resin was recovered by vacuum filtration and dried for 72 h in a ventilated oven at 70 °C under atmospheric pressure. The resulting yellowish solid was mechanically ground with an impact grinder and sieved to collect the 180–400 μm fraction. The solid was then washed on a column with 1.0 M H_2SO_4 (250 mL) and subsequently washed with distilled water until the filtrate pH was neutral. After vacuum filtration, the solid was dried for 6 days in a ventilated oven at 110 °C under atmospheric pressure. The AHE materials were characterized by elemental analysis, back-titration of the sulfonic groups and FTIR spectroscopy.

3.3. Synthesis of the AME Copolymers

Four copolymers containing AMPS, MAA and EDMA were prepared as reported for the AHE materials, except HEMA was replaced with freshly distilled MAA. The polymers were labeled as AME Y-X, where Y and X are the nominal molar ratios of AMPS and the cross-linker (EDMA). AME 64-4 and AME 72-4 are characterized by a 4% nominal EDMA molar fraction and nominal AMPS:MAA molar ratio of 2:1 and 3:1, respectively. AME 40-40 and AME 45-40 are characterized by a 40% nominal EDMA molar fraction and a nominal AMPS:MAA molar ratio of 2:1 and 3:1, respectively. The AME materials were characterized by elemental analysis, back-titration of the acidic groups and FTIR spectroscopy.

3.4. Characterization of the Catalysts

The elemental analysis of the catalysts was carried out with a Leco CHNS-932 analyzer. For the back-titration, approximately 100 mg of resin (previously dried at 110 °C in an oven until the weight was constant) was exactly weighed and transferred into a 50 mL Erlenmeyer flask. Then, 10 mL of a 0.1 M NaOH aqueous solution was added. The flask was stoppered and swirled overnight with a mechanical orbiting plate. The next day, the suspension was titrated under magnetic stirring with a standard 0.1 M HCl aqueous solution using a pH-meter equipped with a glass electrode. To assess the potential reaction between NaOH and CO_2 from air, a blank mixture made with 10 mL of the abovementioned 0.1 M NaOH solution was swirled and titrated using the same procedure. The apparent specific content of the sulfonic groups was determined to be the difference of the initial and final millimoles of NaOH (corrected for any overconsumption due to carbonation of the base) divided by the resin mass. Infrared spectra of the catalysts, in absorption mode, were collected with a Bruker Tensor 27 spectrophotometer with KBr pellets. A total of 128 cumulative scans were performed with a resolution of 4 cm^{-1} in the wavenumber range of 4000–400 cm^{-1} in absorption mode.

3.5. Characterization of the Wheat Straw Samples

Raw material (approximately 250 μm particle size) was provided by the Biofuels Unit from the Centre of Energy, Environmental and Technological Research (CIEMAT, Madrid, Spain) and characterized using the National Renewable Energy Laboratory (NREL) standard methods for determining structural carbohydrates and lignin in biomass [21,56] at CIEMAT Biofuels Unit laboratories (Table 3).

Table 3. Composition of the wheat straw samples with respect to the dry weight.

Original Wheat Straw	
Extractives	**12.2 ± 0.7**
Aqueous	9.9 ± 0.4
Organic	1.7 ± 1.22
Monomeric glucose [1]	0.23 ± 0.08
Total glucose [2]	1.04 ± 0.13
Cellulose	**40.1 ± 0.2**
Hemicellulose	**28.6 ± 0.3**
Xylose	24.2 ± 0.2
Galactose	1.6 ± 0.01
Arabinose + Mannose	2.8 ± 0.14
Acid insoluble lignin	**15.4 ± 0.3**
Ashes	**3.6 ± 0.1**
Acetyl groups	**1.6 ± 0.02**

[1] The fraction of monomeric glucose determined in the aqueous extracts after hydrolysis (4% sulfuric acid, 120 °C, 30 min), which was obtained with respect to the dry weight (105 °C). [2] The fraction of monomeric and oligomeric glucose determined in the aqueous extracts after hydrolysis (4% sulfuric acid, 120 °C, 30 min), which was determined with respect to the dry weight (105 °C).

The wheat straw samples before and after pretreatment with [EMIM]OAc were analyzed by X-ray diffraction (XRD, Kassel, Germany), scanning electron microscopy (SEM, Tokyo, Japan) and FTIR spectroscopy (Tokyo, Japan). The XRD diffractograms were recorded using an X'Pert Pro PANalytical diffractometer equipped with a Cu Kα radiation source (λ = 0.15418 nm) and X'Celerator detector that recorded data by Real-Time Multiple Strip (RTMS). After grinding, the samples were placed on a stainless-steel plate. The diffraction patterns were recorded in steps over a range of Bragg angles (2θ) between 4° and 90° at a scanning rate of 0.02° per step and an accumulation time of 50 s. The diffractograms were analyzed with X'Pert HighScore Plus software (Kassel, Germany).

The scanning electron microscopy (SEM) images of the samples were recorded using a Hitachi S-3000 N microscope. The samples were treated with increasing concentrations of ethanol to fix the structure and dehydrate the samples. After critical-point drying with a Polaron CPD7501 critical point drier, the samples were sputter-coated with a thin layer of gold in a Balzers SCD 004 gold-sputter coater.

The infrared spectra of the wheat straw samples were obtained using Jasco FT/IR-6300 spectrophotometers. The FT-IR spectra were collected in an attenuated total reflectance geometry using a diamond crystal. A total of 180 cumulative scans from 4000–800 cm^{-1} were taken in absorption mode with a resolution of 4 cm^{-1}.

3.6. Wheat Straw Fragmentation

The wheat straw pretreatment was performed in a Mettler Toledo Easy Max 102®reactor equipped with a 100 mL glass vessel, and the samples were mechanically stirred at atmospheric pressure. In a typical pretreatment, 0.5 g of wheat straw was dissolved in 9.5 g of [EMIM]OAc, which had been previously heated at 105 °C under mechanical stirring (500 rpm). The reactor was stoppered, and the mixture was maintained at 105 °C under vigorous mechanical stirring (950 rpm) for 25 min, according to a previous study [6]. A viscous dark brown suspension was obtained. A partially dissolved biomass was regenerated by adding 50 mL of a 1:1 (v/v) water:acetone solution previously cooled in an ice-water bath. The cellulose- and hemicellulose-enriched solid was collected by filtration using a glass fiber filter (pore diameter: 1.6 μm). The wet solid was weighed, and the amount of dry material was determined from the weight loss of an aliquot after drying overnight in an oven at 100 °C.

3.7. Catalytic Tests

The hydrolytic tests were carried out by adapting the optimized procedure for hydrolyzing cellulose as reported by Morales-delaRosa et al. [17]. The tests were performed under magnetic stirring in a 100 mL Teflon-lined steel Berghof reactor equipped with a thermostat, a pressure addition funnel and a tap fitted with a filter, which allowed aliquots to be collected from the reaction mixture free from the solid material.

All the catalytic tests were performed on the fraction that was enriched with polysaccharides, obtainedby precipitating the wheat straw treated with [EMIM]OAc using a water:acetone (1:1) solution for 25 min at 110 °C.

For all the experiments, the amount of catalyst added to the reaction mixture was varied to maintain the sulfonic group concentration at 0.1 M. In a typical run, the heterogeneous catalyst, 0.5 g of wheat straw and 50 mL of water were added, and the suspension was heated to 140 °C. The stirring (500 rpm) and reaction time recording was started when a temperature of 140 °C was attained. Aliquots were periodically withdrawn from the reactor. After 5 h, the reaction was stopped by rapidly cooling the reactor in an ice-water bath (5 min).

3.8. HPLC Analysis

The samples from the reaction mixture were analyzed by HPLC. The monosaccharides (glucose and xylose) were analyzed with an Agilent Technologies 1260 Infinity HPLC (Santa Clara, CA, USA) equipped with an Agilent Technologies Hi-Plex Pb capillary column (300 mm × 7.7 mm) held at 70 °C, and water was used as the mobile phase (0.6 mL/min). Levulinic acid, 5-HMF and furfural were analyzed with an Agilent Technologies 1200 Series HPLC equipped with an Aminex HPX-87H capillary column (300 mm × 7.8 mm) held at 65 °C, and a 0.01 M H_2SO_4 aqueous solution was used as the mobile phase (0.6 mL/min). For the analysis of 5-HMF and furfural, a UV-vis detector (λ = 285 nm) was employed, while the other compounds were analyzed using a refractive index detector. The samples used for the analysis in the 1260 Infinity HPLC were diluted with distilled water (reaction slurry/water

1:1, v:v). The products were identified by comparing their retention times with those of pure reference analytes. The identified compounds were quantified using internal standard calibration curves.

4. Conclusions

The dissolution of wheat straw in [EMIM]OAc and subsequent fractional precipitation with a mixture of acetone:water (1:1) lead to the enrichment of the solid in the polysaccharide fraction, checked by FTIR Spectroscopy. This pretreatment method allows to recover the IL in the filtrate liquid avoiding the presence of [EMIM]OAc in the following steps and rendering the process more economically attractive.

This methodology allows to make the wheat straw surface much more accessible to an acid catalyst. The XRD pattern of the treated wheat straw indicated that it was noncrystalline, and SEM micrographs showed that the original structure disappeared after IL-pretreatment.

Sulfonic resins functionalized with polar groups in addition to -SO_3H groups (AHE and AME) are more effective catalysts in the hydrolytic process than Amberlyst 70. In the case of the gel-type catalysts (AHE 72-4, AHE 64-4, AME 72-4 and AME 64-4) this could be simply the result of their higher swelling degree, which makes a larger fraction of active sites accessible to the molecules of the reagents. This highlights the importance of the morphology of the catalyst and its swelling ability when it is a cross-linked organic resin and is used under liquid–solid conditions. However, it can be also argued that the alcoholic hydroxyl groups and carboxylic groups in the polymer framework of the AHE and AME catalysts, respectively, speed up the breakdown of the β-(1,4) glycosidic bonds by favoring the adsorption of glucans ("assisted adsorption") through hydrogen bonding. This effect seems particularly relevant in the macroreticular catalysts (AME 40-40, AME 45-40).

In addition, the morphology of the catalysts appears to be crucial in terms of glucose production. Gel-type catalysts had the highest glucose concentration, as a larger number of the catalytic sites were accessible to the substrate.

The use of heterogeneous catalysts in the hydrolysis of lignocellulosic materials is a difficult topic because it involves an interaction between two solids. In this way, some different approaches have been studied, firstly the modification of the lignocellulosic materials increasing their accessibility and favoring the contact with catalysts, and the second approach beingthe use of the catalysts supports with polar groups that favor the adsorption of cellulose chains increasing the interaction with the acid site. In this work, we add a step forward, using catalysts functionalized with some polar groups, but also showing the importance of the morphology of the catalysts (gel-type or macroreticular), to enhance the interaction with carbohydrate substrates, mainly cellulose chains. Summing up, this manuscript combines all three approaches, pretreatment of lignocellulosic biomass, catalysts with polar groups and morphology of the catalysts, and the results showed opens the way to design new and efficient heterogeneous catalysts for the hydrolysis of lignocellulosic biomass.

Author Contributions: S.M.-d. designed the experiments. G.L. and S.M.-d. wrote the manuscript. J.M.C.-M. and P.C. designed the experiments and reviewed the manuscript. G.L. and S.M.-d. performed the experimental analyses. J.L.G.F. and M.Z. critically reviewed the manuscript.

Funding: This research was funded by Comunidad de Madrid (Spain), grant numbers S2013/MAE-2882 (RESTOENE-2-CM) and P2018/EMT-4344 (BIOTRES-CM), and the University of Padova under project CPDA123471.

Conflicts of Interest: The authors declare that the research was conducted in the absence of any commercial or financial relationships that could be construed as a potential conflict of interest.

References

1. García-Condado, S.; López-Lozano, R.; Panarello, L.; Cerrani, I.; Nisini, L.; Zucchini, A.; Van der Velde, M.; Baruth, B. Assessing lignocellulosic biomass production from crop residues in the European Union: Modelling, analysis of the current scenario and drivers of interannual variability. *GCB Bioenergy* **2019**. [CrossRef]

2. Scarlat, N.; Dallemand, J.-F.; Monforti-Ferrario, F.; Nita, V. The role of biomass and bioenergy in a future bioeconomy: Policies and facts. *Environ. Dev.* **2015**, *15*, 3–34. [CrossRef]

3. Instituto para la Diversificación y ahorro de la Energía. *Plan de Energías Renovables 2011–2020*; IDAE: Madrid, Spain, 2011.

4. Malins, S.S.A.C. *Availability of Cellulosic Residues and Wastes in the Eu*; The International Council on Clean Transportation: Washington, DC, USA, 2013; White Paper.

5. Alvira, P.; Tomas-Pejo, E.; Ballesteros, M.; Negro, M.J. Pretreatment technologies for an efficient bioethanol production process based on enzymatic hydrolysis: A review. *Bioresour. Technol.* **2010**, *101*, 4851–4861. [CrossRef] [PubMed]

6. Lara-Serrano, M.; Morales-delaRosa, S.; Campos-Martín, M.J.; Fierro, L.J. Fractionation of Lignocellulosic Biomass by Selective Precipitation from Ionic Liquid Dissolution. *Appl. Sci.* **2019**, *9*, 1862. [CrossRef]

7. Zhang, Y.H.P.; Ding, S.Y.; Mielenz, J.R.; Cui, J.B.; Elander, R.T.; Laser, M.; Himmel, M.E.; McMillan, J.R.; Lynd, L.R. Fractionating recalcitrant lignocellulose at modest reaction conditions. *Biotechnol. Bioeng.* **2007**, *97*, 214–223. [CrossRef]

8. Lan, W.; Liu, C.F.; Sun, R.C. Fractionation of bagasse into cellulose, hemicelluloses, and lignin with ionic liquid treatment followed by alkaline extraction. *J. Agric. Food. Chem.* **2011**, *59*, 8691–8701. [CrossRef] [PubMed]

9. da Costa Lopes, A.M.; João, K.G.; Rubik, D.F.; Bogel-Łukasik, E.; Duarte, L.C.; Andreaus, J.; Bogel-Łukasik, R. Pre-treatment of lignocellulosic biomass using ionic liquids: Wheat straw fractionation. *Bioresour. Technol.* **2013**, *142*, 198–208. [CrossRef]

10. Yang, D.; Zhong, L.-X.; Yuan, T.-Q.; Peng, X.-W.; Sun, R.-C. Studies on the structural characterization of lignin, hemicelluloses and cellulose fractionated by ionic liquid followed by alkaline extraction from bamboo. *Ind. Crop. Prod.* **2013**, *43*, 141–149. [CrossRef]

11. Yang, X.; Wang, Q.; Yu, H. Dissolution and regeneration of biopolymers in ionic liquids. *Russ. Chem. Bull.* **2014**, *63*, 555–559. [CrossRef]

12. Zhang, P.; Dong, S.J.; Ma, H.H.; Zhang, B.X.; Wang, Y.F.; Hu, X.M. Fractionation of corn stover into cellulose, hemicellulose and lignin using a series of ionic liquids. *Ind. Crop. Prod.* **2015**, *76*, 688–696. [CrossRef]

13. Morais, A.R.C.; Pinto, J.V.; Nunes, D.; Roseiro, L.B.; Oliveira, M.C.; Fortunato, E.; Bogel-Łukasik, R. Imidazole: Prospect Solvent for Lignocellulosic Biomass Fractionation and Delignification. *ACS Sustain. Chem. Eng.* **2016**, *4*, 1643–1652. [CrossRef]

14. Mohtar, S.S.; Tengku Malim Busu, T.N.; Md Noor, A.M.; Shaari, N.; Mat, H. An ionic liquid treatment and fractionation of cellulose, hemicellulose and lignin from oil palm empty fruit bunch. *Carbohydr. Polym.* **2017**, *166*, 291–299. [CrossRef] [PubMed]

15. Zhou, C.-H.; Xia, X.; Lin, C.-X.; Tong, D.-S.; Beltramini, J. Catalytic conversion of lignocellulosic biomass to fine chemicals and fuels. *Chem. Soc. Rev.* **2011**, *40*, 5588–5617. [CrossRef] [PubMed]

16. Geboers, J.A.; Van de Vyver, S.; Ooms, R.; Op de Beeck, B.; Jacobs, P.A.; Sels, B.F. Chemocatalytic conversion of cellulose: Opportunities, advances and pitfalls. *Catal. Sci. Tech.* **2011**, *1*, 714. [CrossRef]

17. Morales-delaRosa, S.; Campos-Martin, J.M.; Fierro, J.L.G. Optimization of the process of chemical hydrolysis of cellulose to glucose. *Cellulose* **2014**, *21*, 2397–2407. [CrossRef]

18. Li, C.; Wang, Q.; Zhao, Z.K. Acid in ionic liquid: An efficient system for hydrolysis of lignocellulose. *Green Chem.* **2008**, *10*, 177. [CrossRef]

19. Morales-delaRosa, S.; Campos-Martin, J.M.; Fierro, J.L.G. High glucose yields from the hydrolysis of cellulose dissolved in ionic liquids. *Chem. Eng. J.* **2012**, *181–182*, 538–541. [CrossRef]

20. Padrino, B.; Lara-Serrano, M.; Morales-delaRosa, S.; Campos-Martín, J.M.; Fierro, J.L.G.; Martínez, F.; Melero, J.A.; Puyol, D. Resource Recovery Potential From Lignocellulosic Feedstock Upon Lysis with Ionic Liquids. *Front. Bioeng. Biotechnol.* **2018**, *6*. [CrossRef]

21. Lara-Serrano, M.; Sáez Angulo, F.; Negro, M.J.; Morales-delaRosa, S.; Campos-Martin, J.M.; Fierro, J.L.G. Second-Generation Bioethanol Production Combining Simultaneous Fermentation and Saccharification of IL-Pretreated Barley Straw. *ACS Sustain. Chem. Eng.* **2018**, *6*, 7086–7095. [CrossRef]

22. Morales-delaRosa, S.; Campos-Martin, J.M.; Fierro, J.L.G. Chemical hydrolysis of cellulose into fermentable sugars through ionic liquids and antisolvent pretreatments using heterogeneous catalysts. *Catal. Today* **2018**, *302*, 87–93. [CrossRef]

23. Fukuoka, A.; Dhepe, P.L. Catalytic Conversion of Cellulose into Sugar Alcohols. *Angew. Chem. Int. Ed.* **2006**, *45*, 5161–5163. [CrossRef] [PubMed]

24. Onda, A.; Ochi, T.; Yanagisawa, K. Selective hydrolysis of cellulose into glucose over solid acid catalysts. *Green Chem.* **2008**, *10*, 1033. [CrossRef]

25. Vilcocq, L.; Castilho, P.C.; Carvalheiro, F.; Duarte, L.C. Hydrolysis of Oligosaccharides Over Solid Acid Catalysts: A Review. *ChemSusChem* **2014**, *7*, 1010–1019. [CrossRef] [PubMed]

26. Biasi, P.; Mikkola, J.P.; Sterchele, S.; Salmi, T.; Gemo, N.; Shchukarev, A.; Centomo, P.; Zecca, M.; Canu, P.; Rautio, A.R.; et al. Revealing the role of bromide in the H2O2 direct synthesis with the catalyst wet pretreatment method (CWPM). *AlChE J.* **2017**, *63*, 32–42. [CrossRef]

27. Centomo, P.; Meneghini, C.; Sterchele, S.; Trapananti, A.; Aquilanti, G.; Zecca, M. EXAFS in situ: The effect of bromide on Pd during the catalytic direct synthesis of hydrogen peroxide in memory of the late Professor Benedetto Corain (July 8th 1941–September 24th 2014), passionate chemist and teacher. *Catal. Today* **2015**, *248*, 138–141. [CrossRef]

28. Centomo, P.; Zecca, M.; Di Noto, V.; Lavina, S.; Bombi, G.G.; Nodari, L.; Salviulo, G.; Ingoglia, R.; Milone, C.; Galvagno, S.; et al. Characterization of Synthetic Iron Oxides and their Performance as Support for Au Catalysts. *ChemCatChem* **2010**, *2*, 1143–1149. [CrossRef]

29. Centomo, P.; Zecca, M.; Zoleo, A.; Maniero, A.L.; Canton, P.; Jeřábek, K.; Corain, B. Cross-linked polyvinyl polymersversus polyureas as designed supports for catalytically active M0 nanoclusters Part III. Nanometer scale structure of the cross-linked polyurea support EnCat 30 and of the PdII/EnCat 30 and Pd0/EnCat 30NP catalysts. *Phys. Chem. Chem. Phys.* **2009**, *11*, 4068–4076. [CrossRef] [PubMed]

30. Sterchele, S.; Centomo, P.; Zecca, M.; Hanková, L.; Jeřábek, K. Dry- and swollen-state morphology of novel high surface area polymers. *Microporous Mesoporous Mater.* **2014**, *185*, 26–29. [CrossRef]

31. Kennedy, F.R. Syntheses and separations using functional polymers. *Br. Polym. J.* **1989**, *21*, 359–360. [CrossRef]

32. Zecca, M.; Centomo, P.; Corain, B. Metal Nanoclusters Supported on Cross-Linked Functional Polymers: A Class of Emerging Metal Catalysts. In *Metal Nanoclusters in Catalysis and Materials Science: The Issue of Size Control*; Elsevier: Amsterdam, The Netherlands, 2008; pp. 201–232. [CrossRef]

33. Jeřábek, K.; Hanková, L.; Holub, L. Working-state morphologies of ion exchange catalysts and their influence on reaction kinetics. *J. Mol. Catal. A Chem.* **2010**, *333*, 109–113. [CrossRef]

34. Suganuma, S.; Nakajima, K.; Kitano, M.; Yamaguchi, D.; Kato, H.; Hayashi, S.; Hara, M. Hydrolysis of Cellulose by Amorphous Carbon Bearing SO3H, COOH, and OH Groups. *J. Am. Chem. Soc.* **2008**, *130*, 12787–12793. [CrossRef] [PubMed]

35. Pang, J.; Wang, A.; Zheng, M.; Zhang, T. Hydrolysis of cellulose into glucose over carbons sulfonated at elevated temperatures. *Chem. Commun.* **2010**, *46*, 6935–6937. [CrossRef] [PubMed]

36. Foo, G.S.; Sievers, C. Synergistic Effect between Defect Sites and Functional Groups on the Hydrolysis of Cellulose over Activated Carbon. *ChemSusChem* **2015**, *8*, 534–543. [CrossRef] [PubMed]

37. Huang, Y.-B.; Fu, Y. Hydrolysis of cellulose to glucose by solid acid catalysts. *Green Chem.* **2013**, *15*, 1095. [CrossRef]

38. Van de Vyver, S.; Peng, L.; Geboers, J.; Schepers, H.; de Clippel, F.; Gommes, C.J.; Goderis, B.; Jacobs, P.A.; Sels, B.F. Sulfonated silica/carbon nanocomposites as novel catalysts for hydrolysis of cellulose to glucose. *Green Chem.* **2010**, *12*, 1560–1563. [CrossRef]

39. Morales-delaRosa, S.; Campos-Martin, J.M.; Fierro, J.L.G. Complete Chemical Hydrolysis of Cellulose into Fermentable Sugars through Ionic Liquids and Antisolvent Pretreatments. *ChemSusChem* **2014**, *7*, 3467–3475. [CrossRef] [PubMed]

40. Li, X.; Jiang, Y.; Shuai, L.; Wang, L.; Meng, L.; Mu, X. Sulfonated copolymers with SO3H and COOH groups for the hydrolysis of polysaccharides. *J. Mater. Chem.* **2012**, *22*, 1283–1289. [CrossRef]

41. Silverstein, R.M.; Webster, F.X.; Kiemle, D.J. *Spectrometric Identification of Organic Compounds*, 7th ed.; Wiley: Hoboken, NJ, USA, 2005.

42. Park, S.; Baker, J.O.; Himmel, M.E.; Parilla, P.A.; Johnson, D.K. Cellulose crystallinity index: Measurement techniques and their impact on interpreting cellulase performance. *Biotechnol. Biofuels* **2010**, *3*, 10. [CrossRef]

43. Wang, H.; Gurau, G.; Pingali, S.V.; O'Neill, H.M.; Evans, B.R.; Urban, V.S.; Heller, W.T.; Rogers, R.D. Physical Insight into Switchgrass Dissolution in Ionic Liquid 1-Ethyl-3-methylimidazolium Acetate. *ACS Sustain. Chem. Eng.* **2014**, *2*, 1264–1269. [CrossRef]

44. Pandey, K.K.; Pitman, A.J. FTIR studies of the changes in wood chemistry following decay by brown-rot and white-rot fungi. *Int. Biodeterior. Biodegrad.* **2003**, *52*, 151–160. [CrossRef]

45. Osatiashtiani, A.; Lee, A.F.; Granollers, M.; Brown, D.R.; Olivi, L.; Morales, G.; Melero, J.A.; Wilson, K. Hydrothermally Stable, Conformal, Sulfated Zirconia Monolayer Catalysts for Glucose Conversion to 5-HMF. *ACS Catal.* **2015**, *5*, 4345–4352. [CrossRef]

46. Hergert, H.L. Infrared Spectra of Lignin and Related Compounds. II. Conifer Lignin and Model Compounds 1,2. *J. Org. Chem.* **1960**, *25*, 405–413. [CrossRef]

47. Sun, R.C.; Fang, J.M.; Tomkinson, J.; Geng, Z.C.; Liu, J.C. Fractional isolation, physico-chemical characterization and homogeneous esterification of hemicelluloses from fast-growing poplar wood. *Carbohydr. Polym.* **2001**, *44*, 29–39. [CrossRef]

48. Buckingham, J.; Cadogan, J.D.G.; Raphael, R.A.; Rees, C.W. *Dictionary of Organic Compounds*, 5th ed.; Chapman and Hall: New York, NY, USA, 1982.

49. Serjeant, E.P.; Dempsey, B. *Ionisation Constants of Organic Acids in Aqueous Solution*; IUPAC Chemical Data Series No. 23; Pergamon Press, Inc.: New York, NY, USA, 1979.

50. Martinuzzi, S.; Cozzula, D.; Centomo, P.; Zecca, M.; Müller, T.E. The distinct role of the flexible polymer matrix in catalytic conversions over immobilised nanoparticles. *Rsc Adv.* **2015**, *5*, 56181–56188. [CrossRef]

51. Centomo, P.; Bonato, I.; Hanková, L.; Holub, L.; Jeřábek, K.; Zecca, M. Novel Ion-Exchange Catalysts for Reactions Involving Lipophilic Reagents: Perspectives in the Reaction of Esterifications of Fatty Acids with Methanol. *Top. Catal.* **2013**, *56*, 611–617. [CrossRef]

52. Jerabek, K.H.L.; Holub, L.; Corain, B.; Zecca, M.; Centomo, P.; Bonato, I. Strongly Acidic Ion Exchanger Catalyst and Method of Preparing The Same. WO2012127414, 27 september 2012.

53. Morales-Delarosa, S.; Campos-Martin, J.M. 6—Catalytic processes and catalyst development in biorefining. In *Advances in Biorefineries*; Waldron, K., Ed.; Woodhead Publishing: Cambridge, UK, 2014; pp. 152–198. [CrossRef]

54. Kumar, P.; Barrett, D.M.; Delwiche, M.J.; Stroeve, P. Methods for Pretreatment of Lignocellulosic Biomass for Efficient Hydrolysis and Biofuel Production. *Ind. Eng. Chem. Res.* **2009**, *48*, 3713–3729. [CrossRef]

55. Centomo, P.; Jeřábek, K.; Canova, D.; Zoleo, A.; Maniero, A.L.; Sassi, A.; Canton, P.; Corain, B.; Zecca, M. Highly Hydrophilic Copolymers of N,N-Dimethylacrylamide, Acrylamido-2-methylpropanesulfonic acid, and Ethylenedimethacrylate: Nanoscale Morphology in the Swollen State and Use as Exotemplates for Synthesis of Nanostructured Ferric Oxide. *Chem. A Eur. J.* **2012**, *18*, 6632–6643. [CrossRef]

56. Sluiter, J.B.; Ruiz, R.O.; Scarlata, C.J.; Sluiter, A.D.; Templeton, D.W. Compositional analysis of lignocellulosic feedstocks. 1. Review and description of methods. *J. Agric. Food. Chem.* **2010**, *58*, 9043–9053. [CrossRef]

 © 2019 by the authors. Licensee MDPI, Basel, Switzerland. This article is an open access article distributed under the terms and conditions of the Creative Commons Attribution (CC BY) license (http://creativecommons.org/licenses/by/4.0/).

Article

Role of Humic Acid Chemical Structure Derived from Different Biomass Feedstocks on Fe(III) Bioreduction Activity: Implication for Sustainable Use of Bioresources

Yuquan Wei [1], Zimin Wei [2], Fang Zhang [1], Xiang Li [3], Wenbing Tan [3] and Beidou Xi [1,3,*]

[1] School of Environment, Tsinghua University, Beijing 100084, China; weiyq2013@gmail.com (Y.W.); fangzhang@tsinghua.edu.cn (F.Z.)

[2] College of Life Science, Northeast Agricultural University, Harbin 150030, China; weizm691120@163.com

[3] State Key Laboratory of Environmental Criteria and Risk Assessment, Chinese Research Academy of Environmental Sciences, Beijing 100012, China; david2013marshall@outlook.com (X.L.); wenbingtan@126.com (W.T.)

* Correspondence: wbtann@126.com; Tel.: +86-10-8491-3133

Received: 24 April 2019; Accepted: 14 May 2019; Published: 15 May 2019

Abstract: Humic acids (HAs) are redox-active components that play a crucial role in catalyzing relevant redox reactions in various ecosystems. However, it is unclear what role the different compost-derived Has play in the dissimilatory Fe(III) bioreduction and which chemical structures could accelerate Fe reduction. In this study, we compared the effect of eighteen HAs from the mesophilic phase, thermophilic phase and mature phase of protein-, lignocellulose- and lignin-rich composting on catalyzing the bioreduction of Fe(III)-citrate by *Shewanella oneidensis* MR-1 in temporarily anoxic laboratory systems. The chemical composition and structure of different compost-derived HAs were analyzed by UV–Vis spectroscopy, excitation-emission matrices of the fluorescence spectra, and ^{13}C-NMR. The results showed that HAs from lignocellulose- and lignin-rich composting, especially in the thermophilic phase, promoted the bioreduction of Fe(III). They also showed that HA from protein-rich materials suppressed significantly the Fe(II) production, which was mainly affected by the amount and structures of functional groups (e.g., quinone groups) and humification degree of the HAs. This study can aid in searching sustainable HA-rich composts for wide-ranging applications to catalyze redox-mediated reactions of pollutants in soils.

Keywords: Humic acid; Bioreduction of dissimilatory Fe(III); *Shewanella oneidensis* MR-1; Redox-active structures; Composts from different sources

1. Introduction

Humic acid (HA) is one of the most important components of humic substances as it plays a crucial role in regulating carbon cycles and catalyzing relevant redox reactions in various ecosystems [1,2]. Under anoxic conditions, both dissolved and particulate HA may accept electrons from anaerobic microbial respiration. Reduced HA can also serve as an electron donor and transfer electrons to poorly soluble Fe(III), mediating dissimilatory Fe(III) reduction and affecting their transformation and speciation [3]. It is widely accepted that quinone moieties, as well as other redox-active functional groups in the HA, have an important role in electron shuttling process, thereby affecting microbial electron transfer onto Fe(III) [4]. Therefore, this process has received a great deal of attention owing to the broad impact of dissimilatory iron reduction on the geochemical cycles.

Composting is a controlled biological transformation process to stabilize different organic solid wastes, which is generally divided into mesophilic phase, thermophilic phase, and mature phase [5].

Numerous studies have found that microbial degradation reduced the soluble organic carbon content and promoted the formation of HA with increasing molecular weight and aromatic characteristics during composting [6]. Composts from different sources may form diverse humic substance precursors (e.g., polyphenols, carboxyl acids, amino acids, reducing sugars, etc.) due to their different organic compositions, causing HA produced by composting to have different molecular composition and chemical structures [2,7]. The HA with the functional groups including phenolic, alcoholic, quinone, ketone, nitrogen and sulfur redox-active moieties as well as organic-metal chelates could mediate electron transfer process and enhance the reduction rate of ferric oxy/hydroxides [8]. Considering different microbial activities at varied composting periods, HA from different stages of composting may also have different composition and functional groups. A study has shown that HA precursors are mainly formed in the mesophilic or thermophilic phase and HAs are polymerized in the mature phase [9]. Therefore, compost-derived HA from different sources and different stages may have diverse redox-active potential for accelerating microbial electrons transfer. Fe(III) is an important terminal electron shuttle and its redox transformation is crucial to stimulate the biodegradation of organic contaminants. Although various papers from the literature have already studied on the relationship between the dissimilatory Fe(III) bioreduction and HA [3,4], few research papers have explored the roles of different compost-derived HA as redox mediators in dissimilatory Fe(III) bioreduction and which chemical structures could accelerate Fe reduction.

In this work, we isolated eighteen HAs from mesophilic phase, thermophilic phase and mature phase of protein-, lignocellulose- and lignin-rich composting to determine their effect as electron shuttles for catalyzing the microbial reduction of Fe(III)-citrate by *Shewanella oneidensis* MR-1 in temporarily anoxic laboratory systems. We applied UV–Vis spectroscopy, 3D excitation-emission matrices of the fluorescence spectra, and ^{13}C-NMR to compare the changes of chemical composition and structure of different compost-derived HA. This study aims to (1) compare the chemical structures of HA from different composting, (2) evaluate their redox capacity of mediating Fe(III)–citrate bioreduction by *Shewanella oneidensis* MR-1, and (3) explore which chemical structure properties of HA are most responsible for accelerating Fe(III) reduction. The results of the present study can promote the application of HA-rich composts in contaminated soils.

2. Results and Discussion

2.1. Bioreduction of Dissimilatory Fe(III) Mediated by Different HAs

Composting HAs from different sources all had relatively high molecular weight and aromatic characteristics, but the ability of HAs to mediate electron transfer process and affect the reduction rate of dissimilatory Fe(III) were different. The changes of Fe(II) in different treatments during the reduction process were shown in Figure 1, which were fitted to the first order pharmacokinetic model. *Shewanella oneidensis* MR-1 is a typical dissimilatory iron reducing bacteria with the capable of respiring anaerobically on Fe(III) as the sole terminal electron acceptor [10]. The increase in the Fe(II) concentration after the inoculation of *Shewanella oneidensis* MR-1 suggested that Fe(III)–citrate was bioreduced and total Fe(II) concentration increased steadily to around 2.254 ± 0.003 mM in 190 h in the absence of HA. With the addition of HA, Fe(III)–citrate was also reduced and gradually leveling-off Fe(II) concentration in 288 h, but the addition of HA from six composts had different reduction extents and Fe(II) production rate. The average level of Fe(II) formation in different composts were ranked in the order WW > CW > GW > SW > CM > DM (Figure 2). A post hoc test for Fe(II) concentration in all the systems showed that there was a significant difference between protein-, lignocellulose- and lignin-rich composting, suggesting that lignocellulose-rich composting might have more redox-active functional groups for transferring electrons than protein- and lignin-rich composts.

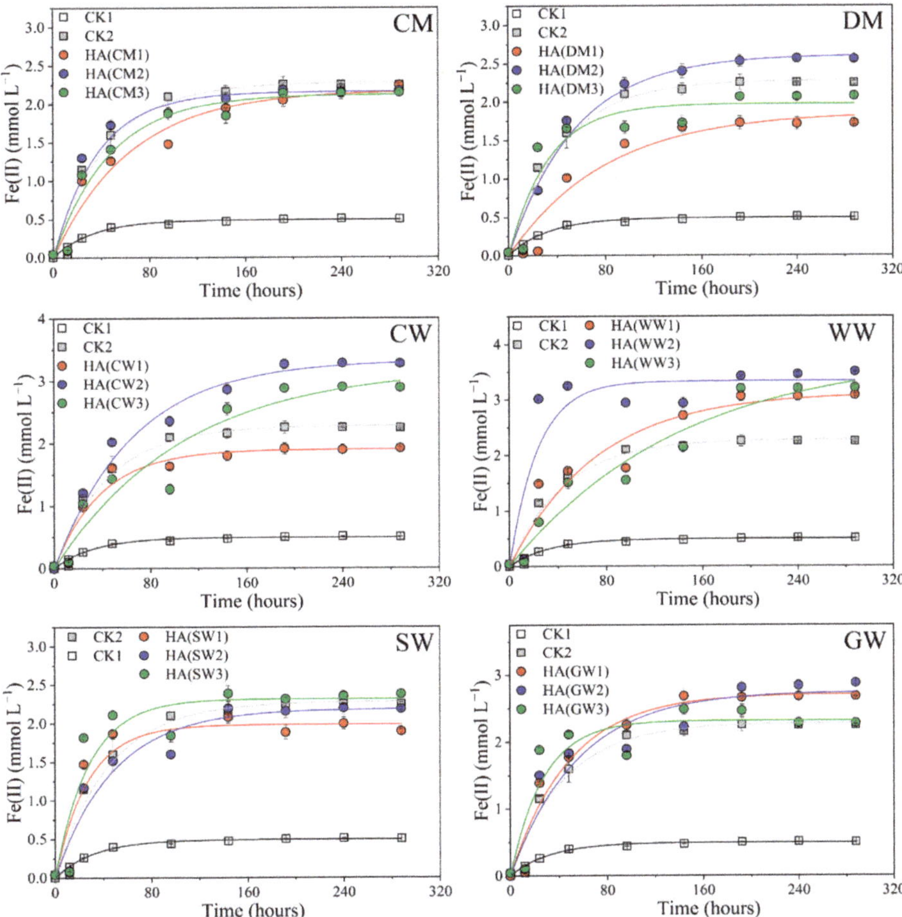

Figure 1. Effect of different compost-derived humic acid (HA) on the bioreduction of Fe(III)-citrate by *Shewanella oneidensis* MR-1 in cultures containing 20 mM Fe(III)-citrate, *Shewanella oneidensis* MR-1 cells (10^7 cells mL^{-1}), and HA (final concentration of 100 mg L^{-1}). Lines showed nonlinear least-squares regression fits of data to an equation describing product accumulation from a first-order reaction. Bars on symbols represent standard deviations of three replicates. CK1 was blank control, CK2 was the control without HA addition and the treatments with HA addition are described in Materials and Methods.

Here, CK1 was blank control and CK2 was the control without HA addition. In the presence of HA from WW2, the reduction extents were generally 55.6% higher than those observed in CK2, which showed the best stimulation capacity among the eighteen tested HAs. The addition of HA species from WW, CW and GW composting resulted in obvious increases of the Fe(II) production (1.2−1.5 times higher than that of control system without HA), especially in WW1-3, CW2-3 and GW2. However, when HA from CM, DM, and SW composting were added, the Fe(II) production of systems were generally lower than those obtained in CK2, especially in CM1-3, DM1, DM3 and SW1, which might have an inhibition effect on the bioreduction of dissimilatory Fe(III). In the composting of protein-rich materials, HA is easier to be formed due to more degradation of easily degradable organics but it may be unstable, which could be served as nutrient for promoting microbial growth [11,12]. For the bioreduction in DM2, more Fe(II) production by HA addition was also observed than those in DM1 and DM3, although the stimulating effects were less significant compared to CK2, suggesting that thermophilic phase of DM composting may benefit the formation of redox functional groups of HA [13]. Though no significant difference could be found with Fe(II) production in systems with the addition of HA from different stages, the formation quantity of Fe(II) with HA obtained from the thermophilic or mature stage were generally higher than that from mesophilic stage for all the composting except CM, demonstrating that the capacity of HA as electron shuttles was enhanced after composting [1,14].

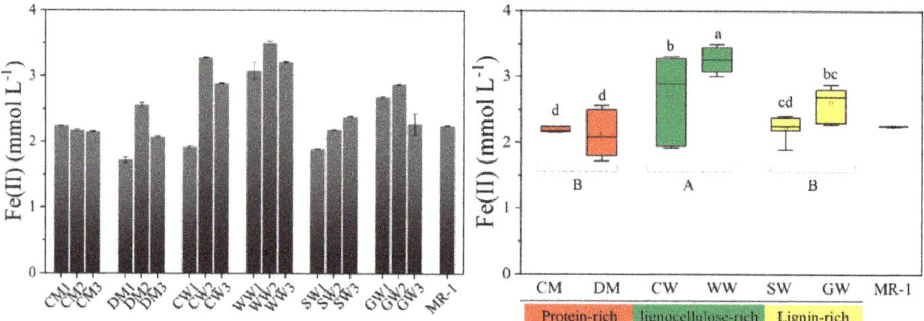

Figure 2. The Fe(II) formation of *Shewanella oneidensis* MR-1 with HA addition for Fe(III)–citrate bioreduction after 288 h in anoxic conditions with 20 mM Fe(III)-citrate addition. Different letters (A-B, a-d) are related to significant differences. Acronyms for different treatments are described in Materials and Methods.

In this study, the bioreduction of Fe(III)–citrate were fitted to first-order kinetic equation by the non-linear least-square curve fitting technique ($P < 0.05$) (Table 1). Results showed that the k_{red} of initial Fe(II) production rates of systems with HA from thermophilic stage of lignocellulose-rich composting higher than that in CK2. The HA from mesophilic and mature stages of lignin-rich composting still demonstrated higher k_{red} and better stimulation capacity than only adding *S. oneidensis* MR-1 at initial Fe(II) production stage (48 h), whereas the total Fe(II) production of HA from mesophilic stages of SW composting was very limited. Overall, there is an obvious stimulation capacity of HA from lignocellulose-rich composting (WW and CW), and HA from protein-rich composting (CM and DM) mainly suppressed the electron transfer between the *Shewanella oneidensis* MR-1 and Fe(III).

Table 1. First-order kinetics equations for describing the dynamics of Fe(III) reduction.

	k_{red} (h^{-1})			R		
	1	2	3	1	2	3
CM	0.0158 ± 0.0032	0.0271 ± 0.0060	0.0213 ± 0.0040	0.9564	0.9384	0.9583
DM	0.0126 ± 0.0037	0.0183 ± 0.0030	0.0309 ± 0.0091	0.9365	0.9705	0.8917
CW	0.0260 ± 0.0056	0.0354 ± 0.0026	0.0094 ± 0.0031	0.9421	0.9709	0.9296
WW	0.0141 ± 0.0039	0.0432 ± 0.0166	0.0072 ± 0.0022	0.9263	0.8293	0.9539
SW	0.0367 ± 0.0108	0.0209 ± 0.0048	0.0356 ± 0.0118	0.8918	0.9378	0.8664
GW	0.0208 ± 0.0041	0.0182 ± 0.0051	0.0368 ± 0.0129	0.9565	0.9120	0.8517
CK1		0.0227 ± 0.0044			0.9568	
CK2		0.0310 ± 0.0030			0.9841	

1, 2, and 3 represented the different stages of composting, i.e., mesophilic phase, thermophilic phase and maturation phase. Fe(III)–citrate reduction experiments were conducted in cultures containing 20 mM Fe(III)-citrate, *Shewanella oneidensis* MR-1 cells (10^7 cells mL^{-1}), and HA (final concentration of 100 mg L^{-1}) at 30 °C.

2.2. Structural Characteristics of Different HAs

In order to investigate the composition and chemical structures of HA from different composting, the HAs at different stages of composting were characterized firstly by the indexes of UV–Vis spectroscopy including $SUVA_{254}$, $SUVA_{290}$, E_4/E_6, $A_{240-440}$, and S_R. A number of studies have shown that $SUVA_{254}$ and $A_{240-440}$ were positively related to the abundance of aromatic C and aromatic rings of polar functional groups of the compost-derived HA, while E_4/E_6 and S_R were negatively correlated with humification degree and molecular weight of humic substances [15,16]. As shown in Figure 3, the values of $SUVA_{254}$, $SUVA_{290}$ and $A_{240-440}$ of HA in different composts were generally all increased from mesophilic stage to mature stage, while E_4/E_6 and S_R were decreased, which suggested the increase of the polymerization of HA structure and the accumulation of the substituent group type on aromatic rings and the electron transfer band. The increase of $SUVA_{254}$ in WW, CW, GW and SW (above 20%) were higher than those in CM and DM, indicating a higher humification degree of HA [17,18]. Notably, the highest values of $SUVA_{290}$ was found in the thermophilic phase for lignocellulose- and lignin-rich composting, suggesting that most of the lower organic matter such as quinones and more complex compound have incorporated in the aromatic structure at the later stage of composting [16,19]. The changes of $SUVA_{290}$ was in agreement with the trends of Fe(II) production, thereby corroborating that the changes in quinone moieties were directly associated with redox properties of HA in composts.

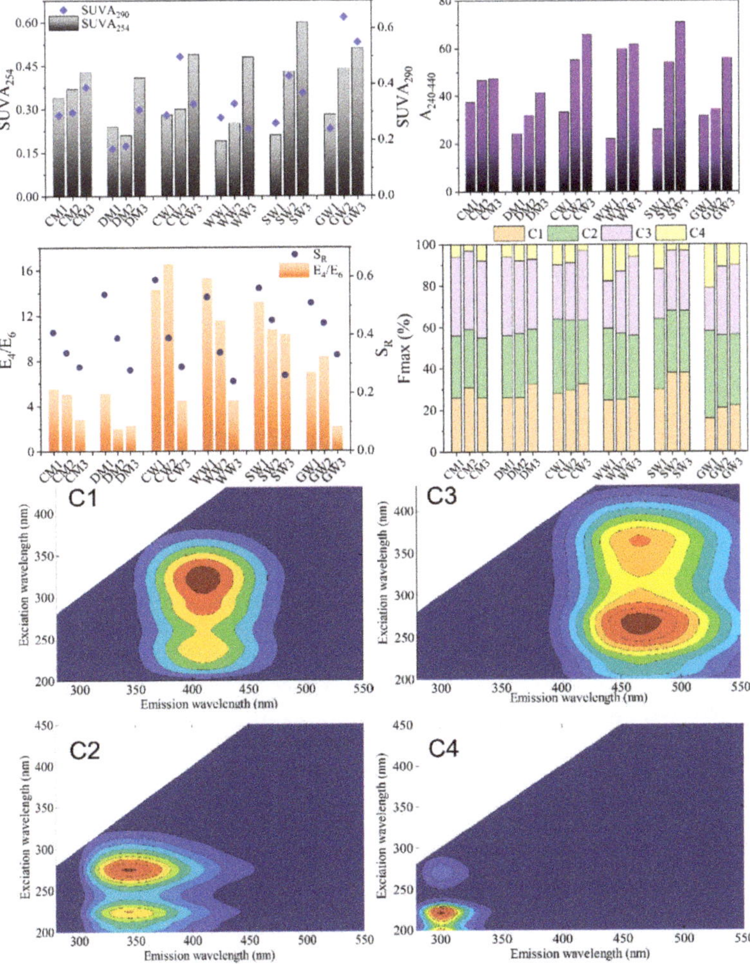

Figure 3. The analysis based on UV–Vis spectroscopy, and fluorescence spectroscopy for the chemical structures of different compost-derived HA. Acronyms definitions can be found in the Abbreviations list.

Parallel factor (PARAFAC) analysis showed that four components were identified to explain most of the HA variation from different composting (Figure 3). Component 1 (C1) and component 3 (C3) was characterized as fulvic-like and humic-like substances, respectively. Component 2 (C2) was attributed to tyrosine-like substances and component 4 (C4) was associated with soluble microbial byproduct-like material with a few tryptophan-like substances [20]. There were obvious differences in Fmax among components of HA from different composts ($P < 0.05$), which decreased as follows: C2 (32.22%), C3 (31.72%) > C1 (27.94%) > C4 (8.78%). However, each fluorescence components of HA presented different content in different composts due to their different organic compositions and humic substance precursors [6,21]. There was no significant increase of fulvic-like and humic-like substances in CM and DM. An obvious increase in the combination of C1 and C3 was identified in HA from WW, CW, GW and SW. This indicated the synthesis of HA that are chemically and biologically more stable in lignin- and lignocellulose-rich composting.

In order to better understand the difference of these chemical structures of HA from diverse composting, ^{13}C-NMR was monitored, which indicated the presence of many types of carbons, including aliphatic, carbohydrate, aromatic, and carboxyl carbons, with a wide variation in the relative abundance of aliphatic vs. aromatic carbons in HA. As depicted in Figure 4, the peak intensity of alkyl C region were obviously decreased from mesophilic stage to mature stage in protein-rich composting, followed by lignocellulose-rich composting. The peak intensity of aryl C region were increased in all the composts. In the mature phase of different composts, there was a significant increase in peak intensity at 148 and 153 ppm, which represent methoxy-/hydroxy-substituted phenyl C and oxygen-substituted aromatic C, respectively [22]. ^{13}C-NMR of CM, DM, and CW exhibited common major peaks at 50–110 ppm, suggesting abundant precursors of hydroxyl and amino acids for the formation of redox-active functional groups of HA [23]. HA from WW, SW and GW exhibited peaks between 170–185 ppm, confirming the presence of carbonyl groups from ester, carboxylic acids, quinone, ketone, etc. An increase of carbonyl groups could be found during composting of CM and DM, suggesting that lignin- and lignocellulose-rich organics were easier for arylation during composting and protein-rich organics were easier for carboxylation to form different HA [9]. The trends of ^{13}C-NMR, UV–Vis spectroscopy and EEM fluorescence plots explained the structural changes of HA during composting and obvious differences of HAs from different materials. These results suggested that HA from lignocellulose- and lignin-rich composting, especially in the thermophilic phase had more polar functional groups with redox activity such as phenolic, alcoholic, quinone and ketone groups, nitrogen and sulfur redox-active moieties, as well as organic-metal chelates.

2.3. The Potential Role of HA in Dissimilatory Fe(III) Bioreduction

The data in Figure 5 gives an indication of the interdependence of two random variables that range in value from −1 to +1. $SUVA_{254}$ was positively correlated with $A_{240\text{-}440}$ ($P < 0.01$). ^{13}C-NMR3 was positively correlated with ^{13}C-NMR2 but they were both negatively related to ^{13}C-NMR1, ^{13}C-NMR4, C2 and E_4/E_6, suggesting that the formation of aryl C and hydroxyl C fractions was directly sourced from degradation of tyrosine-like substances and alkyl C for the condensation of C skeletons in different composts. Correlation analysis showed that $SUVA_{290}$ was positively correlated with C1, but negatively with S_R. This result supported that quinones might act as main fractions of the fulvic-like substances with high molecular weight structure [24]. A strong positive correlation between ^{13}C-NMR4 and C2, C4 as well as E_4/E_6 was observed, indicating that carboxyl and protein-like matter could be used by the microorganisms and obviously promote the formation of HA with a high degree of humification [25].

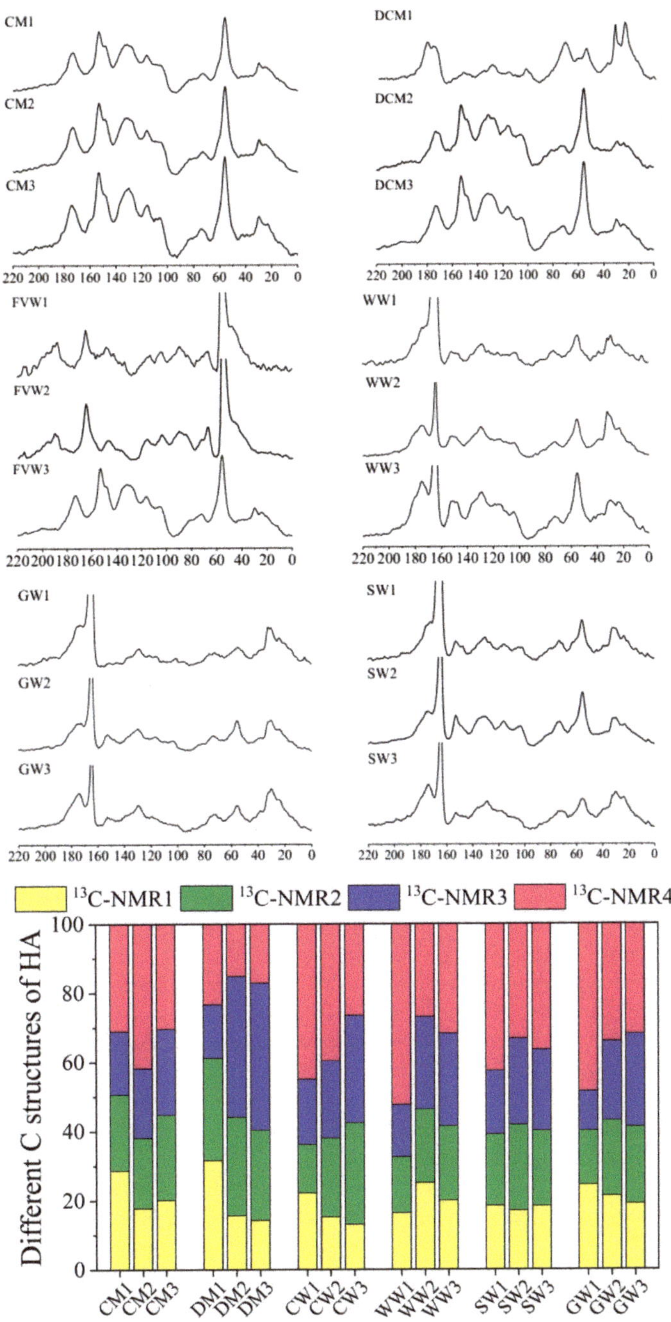

Figure 4. The solid state ^{13}C-NMR spectrum of HA during composting. Acronyms definitions can be found in the Abbreviations list.

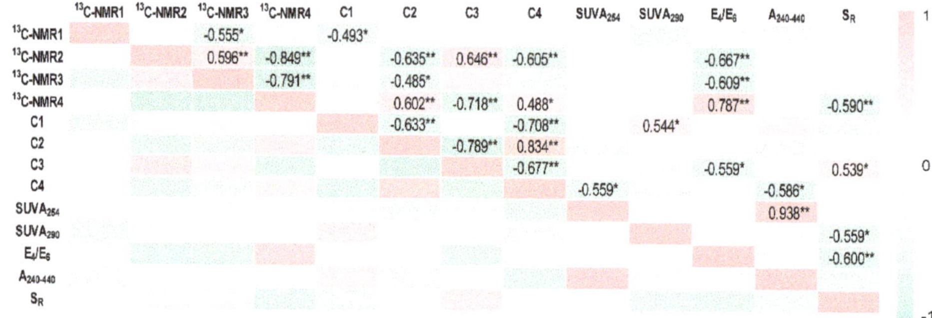

	^{13}C-NMR1	^{13}C-NMR2	^{13}C-NMR3	^{13}C-NMR4	C1	C2	C3	C4	SUVA$_{254}$	SUVA$_{290}$	E$_4$/E$_6$	A$_{240-440}$	S$_R$
^{13}C-NMR1			-0.555*		-0.493*								
^{13}C-NMR2			0.596**	-0.849**		-0.635**	0.646**	-0.605**			-0.667**		
^{13}C-NMR3				-0.791**		-0.485*					-0.609**		
^{13}C-NMR4						0.602**	-0.718**	0.488*			0.787**		-0.590**
C1						-0.633**		-0.708**	0.544*				
C2							-0.789**	0.834**					
C3								-0.677**			-0.559*		0.539*
C4									-0.559*			-0.586*	
SUVA$_{254}$												0.938**	
SUVA$_{290}$													-0.559*
E$_4$/E$_6$													-0.600**
A$_{240-440}$													
S$_R$													

Figure 5. Pearson correlation matrix between the chemical structure parameters of HA from different composting. * Significant at $P < 0.05$. ** Significant at $P < 0.01$. Colors are proportional to values in table (correlation coefficient). Acronyms definitions can be found in Materials and Methods.

In order to determine to what extent different chemical structures of HA from different composts affected the dissimilatory Fe(III) bioreduction, RDA was performed based on the spectral dataset and Fe(II) production indexes of different HA treatments. Figure 6 indicated an ordination graph where the first and all canonical axes were highly significant based on Monte Carlo tests ($P < 0.01$). The HA samples of different composting could be divided into three groups. The humification degree and electron transfer capacity of HA were different at each stage of diverse composting. The HA from mesophilic phase of composting were similar in most of the materials that clustered together, which had more aliphatic compounds and protein-like matter with an inhibition effect on the bioreduction of dissimilatory Fe(III). The HA from different lignocellulose-rich composts and the thermophilic phase of lignin-rich composts mainly clustered together except SW2, which had higher projection value on the arrow of Fe(II) formation, and SUVA$_{290}$. The findings highlighted the role of HA from thermophilic phase of lignocellulose- and lignin-rich composting on stimulating the bioreduction of Fe(III) mainly based on the quinone moieties as well as other redox-active functional groups [3]. HA from the mature stage of composting, especially protein-rich materials, were grouped together but had shorter projection value on the Fe(II) formation. The shorter distance of these HA from the mature stage of composting to C3, ^{13}C-NMR3, ^{13}C-NMR2, and SUVA$_{254}$ suggested that the high abundance of aromatic C structure with hydroxyl in humic-like substances would not obviously affect the electron transfer process of Fe(III) reduction by *Shewanella oneidensis* MR-1.

Taking the aforementioned information into account, our results suggested that the HA from lignocellulose- and lignin-rich composting, especially in the thermophilic phase promoted obviously the bioreduction of Fe(III), and HA from protein-rich materials suppressed significantly the Fe(II) production. Aromaticity of HA is not a reason for the electron transfer, while quinone/hydroquinone groups in HA are the major contributors to redox reaction between Fe(III), HA and microorganisms. The HA from diverse composts has different redox-active potential for Fe reduction, mainly because the origin and stages of composting affects the humification degree and chemical structures of HA, leading to different amount, structures and the redox activity of functional groups such as phenolic, alcoholic, quinone and ketone groups, nitrogen and sulfur redox-active moieties, as well as organic-metal chelates [8].

Fe(III) is an important terminal electron shuttles to stimulate the biodegradation of organic contaminants [26]. Dissimilatory iron reduction often occurs in soils, which caused the release of inorganic compounds that are bound to Fe(III) oxides and the soluble Fe(II) to be leached from the soil [27]. Considering the different stimulation capacity of HA from diverse composting, it can be reasonably concluded that the HA of composts will play an important role on heavy metal transformation and organic pollutant degradation in the contaminated soils. When composts are applied to soils, composts can be used in different ways based on the chemical structures and electron transfer capacity of HA. HA-rich composts from lignocellulose- and lignin-rich materials are more suitable bioresources for wide-ranging applications to catalyze the redox-mediated reactions of pollutants for contaminated soils remediation.

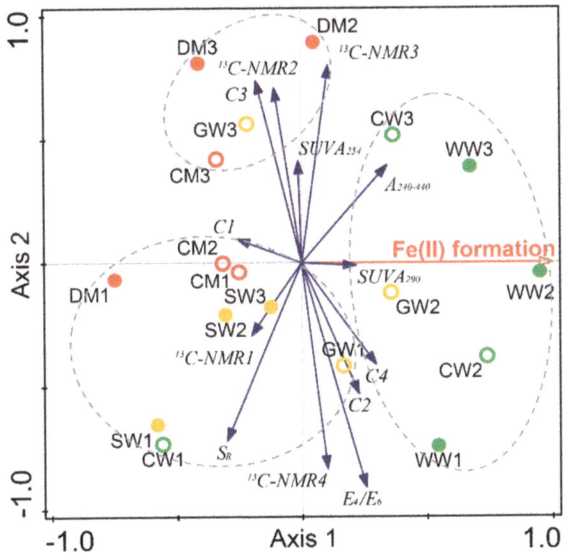

Figure 6. Redundancy analysis of the correlation between the redox-active functional groups of HA and Fe(III)-citrate bioreduction for HA samples from different composting. The red, green and yellow circles represented the treatments with adding HA from protein-, lignocellulose- and lignin-rich composting, respectively. Acronyms definitions can be found in Materials and Methods.

3. Materials and Methods

3.1. Preparation of Shewanella oneidensis MR-1

Shewanella oneidensis MR-1 was grown aerobically from frozen (−80 °C) stock to lag-log phase in a Luria-Bertani (LB) broth at 30 °C. The cells were harvested by centrifugation at 3600 g for 10 min at 4 °C when it reached the exponential phase, washed twice with sterile 30 mM of anoxic bicarbonate buffer (pH = 7) and suspended before use in the following reduction studies. The concentrations of the strains were 1×10^7 CFU mL^{-1}. All growth media and containers used in this study were autoclaved (121 °C, 20 min) prior to incubation.

3.2. Extraction and analysis of HA from Different Composting

Humic acid (HA) samples were extracted from three main stages (i.e., mesophilic phase, thermophilic phase and mature phase) of different composting of six raw materials, including chicken manure (CM), dairy manure (DM), cabbage wastes (CW), weeds wastes (WW), straw wastes (SW) and green wastes (GW). Details of composting process and relevant properties have been described previously in Zhao et al. [17]. Each compost pile was more than 2 ton, and the piles were turned mechanically every 7 days at Shanghai Songjiang Composting plant, China. The extraction of HA followed the procedure of Wu et al. [9]. Compost samples (10 g) were mixed with 0.1 M (NaOH + Na$_4$P$_2$O$_7$) at 1:10 (w: v) ratio for shaking 24 h with rotational speed 150 rpm. After centrifugation at 10,000 rpm for 15 min, the supernatant was filtered through a 0.45 μm membrane and then acidified to pH 1 with 6 M hydrochloric acid (HCl). HA was separated by centrifugation at 11,000 rpm for 15 min and suspended in a solution of 0.1 M HCl/0.3 M hydrogen fluoride to remove mineral impurities and dialyzed until the elimination of chloride ions. All chemicals were analytical reagent grade or higher and were purchased from Sinopharm Chemical Reagent Co., Ltd. (Shanghai, China).

UV–Vis spectroscopy, fluorescence spectroscopy, and ^{13}C-NMR spectra were carried out to analyze the chemical structures of HA. Prior to fluorescence and UV–Vis analysis, total organic carbon (TOC) of all the samples was measured with an Analytik Jena model Multi N/C 2100 TOC analyzer (Jena, Germany) and the concentrations of HA were adjusted to 10 mg L^{-1}.

UV–Vis spectroscopy was performed with a Shimadzu model UV-1700 PC spectrophotometer. Specific ultraviolet absorbance at 254 nm (SUVA$_{254}$) and 290 nm (SUVA$_{290}$) was calculated as the absorbance divided by the TOC concentration, and the absorbance ratio at 465 and 665 nm (E$_4$/E$_6$) was used to characterize the humification degree of HA. The integral area was calculated from 240 to 400 nm, designated as A$_{240-400}$, which was referred to the presence of the aromatic rings of polar functional groups [28]. The slope ratio or S$_R$, the ratio of the slope of the shorter ultraviolet absorbance wavelength region (275–295 nm) to that of the longer wavelength region (350–400 nm), was selected to determine the molecular weight [15].

Fluorescence spectroscopy was recorded using a Perkin-Elmer model LS50B fluorescence spectrophotometer in a clear quartz cuvette. Excitation–emission matrices (EEMs) spectra were recorded for excitation (200–450 nm) and emission (280–550 nm) wavelengths at intervals of 5 nm with a scan speed of 2400 nm min^{-1}. The EEMs were blank subtracted, corrected for inner-filter effects and instrument-specific biases, and normalized to the Raman area [29]. Parallel factor (PARAFAC) analysis of the EEMs spectral data was carried out in MATLAB 2013a (Mathworks, Natick, MA) with the DOMFluor toolbox (www.models.life.ku.dk) to identify the co-varied independently components. The concentration scores of the PARAFAC components were expressed as maximum fluorescence intensity (Fmax) (R.U.) for each modeled component [30].

Solid state ^{13}C-NMR data of HA were acquired with a Bruker AV-300 spectrometer equipped with a direct polarization magic angle spinning (DP-MAS) probe, which were performed according to the method of Amir et al. [31]. The solutions for NMR were prepared by dissolving 100 mg of the freeze-dried HA in 1 mL of 0.5 M NaOD/D$_2$O. On the basis of qualitative analysis, the ^{13}C-NMR spectra are subdivided into four main resonance regions (i.e., ^{13}C-NMR1, ^{13}C-NMR2, ^{13}C-NMR3, and ^{13}C-NMR4), which were 0–55 ppm (aliphatic carbon), 55–110 ppm (aliphatic carbon substituted with O or N), 110–165 ppm (aromatic carbon), and 165–200 ppm (carboxylic carbon) [32]. The proportion of each type of carbon was calculated by integrating the spectral regions.

3.3. Batch Experiments of Dissimilatory Fe(III) Reduction

Dissimilatory Fe(III) reduction assays were set up in triplicate in serum bottles containing 50 mL of basal medium with sodium lactate (5 mM) as an electron donor, 20 mM Fe(III)-citrate as an electron acceptor, *Shewanella oneidensis* MR-1 cells (10^7 cells mL^{-1}), and HA (final concentration of 100 mg L^{-1}). Only sodium lactate and Fe(III)-citrate were added as a blank control (CK1). Controls (CK2) were prepared without HA addition. Various treatments were applied as shown in Table 2. The pH values were adjusted to 7.0–7.5 with trace metal grade 1.0 M NaOH or HCl. The reaction solution purged with 100% N$_2$ for 30 min and static incubation in anoxic glovebox (N$_2$ atmosphere, at 30 °C, O$_2$ < 0.1 ppm). Samples were periodically extracted with 1 M HCl at 1:1 (v: v) ratio, then removed with sterile needles and syringes, and filtered through a 0.22 µm membrane filter to analyze the total Fe(II) concentration with phenanthroline assay in an anaerobic chamber until Fe(II) became stable [33]. Absorbance was measured immediately with an ultraviolet-visible spectrophotometer (Shimadzu UV-1800, Tokyo, Japan) at 510 nm.

Table 2. Different treatments of batch experiments for dissimilatory Fe(III) reduction.

Treatment	Sodium Lactate and Fe(III)-citrate	*Shewanella oneidensis* MR-1	Humic Acids		
			Mesophilic Phase	Thermophilic Phase	Mature Phase
CK1	+	−	−	−	−
CK2	+	+	−	−	−
HA(X1)	+	+	+	−	−
HA(X2)	+	+	−	+	−
HA(X3)	+	+	−	−	+

X represented the different materials composting, including chicken manure (CM), dairy manure (DM), cabbage wastes (CW), weeds wastes (WW), straw wastes (SW) and green wastes (GW).

3.4. Statistical Analyses

Figure processing was generated by Origin 9.1 software (OriginLab, Los Angeles, USA). SPSS version 22.0 was used for the analysis of variance (ANOVA), and Pearson correlation analysis. The potential reduction of Fe(III)–citrate was fitted to mathematical equations by the non-linear least-square curve fitting technique (Marquardt-Levenberg algorithm) using MATLAB v.19.0 software. $Fe(II)_t = Fe(II)_0 + Fe(III)_0 [1\text{-}exp(\text{-}k_{red}t)]$, where $Fe(II)_t$ is the Fe(II) concentration at time t, $Fe(II)_0$ is the initial Fe (II) concentration, $Fe(III)_0$ is the initial Fe(III) measured values. Redundancy analysis (RDA) was performed to analyze the multivariate relationships between the chemical structures of HA and Fe(III)-citrate bioreduction by Canoco (Version 5.0, Centre for Biometry, Wageningen, The Netherlands). Monte Carlo reduced model tests with 499 unrestricted permutations were also used to evaluate the significance of the first canonical axis and of all canonical axes together. Statistical significance was kept at $P < 0.05$ for all analyses.

4. Conclusions

This study provided comprehensive information on the chemical structures of HA from different stages of composting derived from protein-, lignocellulose- and lignin-rich materials. Quinone/hydroquinone groups in HA are the major contributors to redox reaction between Fe(III), HA, and microorganisms, but HA from diverse origins and stages of composting has different humification degree and different amount, structures, and the redox activity of functional groups, leading to different redox-active potential for catalyzing the bioreduction of Fe(III)-citrate by *Shewanella oneidensis* MR-1. The HA from lignocellulose- and lignin-rich composting, especially in the thermophilic phase promoted obviously the bioreduction of Fe(III), and HA from protein-rich materials suppressed significantly the Fe(II) production.

Author Contributions: Y.W. and B.X. conceived and designed the experiments; Y.W., W.T. and X.L. performed the experiments; Y.W., and X.L. analyzed the data with suggestions by F.Z., W.T. and Z.W.; Y.W. and W.T. wrote the paper; Z.W., F.Z. and B.X. proofed the paper.

Funding: This work was funded by the China Postdoctoral Science Foundation (No. 2017M620801), and National Natural Science Foundation of China (No. 51808519).

Acknowledgments: We thank the editor and two anonymous reviewers for their insightful comments that greatly improved this manuscript.

Conflicts of Interest: The authors declare that they have no conflict of interest.

Abbreviations

CM1; CM2; CM3: Chicken manure composting at the mesophilic phase; Chicken manure composting at the thermophilic phase; Chicken manure composting at the maturation phase, respectively. DM1; DM2; DM3: Dairy manure composting at the mesophilic phase; Dairy manure composting at the thermophilic phase; Dairy manure composting at the maturation phase, respectively. CW1; CW2; CW3: Cabbage wastes composting at the mesophilic phase; Cabbage wastes composting at the thermophilic phase; Cabbage wastes composting at the maturation phase, respectively. WW1; WW2; WW3: Weeds wastes composting at the mesophilic phase; Weeds wastes composting at the thermophilic phase; Weeds wastes composting at the maturation phase, respectively. SW1; SW2; SW3: Straw wastes composting at the mesophilic phase; Straw wastes composting at the thermophilic phase; Straw wastes composting at the maturation phase, respectively. GW1; GW2; GW3: Green wastes composting at the mesophilic phase; Green wastes composting at the thermophilic phase; Green wastes composting at the maturation phase, respectively. HA: Humic acids. ^{13}C NMR: Carbon 13 nuclear magnetic resonance.

References

1. Liu, G.; Qiu, S.; Liu, B.; Pu, Y.; Gao, Z.; Wang, J.; Jin, R.; Zhou, J. Microbial reduction of Fe (III)-bearing clay minerals in the presence of humic acids. *Sci. Rep.* **2017**, *7*, 45354. [CrossRef] [PubMed]
2. Zhao, X.; Tan, W.; Dang, Q.; Li, R.; Xi, B. Enhanced biotic contributions to the dechlorination of pentachlorophenol by humus respiration from different compostable environments. *Chem. Eng. J.* **2019**, *361*, 1565–1575. [CrossRef]
3. Klüpfel, L.; Piepenbrock, A.; Kappler, A.; Sander, M. Humic substances as fully regenerable electron acceptors in recurrently anoxic environments. *Nat. Geosci.* **2014**, *7*, 195. [CrossRef]
4. Stern, N.; Mejia, J.; He, S.; Yang, Y.; Ginder-Vogel, M.; Roden, E.E. Dual role of humic substances as electron donor and shuttle for dissimilatory iron reduction. *Environ. Sci. Technol.* **2018**, *52*, 5691–5699. [CrossRef]
5. Wei, Y.; Zhao, Y.; Shi, M.; Cao, Z.; Lu, Q.; Yang, T.; Fan, Y.; Wei, Z. Effect of organic acids production and bacterial community on the possible mechanism of phosphorus solubilization during composting with enriched phosphate-solubilizing bacteria inoculation. *Bioresour. Technol.* **2018**, *247*, 190–199. [CrossRef] [PubMed]
6. Wang, C.; Tu, Q.; Dong, D.; Strong, P.; Wang, H.; Sun, B.; Wu, W. Spectroscopic evidence for biochar amendment promoting humic acid synthesis and intensifying humification during composting. *J. Hazard. Mater.* **2014**, *280*, 409–416. [CrossRef]
7. Gao, X.; Tan, W.; Zhao, Y.; Wu, J.; Sun, Q.; Qi, H.; Xie, X.; Wei, Z. Diversity in the Mechanisms of Humin Formation during Composting with Different Materials. *Environ. Sci. Technol.* **2019**, *53*, 3653–3662. [CrossRef]
8. Lipczynska-Kochany, E. Humic substances, their microbial interactions and effects on biological transformations of organic pollutants in water and soil: A review. *Chemosphere* **2018**, *202*, 420–437. [CrossRef] [PubMed]
9. Wu, J.; Zhao, Y.; Zhao, W.; Yang, T.; Zhang, X.; Xie, X.; Cui, H.; Wei, Z. Effect of precursors combined with bacteria communities on the formation of humic substances during different materials composting. *Bioresour. Technol.* **2017**, *226*, 191–199. [CrossRef]
10. Lovley, D.R.; Coates, J.D.; Blunt-Harris, E.L.; Phillips, E.J.; Woodward, J.C. Humic substances as electron acceptors for microbial respiration. *Nature* **1996**, *382*, 445–448. [CrossRef]
11. Wu, J.; Zhao, Y.; Yu, H.; Wei, D.; Yang, T.; Wei, Z.; Lu, Q.; Zhang, X. Effects of aeration rates on the structural changes in humic substance during co-composting of digestates and chicken manure. *Sci. Total Environ.* **2019**, *658*, 510–520. [CrossRef] [PubMed]
12. Xi, B.; Zhao, X.; He, X.; Huang, C.; Tan, W.; Gao, R.; Zhang, H.; Li, D. Successions and diversity of humic-reducing microorganisms and their association with physical-chemical parameters during composting. *Bioresour. Technol.* **2016**, *219*, 204–211. [CrossRef] [PubMed]
13. Nakasaki, K.; Le, T.H.T.; Idemoto, Y.; Abe, M.; Rollon, A.P. Comparison of organic matter degradation and microbial community during thermophilic composting of two different types of anaerobic sludge. *Bioresour. Technol.* **2009**, *100*, 676–682. [CrossRef] [PubMed]
14. Kulkarni, H.V.; Mladenov, N.; McKnight, D.M.; Zheng, Y.; Kirk, M.F.; Nemergut, D.R. Dissolved fulvic acids from a high arsenic aquifer shuttle electrons to enhance microbial iron reduction. *Sci. Total Environ.* **2018**, *615*, 1390–1395. [CrossRef]
15. Helms, J.R.; Stubbins, A.; Ritchie, J.D.; Minor, E.C.; Kieber, D.J.; Mopper, K. Absorption spectral slopes and slope ratios as indicators of molecular weight, source, and photobleaching of chromophoric dissolved organic matter. *Limnol. Oceanogr.* **2008**, *53*, 955–969. [CrossRef]
16. Yuan, Y.; He, X.; Xi, B.; Li, D.; Gao, R.; Tan, W.; Zhang, H.; Yang, C.; Zhao, X. Polarity and molecular weight of compost-derived humic acid affect Fe(III) oxides reduction. *Chemosphere* **2018**, *208*, 77–83. [CrossRef]
17. Zhao, X.; He, X.; Xi, B.; Gao, R.; Tan, W.; Zhang, H.; Huang, C.; Li, D.; Li, M. Response of humic-reducing microorganisms to the redox properties of humic substance during composting. *Waste Manag.* **2017**, *70*, 37–44. [CrossRef]

18. Tan, W.; Xi, B.; Wang, G.; Jiang, J.; He, X.; Mao, X.; Gao, R.; Huang, C.; Zhang, H.; Li, D. Increased electron-accepting and decreased electron-donating capacities of soil humic substances in response to increasing temperature. *Environ. Sci. Technol.* **2017**, *51*, 3176–3186. [CrossRef]

19. Ratasuk, N.; Nanny, M.A. Characterization and Quantification of Reversible Redox Sites in Humic Substances. *Environ. Sci. Technol.* **2007**, *41*, 7844–7850. [CrossRef]

20. Stedmon, C.A.; Bro, R. Characterizing dissolved organic matter fluorescence with parallel factor analysis: A tutorial. *Limnol. Oceanogr. Meth.* **2008**, *6*, 572–579. [CrossRef]

21. Chen, W.; Westerhoff, P.; Leenheer, J.A.; Booksh, K. Fluorescence Excitation–Emission Matrix Regional Integration to Quantify Spectra for Dissolved Organic Matter. *Environ. Sci. Technol.* **2003**, *37*, 5701–5710. [CrossRef] [PubMed]

22. Pérez, M.G.; Martin-Neto, L.; Saab, S.C.; Novotny, E.H.; Milori, D.M.; Bagnato, V.S.; Colnago, L.A.; Melo, W.J.; Knicker, H. Characterization of humic acids from a Brazilian Oxisol under different tillage systems by EPR, ^{13}C NMR, FTIR and fluorescence spectroscopy. *Geoderma* **2004**, *118*, 181–190. [CrossRef]

23. Tan, K.H. *Humic Matter in Soil and the Environment: Principles and Controversies*; CRC Press: Boca Raton, FL, USA, 2014.

24. Zhou, Y.; Selvam, A.; Wong, J.W.C. Evaluation of humic substances during cocomposting of food waste, sawdust and Chinese medicinal herbal residues. *Bioresour. Technol.* **2014**, *168*, 229–234. [CrossRef] [PubMed]

25. Smilek, J.; Sedláček, P.; Kalina, M.; Klucáková, M. On the role of humic acids' carboxyl groups in the binding of charged organic compounds. *Chemosphere* **2015**, *138*, 503–510. [CrossRef]

26. Lovley, D.R.; Holmes, D.E.; Nevin, K.P. Dissimilatory Fe(III) and Mn(IV) reduction. *Adv. Microb. Physiol.* **2004**, *49*, 219–286. [PubMed]

27. Xu, Y.; He, Y.; Feng, X.; Liang, L.; Xu, J.; Brookes, P.C.; Wu, J. Enhanced abiotic and biotic contributions to dechlorination of pentachlorophenol during Fe(III) reduction by an iron-reducing bacterium Clostridium beijerinckii Z. *Sci. Total Environ.* **2014**, *473–474*, 215–223. [CrossRef]

28. Zhao, X.; Wei, Y.; Fan, Y.; Zhang, F.; Tan, W.; He, X.; Xi, B. Roles of bacterial community in the transformation of dissolved organic matter for the stability and safety of material during sludge composting. *Bioresour. Technol.* **2018**, *267*, 378–385. [CrossRef] [PubMed]

29. Cui, H.Y.; Zhao, Y.; Chen, Y.N.; Zhang, X.; Wang, X.Q.; Lu, Q.; Jia, L.M.; Wei, Z.M. Assessment of phytotoxicity grade during composting based on EEM/PARAFAC combined with projection pursuit regression. *J. Hazard. Mater.* **2017**, *326*, 10–17. [CrossRef] [PubMed]

30. Wei, Z.; Zhao, X.; Zhu, C.; Xi, B.; Zhao, Y.; Yu, X. Assessment of humification degree of dissolved organic matter from different composts using fluorescence spectroscopy technology. *Chemosphere* **2014**, *95*, 261–267. [CrossRef] [PubMed]

31. Amir, S.; Jouraiphy, A.; Meddich, A.; El Gharous, M.; Winterton, P.; Hafidi, M. Structural study of humic acids during composting of activated sludge-green waste: Elemental analysis, FTIR and ^{13}C NMR. *J. Hazard. Mater.* **2010**, *177*, 524–529. [CrossRef]

32. Trubetskoj, O.A.; Hatcher, P.G.; Trubetskaya, O.E. ^{1}H-NMR and ^{13}C-NMR spectroscopy of chernozem soil humic acid fractionated by combined size-exclusion chromatography and electrophoresis. *Chem. Ecol.* **2010**, *26*, 315–325. [CrossRef]

33. Tan, W.; Li, R.; Yu, H.; Zhao, X.; Dang, Q.; Jiang, J.; Wang, L.; Xi, B. Prominent Conductor Mechanism-Induced Electron Transfer of Biochar Produced by Pyrolysis of Nickel-Enriched Biomass. *Catalysts* **2018**, *8*, 573. [CrossRef]

© 2019 by the authors. Licensee MDPI, Basel, Switzerland. This article is an open access article distributed under the terms and conditions of the Creative Commons Attribution (CC BY) license (http://creativecommons.org/licenses/by/4.0/).

Article

Selective Production of Terephthalonitrile and Benzonitrile via Pyrolysis of Polyethylene Terephthalate (PET) with Ammonia over Ca(OH)$_2$/Al$_2$O$_3$ Catalysts

Lujiang Xu [1,2], Xin-wen Na [1], Le-yao Zhang [1], Qian Dong [1], Guo-hua Dong [1], Yi-tong Wang [3] and Zhen Fang [1,*]

[1]   Biomass Group, College of Engineering, Nanjing Agricultural University, 40 Dianjiangtai Road, Nanjing 210031, Jiangsu, China; lujiangxu@njau.edu.cn (L.X.); Na_Xinwen@163.com (X.-w.N.); leyaozhang1003@163.com(L.-y.Z.); dongqianbiomass@163.com (Q.D.); dong_guohua@126.com (G.-h.D.)

[2]   Key Laboratory of Energy Thermal Conversion and Control of Ministry of Education, Southeast University, Nanjing 210096, China

[3]   College of Metallurgy and Energy, North China University of Science and Technology, 21 Bohai Street, Tangshan 063210, China; wangyt@ncst.edu.cn

*   Correspondence: zhenfang@njau.edu.cn; Tel.: +86-025-58606657

Received: 31 March 2019; Accepted: 28 April 2019; Published: 9 May 2019

Abstract: A series of Ca(OH)$_2$/Al$_2$O$_3$ catalysts were synthesized for selectively producing N-containing chemicals from polyethylene terephthalate (PET) via catalytic fast pyrolysis with ammonia (CFP-A) process. During the CFP-A process, the carboxyl group in PET plastic was efficiently utilized for the selective production of terephthalonitrile and benzonitrile by controlling the catalysts and pyrolysis parameters (e.g. temperature, residence time, ammonia content). The best conditions were selected as 2% Ca(OH)$_2$/γ-Al$_2$O$_3$ (0.8 g), 500 °C under pure ammonia with 58.3 C% terephthalonitrile yield and 92.3% selectivity in nitriles. In addition, 4% Ca(OH)$_2$/ Al$_2$O$_3$ was suitable for producing benzonitrile. With catalyst dosage of 1.2 g, residence time of 1.87 s, pyrolysis temperature of 650 °C and pure ammonia (160 mL/min carrier gas flow rate), the yield and selectivity of benzonitrile were 30.4 C% and 82.6%, respectively. The catalysts deactivated slightly after 4 cycles.

Keywords: benzonitrile; terephthalonitrile; polyethylene terephthalate (PET); catalytic pyrolysis; ammonia; Ca(OH)$_2$/Al$_2$O$_3$

1. Introduction

With the rapid development of society and the improvement of people's quality of life, more and more petroleum-based plastics are being consumed to meet people's life needs [1]. Due to its advantages of anti-acid, stability and safety, polyethylene terephthalate (PET) has been widely used as bottle materials, fibers, films, sheets, households and carpet [2]. The worldwide consumption of PET plastic was 60 million tons in 2011 and keeps growing year by year [3]. However, the high stable properties of PET make it difficult to degrade naturally and causes serious environmental problems [4,5]. Thus, the recycling of PET plastics has become urgent [6].

Catalytic fast pyrolysis (CFP) is a promising technology to convert plastics, biomass and other organic wastes to high quality bio-oil or targeted compounds within a few seconds with the help of catalyst [7–9]. Currently, different plastics (e.g. high-density polyethylene, polyvinyl chloride, polypropylene, PET, and polystyrene, etc.) are recycled efficiently by the catalytic pyrolysis process [10]. In addition, numerous kinds of reactors (e.g., batch, semi-batch, fixed bed, fluidized bed, conical spouted bed and microwave-assisted reactors, etc.) were designed for waste plastics recycling [11,12]. Studies on catalytic pyrolysis of PET plastic were mainly carried out under inert atmosphere, and benzene-rich bio-oil was the main pyrolytic product. Yoshioka et al. [13] screened $Ca(OH)_2$, NiO, TiO_2 and Fe_2O_3 on the catalytic fast pyrolysis of PET under inert atmosphere, and found $Ca(OH)_2$ showed best decarboxylation performance, the yield of benzene-rich bio-oil reached 31% at 700 °C. Besides $Ca(OH)_2$, CaO and zeolites (HZSM-5) also showed good catalytic performance to produce benzene-rich bio-oil. Kumagai et al. used the tandem micro (μ)-reactor-gas chromatography/mass spectrometry (TR-GC/MS) system to investigate the CaO catalytic pyrolysis of the PET process and elucidated the relationship between CaO deterioration and the aromatic hydrocarbon selectivity, and the nature of PET catalytic pyrolysis process [14,15]. Du et al. [16] used ZSM-5 zeolite and CaO to catalyze PET via pyrolysis for producing benzene-rich oil. Both catalysts showed good performance on the deoxygenation of pyrolysis products, and CaO presented better performance than ZSM-5. In addition, Xue et al. [17] used HZSM-5 to catalyze PET via pyrolysis for producing benzene-rich oil, elucidated the effect of HZSM-5 and the PET contact mode, and found that in situ catalytic pyrolysis produced more coke and aromatics than ex situ catalytic pyrolysis. Besides the benzene-rich bio-oil, benzoic acid could also be selectively produced from PET via catalytic pyrolysis over sulphated zirconia catalysts [18].

In order to make full use of the oxygen-containing functional groups in biomass and other oxygen-containing wastes, catalytic pyrolysis under ammonia process was proposed to selectively converted biomass and oxygen plastic to produce N-containing chemicals [19]. Different than the traditional pyrolysis process, the CFP-A process consisted of pyrolysis and ammonization, and ammonia was introduced as a reagent and nitrogen source. N-containing chemicals (e.g. acetonitrile, pyrrole, pyridines, indoles, and anilines) could be selectively produced from cellulose, bio-derived furans, glycerol, polylactic acid and lignin via catalytic pyrolysis with the ammonia process [20–24]. The carboxyl group in PET was not used effectively, which could be converted to amides, amines, nitriles, ketones and other value-added compounds by suitable processes.

Aromatic nitriles (e.g., terephthalonitrile, benzonitrile) have been widely applied to pharmaceuticals, pesticides, dyes and polymers [25–27]. Terephthalonitrile is a high value-added fine chemical with a price reaching over $ 6000 per ton. It can be converted to *p*-phenylenediamine (raw material of epoxy resin and polyurethane), terephthalic acid (monomer of PET), pyrethroids pesticide, terephthalonitrile-derived nitrogen-rich network (supercapacitors), polyamidines polymers (light sensitive material) [28–30]. Benzonitrile has been widely used as solvent in the synthesis of nitrile-based rubber, resin, polymer and coatings [31–33]. It also has been used as building block for the synthesis of agrochemicals and pharmaceuticals [34,35]. The price of benzonitrile is more than $ 4000 per ton, with over 10 thousand tons of benzonitrile being produced per year. Currently, terephthalonitrile and benzonitrile are produced by the ammoxidation of *p*-xylene, toluene industrially [36,37]. In industry, toluene is produced by the catalytic reforming gasoline or benzene alkylation reaction. *p*-Xylene is produced by separation from fossil fuel or alkyl transfer reactions of toluene [38]. However, *p*-xylene and toluene are mainly derived from non-renewable fossil-based resources. Therefore, finding a renewable or environment friendly feedstock to produce terephthalonitrile and benzonitrile is highly desirable for the green production of aromatic nitriles.

In this work, the carboxyl group in PET was utilized efficiently via pyrolysis with synthesized $Ca(OH)_2/Al_2O_3$ catalysts. Terephthalonitrile and benzonitrile could be selectively produced from PET by controlling the catalysts and pyrolysis parameters in a fixed bed reactor. The parameters (pyrolysis temperature, $Ca(OH)_2$ loading, catalyst usage, ammonia usage, etc), which affected the production of terephthalonitrile and benzonitrile, were investigated systematically. Furthermore, the possible reaction pathways from PET to different aromatic nitriles were investigated. Finally, catalysts stability was also studied by 4 cycle experiments. The fresh and reused catalysts were characterized by X-Ray Diffraction (XRD), N_2 adsorption/desorption, temperature programmed desorption of ammonia (NH_3-TPD) and temperature programmed desorption of carbon dioxide (CO_2-TPD) analyses.

2. Results and Discussions

2.1. Effect of Pyrolysis Temperature

The effect of temperature (500–700 °C) on the aromatic nitriles was investigated by using 2% Ca(OH)$_2$-Al$_2$O$_3$ (1 g) at a constant flow rate of 160 mL/min carrier gas (50% ammonia and 50% nitrogen). Figure 1 shows the effect of temperature on overall yield of products and nitriles selectivity. Nitriles (the main product) and aromatics were collected by a cold trap. As temperature increased from 500 to 700 °C, the carbon yields of char, gases, nitriles and aromatics, and nitriles selectivity changed significantly. In Figure 1a, the carbon yield of char decreased from 15.85 C% at 500 °C to 6.03 C% at 700 °C, while the carbon yield of gases increased from 11.78 C% at 500 °C to 36.04 C% at 700 °C. In addition, the variation trend of nitriles carbon yield with temperature was consistent with that of char, and aromatics with that of gases. With the temperature increasing from 500 to 700 °C, the carbon yield of nitriles decreased from 53.98 to 20.61 C%. Meanwhile, the carbon yield of aromatics increased from 1.78 C% to 16.46 C%. Figure 1b shows the variation trends of different nitriles selectivity vs. temperature and the selectivity of other nitriles was very low (less than 3%). Nevertheless, the variation trends of terephthalonitrile and benzonitrile were completely opposite with the increase of temperature. At 500 °C, the selectivity of terephthalonitrile and benzonitrile was 88.88% and 8.43%, respectively. However, at 700 °C, the selectivity of terephthalonitrile decreased to 7.91%, and the selectivity of benzonitrile was up to 92.09%. Higher temperature would promote the cracking and decarboxylation reactions. Pyrolytic temperature affected the production of terephthalonitrile and benzonitrile significantly. Together with Figure 1a, the carbon yield of terephthalonitrile obtained at 500 °C reached 47.98 C%, and the carbon yield of benzonitrile obtained at 700 °C was 18.98 C%. At 650 °C, the carbon yield of benzonitrile was 24.53 C%, the highest carbon from 500–700 °C with the selectivity of benzonitrile in nitriles above 70%. Therefore, 500 °C and 650 °C were the optimal pyrolysis temperatures for selectively producing terephthalonitrile and benzonitrile. In the following tests, the effects of catalyst, residence time (between pyrolytic vapors and catalyst) and ammonia fraction in the carrier gas on the production of terephthalonitrile and benzonitrile were investigated at 500 °C and 650 °C, respectively.

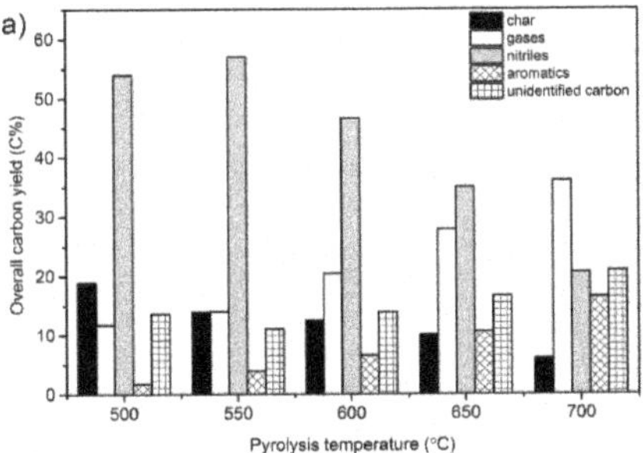

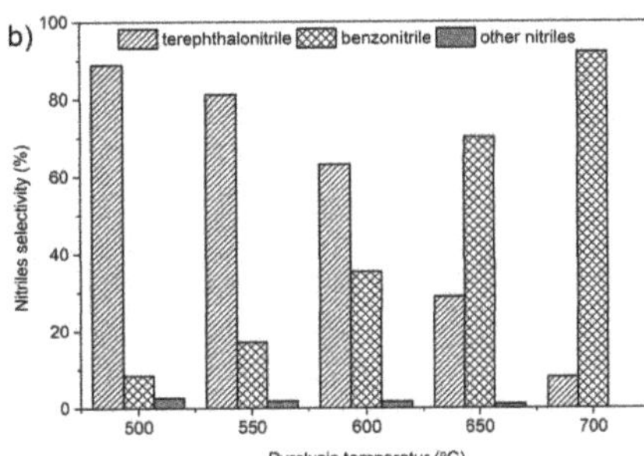

Figure 1. Effect of temperature on (**a**) overall yield and (**b**) nitriles selectivity (2% Ca(OH)$_2$/Al$_2$O$_3$ (1 g); 80 mL/min N$_2$ and NH$_3$ flow rate, 0.5 g polyethylene terephthalate (PET) /batch).

2.2. Optimizing Conditions for Producing Terephthalonitrile at 500 °C

The effects of catalyst, residence time and ammonia content in the carrier gas on the production of terephthalonitrile were studied at 500 °C. Figure 2 shows the effect of Ca(OH)$_2$ loading on Ca(OH)$_2$/Al$_2$O$_3$ catalysts. Tables 1 and 2 give the effects of residence time and ammonia content, respectively.

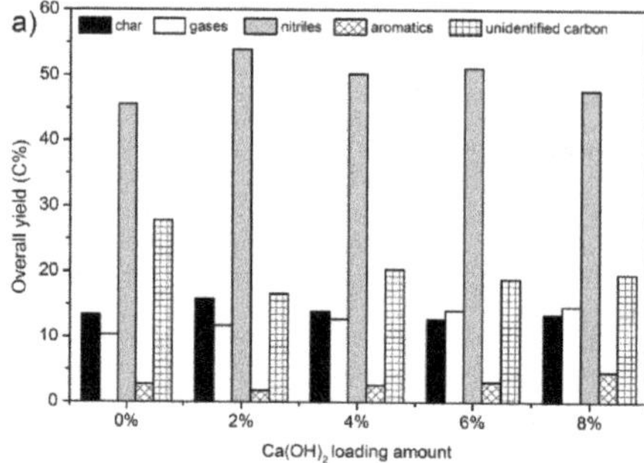

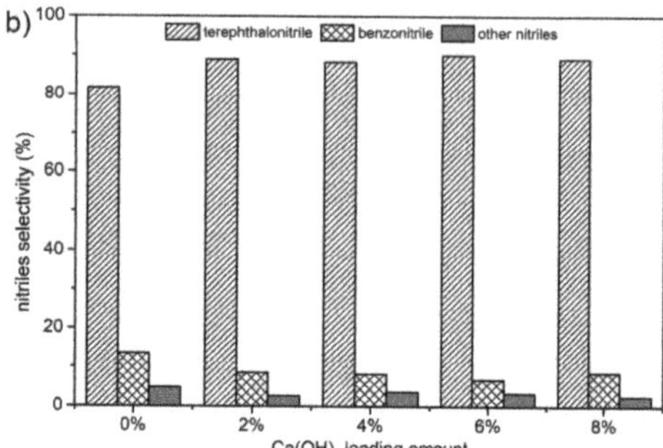

Figure 2. Effect of Ca(OH)$_2$ loading on Ca(OH)$_2$-Al$_2$O$_3$ catalysts on (**a**) overall yield and (**b**) nitriles selectivity for selectively producing terephthalonitrile via the catalytic fast pyrolysis of PET with ammonia at 500 °C. (1 g catalyst; N$_2$ and NH$_3$ flow rate, 80 mL/min; 0.5 g/batch PET feeding).

Table 1. Effect of residence time between pyrolytic vapors and catalyst.[1]

Entry	Catalyst Dosage (g)	Residence Time (s)	Nitriles (C%)	Aromatics[2] (C%)	Nitriles Selectivity (%)		
					Terephthalonitrile	Benzonitrile	Other Nitriles[3]
1	0.6	0.94	57.05	1.44	90.48	6.60	2.52
2	0.8	1.25	60.53	1.51	89.91	6.81	3.29
3	1	1.56	53.98	1.78	88.88	8.43	2.69
4	1.2	1.87	54.96	3.83	86.43	8.26	5.31
5	1.5	2.34	54.52	4.69	82.89	11.48	5.63

[1] Reaction conditions: Pyrolysis temperature 500 °C; catalyst 2% Ca(OH)$_2$/Al$_2$O$_3$; N$_2$ and NH$_3$ flow rate, 80 mL/min; PET feeding at 0.5 g/batch;. [2] Aromatics: Benzene, toluene, xylenes, etc.; [3] Other nitriles: Acetonitrile, alkyl aromatic nitriles, etc.

Table 2. Effect of ammonia content in the carrier gas on selectively producing terephthalonitrile.[1]

Entry	Ammonia Content (%)	Nitriles (C%)	Aromatics[2] (C%)	Nitriles Selectivity (%)		
				Terephthalonitrile	Benzonitrile	Other Nitriles[3]
1	25%	55.93	2.08	88.78	7.34	3.88
2	50%	60.53	1.51	89.90	6.81	3.29
3	75%	62.93	1.49	89.51	6.63	3.86
4	100%	63.18	1.73	92.28	4.11	3.61

[1] Reaction conditions: Pyrolysis temperature 500 °C; catalyst 2% Ca(OH)$_2$/Al$_2$O$_3$; catalyst usage 0.8 g; flow rate of carrier gas: 160 mL/min; PET feeding at 0.5 g/batch;. [2] Aromatics: Benzene, toluene, xylenes, etc.; [3] Other nitriles: Acetonitrile, alkyl aromatic nitriles, etc.

2.2.1. Effect of Catalyst

The effect of catalyst on selectively producing terephthalonitrile by changing Ca(OH)$_2$ loading on Al$_2$O$_3$ at the range of 0–8% at 500 °C was studied with catalyst dosage and flow rate of carrier gas fixed at 1 g and 160 mL/min, as well as ammonia content in carrier gas of 50%. The effect of Ca(OH)$_2$ loading on Al$_2$O$_3$ affected the nitriles and terephthalonitrile production obviously (Figure 2). In Figure 1a, the carbon yield of char, gases and aromatics changed slightly with Ca(OH)$_2$ increasing from 0–8.0%, and kept around 14 C%, 12 C% and 3%, respectively. When Al$_2$O$_3$ served as catalyst, less nitriles and more unidentified compounds were produced with yields of 45.61 C% and 27.80 C%, respectively. The unidentified compounds were much more than those catalyzed by Ca(OH)$_2$/Al$_2$O$_3$ catalysts. At 2% Ca(OH)$_2$ loading, nitriles reached the highest yield of 53.98 C% and decreased with Ca(OH)$_2$ loading increasing. At 8% Ca(OH)$_2$ loading, the carbon yield of nitriles decreased to 47.69 C%. In Figure 2b, the selectivity of other nitriles was very low (around 3%). The lowest selectivity of terephthalonitrile (81.71%) and the highest selectivity of benzonitrile (13.53%) in nitriles were obtained by using neat Al$_2$O$_3$. The highest selectivity of terephthalonitrile was 90.0% by using 6% Ca(OH)$_2$/Al$_2$O$_3$. However, combined with Figure 2a, the carbon yield of terephthalonitrile was only 45.05%, which was less than that (47.98%) with 6% Ca(OH)$_2$/Al$_2$O$_3$. Meanwhile, the terephthalonitrile selectivity in nitriles in the presence of 2% Ca(OH)$_2$/Al$_2$O$_3$ was 88.9%, which was slightly less than that (90.0%) with 6% $Ca(OH)_2/Al_2O_3$. Therefore, 2% $Ca(OH)_2$ was selected as the optimal loading for producing terephthalonitrile in the following tests.

2.2.2. Effect of Residence Time Between the Pyrolytic Vapors and Catalyst

Besides temperature and catalyst, residence time between the pyrolytic vapor and catalyst also influenced the product distributions. Herein, the effect of catalyst dosage was investigated through changing catalyst dosage (from 0.6 to 1.5 g) in the presence of 2% $Ca(OH)_2/Al_2O_3$ at 500 °C and flowing carrier gas (50% ammonia and 50% nitrogen) at 160 mL/min. Table 2 shows that residence time and catalyst also affected the terephthalonitrile production. When catalyst dosage was 0.8 g and residence time was 1.25 s, the carbon yields of nitriles and terephthalonitrile were the highest of 60.53 C% and 54.42%, respectively. If residence time was extended further, the carbon yield of nitriles and terephthalonitrile decreased. Longer contacting time between pyrolytic vapor and catalyst would promote adequate decarboxylation and alkylation reactions during CFP-A process. At 2.34 s, the carbon yield of nitriles and terephthalonitrile was only 54.52 C% and 45.19 C%, respectively. In addition, the selectivity of terephthalonitrile in nitriles also decreased with the increase of residence time from above 90% at 0.94 s to 82.9% at 2.34 s, respectively. While, the selectivity of benzonitrile and other nitriles increased from 6.60% and 2.52% at 0.94 s to 11.48% and 5.63% at 2.34 s, respectively. By comprehensively considering the carbon yield and selectivity of terephthalonitrile, 1.25 s was the optimal residence time for selective producing terephthalonitrile, and catalyst dosage was selected as 0.8 g in the following tests.

2.2.3. Effect of Ammonia Content in the Carrier Gas

Besides decomposition reactions, ammonolysis reaction also occurred. Ammonia acted both as carrier gas and reactant for producing nitriles, and is as important as PET and catalyst in this process. The effect of NH_3 content (from 25% to 100%) in the carrier gas was investigated by fixing pyrolysis temperature (500 °C), catalyst dosage (2% $Ca(OH)_2/Al_2O_3$, 0.8 g), and carrier gas flow rate (160 mL/min). In Table 3, the higher ammonia content yielded more nitriles and terephthalonitrile. As ammonia content increased from 25% to 100%, the carbon yield of nitriles and terephthalonitrile increased from 55.93 C% and 49.66 C% to 63.18 C% and 58.30 C%, respectively. Meanwhile, the selectivity of terephthalonitrile in nitriles also increased from 88.78% (25% NH_3) to 92.28% (100% NH_3). Therefore, as compared with the mixture of N_2 and NH_3, pure ammonia was more suitable and used as carrier gas in the terephthalonitrile production process. The best conditions for the production of terephthalonitrile are selected as 0.8 g of 2% $Ca(OH)_2/\gamma\text{-}Al_2O_3$, 500 °C under pure ammonia with 58.30 C% terephthalonitrile yield and 92.28% selectivity in nitriles.

Table 3. Effect of residence time between pyrolytic vapors and catalyst.[1]

Entry	Catalyst Dosage (g)	Residence Time (s)	Nitriles (C%)	Aromatics[2] (C%)	Nitriles Selectivity (%)		
					Terephthalonitrile	Benzonitrile	Other Nitriles[3]
1	0.6	0.94	32.93	6.33	28.36	64.74	6.89
2	0.8	1.25	31.14	7.85	15.35	77.91	6.74
3	1	1.56	33.1	10.11	12.36	81.12	6.53
4	1.2	1.87	32.96	13.94	6.43	84.98	8.59
5	1.5	2.34	32.05	15.87	4.21	84.21	11.58

[1] Reaction conditions: Pyrolysis temperature 650 °C; catalyst 4% $Ca(OH)_2/Al_2O_3$; N_2 and NH_3 flow rate, 80 mL/min; PET feeding at 0.5 g/batch;. [2] Armatics: Benzene, toluene, xylenes, etc.; [3] Other nitriles: Acetonitrile, alkyl aromatic nitriles, etc.

2.3. *Optimizing Benzonitrile Production at 650 °C*

Similarly, the effects of catalyst, residence time and ammonia content on the production of benzonitrile were studied at 650 °C. Figure 3 shows the effect of $Ca(OH)_2$ loading of $Ca(OH)_2/Al_2O_3$ catalysts, Tables 4 and 5 give the effect of residence time and effect of ammonia content in the carrier gas on the production of benzonitrile.

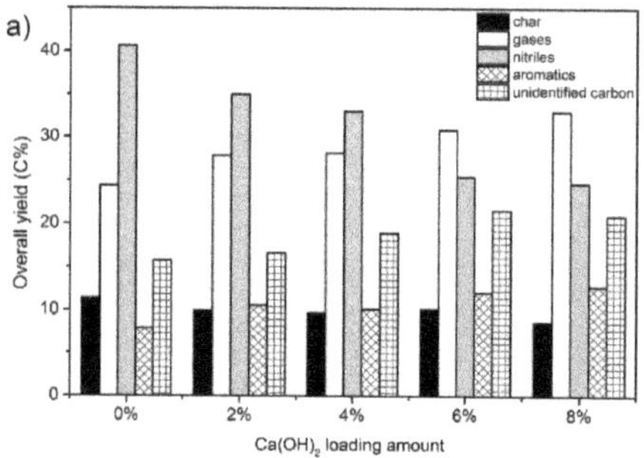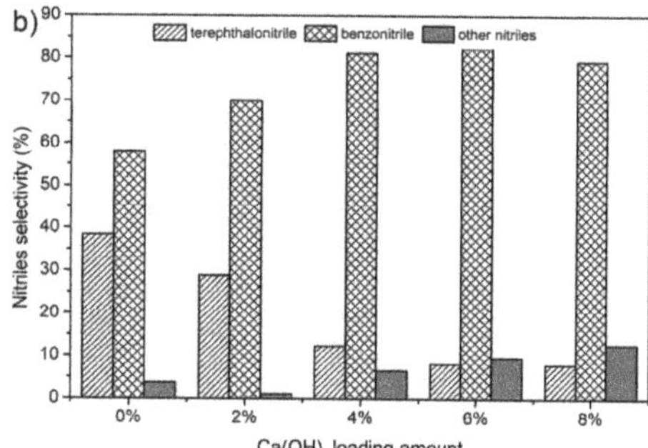

Figure 3. Effect of Ca(OH)$_2$ loading on Ca(OH)$_2$-Al$_2$O$_3$ catalysts on (**a**) overall yield and (**b**) nitriles selectivity for selectively producing terephthalonitrile via the catalytic fast pyrolysis of PET with ammonia at 650 °C. (1 g catalyst; N$_2$ and NH$_3$ flow rate, 80 mL/min; PET feeding, 0.5 g/batch)

Table 4. Effect of ammonia content in the carrier gas on selectively producing benzonitrile.[1]

Entry	Ammonia Content (%)	Nitriles (C%)	Aromatics[2] (C%)	Nitriles Selectivity (%)		
				Terephthalonitrile	Benzonitrile	Other Nitriles[3]
1	25%	25.53	15.81	7.46	84.04	8.5
2	50%	32.96	13.94	6.43	84.98	8.59
3	75%	34.14	13.5	6.85	84.33	8.82
4	100%	36.83	12.56	8.12	82.60	9.28

[1] Reaction conditions: Pyrolysis temperature 650 °C; catalyst: 4% Ca(OH)$_2$/Al$_2$O$_3$; catalyst usage 1.2 g; flow rate of carrier gas: 160 mL/min; PET feeding at 0.5 g/batch;. [2] Aromatics: Benzene, toluene, xylenes, etc.; [3] Other nitriles: Acetonitrile, alkyl aromatic nitriles, etc.

Table 5. Summary of catalytic fast pyrolysis with ammonia (CFP-A) of model chemicals over 4% Ca(OH)$_2$/Al$_2$O$_3$ at 650 °C.[1]

Entry	Feedstock	Nitriles (C%)	Aromatics[2] (C%)	Nitriles Selectivity (%)		
				Terephthalonitrile	Benzonitrile	Other Nitriles[3]
1	Benzoic acid	68.75	3.02	N.D	100	N.D
2	Methyl benzoate	64.34	3.46	N.D	98.78	1.22
3	Benzamide	86.56	5.68	N.D	100	N.D
4	Terephthalic acid	56.05	2.89	53.02	45.43	1.55
5	Dimethyl terephthalate	59.52	3.27	49.59	47.54	2.86
6	Benzonitrile	95.74	0.45	N.D	100	N.D

[1] Reaction conditions: Pyrolysis temperature 650 °C; catalyst: 4% Ca(OH)$_2$/Al$_2$O$_3$; catalyst usage 1.2 g; flow rate of carrier gas: 160 mL/min; PET feeding at 0.5 g/batch;. [2] Aromatics: Benzene, toluene, xylenes, etc.; [3] Other nitriles: Acetonitrile, alkyl aromatic nitriles, etc.

2.3.1. Effect of Catalyst

The effect of Ca(OH)$_2$ loading on Al$_2$O$_3$ on selectively producing benzonitrile was investigated. The Ca(OH)$_2$ loading on Al$_2$O$_3$ was in the range of 0–8%. Compared to the production of terephthalonitrile, Ca(OH)$_2$ loading on Al$_2$O$_3$ had a more obvious influence (especially on the catalytic cracking and decarboxylation) on the producing benzonitrile. In Figure 3a, the more Ca(OH)$_2$ loading on Al$_2$O$_3$, the more gases, aromatics and unidentified carbon were produced with less nitriles. When 8% Ca(OH)$_2$/Al$_2$O$_3$ served as catalyst, the carbon yield of nitriles was only 24.67 C%, much less than that (40.65 C%) with neat Al$_2$O$_3$. Figure 3b shows that Ca(OH)$_2$ loading on Al$_2$O$_3$ could promote the formation of benzonitrile but inhibit terephthalonitrile production. In the presence of 6% Ca(OH)$_2$/Al$_2$O$_3$, the highest selectivity of benzonitrile in nitriles (82.24%) was obtained. In addition, the selectivity of benzonitrile with 4% Ca(OH)$_2$/Al$_2$O$_3$ was 81.12 %, which was similar to that of 6% Ca(OH)$_2$/Al$_2$O$_3$. Therefore, 4% Ca(OH)$_2$/Al$_2$O$_3$ was selected for producing benzonitrile in the following tests.

2.3.2. Effect of Residence Time

The effect of residence time on benzonitrile production was studied by changing catalyst (4% Ca(OH)$_2$/Al$_2$O$_3$) dosage (from 0.6 to 1.5 g) at 650 °C and carrier gas (50% NH$_3$ and 50% N$_2$) flow rate of 160 mL/min. In Table 2, residence time affected the carbon yield of nitriles slightly, but affected the carbon yield of aromatics and the nitriles selectivity significantly. The effect of residence time on the production of benzonitrile at 650 °C was similar to that of the production of terephthalonitrile at 500 °C. Longer residence time could promote PET to form more benzonitrile and aromatics and less terephthalonitrile. The carbon yield and selectivity of benzonitrile in nitriles were the highest of 28.01 C% and 84.98 %, respectively at residence time of 1.87 s (catalyst dosage of 1.2 g). Therefore, 1.87 s was selected as optimal residence time for producing benzonitrile with catalyst dosage of 1.2 g in the following tests.

2.3.3. Effect of Ammonia Content

The effect of NH$_3$ content (25% to 100%) on the production of benzonitrile was investigated by fixing temperature (650 °C), catalyst dosage (4% Ca(OH)$_2$/Al$_2$O$_3$, 1.2 g) and carrier gas flow rate (160 mL/min). The effect of ammonia content on benzonitrile production (Table 4) was the same as that on terephthalonitrile production (Table 3). The higher ammonia content could cause more nitriles and benzonitrile production. When pure ammonia served as carrier gas, the carbon yields of nitriles and benzonitrile were 36.83 C% and 30.42 C%, which were much higher than those (25.53 C% and 21.45%) with 25% ammonia. Meanwhile, the selectivity of benzonitrile in nitriles was up to 82.60%. Therefore, pure ammonia was more suitable and used as carrier gas in the benzonitrile production process. The optimal conditions for the production of benzonitrile were selected as 1.2 g of 4% Ca(OH)$_2$/γ-Al$_2$O$_3$, 650 °C under pure ammonia with 30.42 C% benzonitrile yield and 82.60% selectivity in nitriles.

2.4. Possible Reaction Pathways from PET to Terephthalonitrile and Benzonitrile

The reaction pathways from PET to terephthalonitrile were investigated in the previous study on selectively producing terephthalonitrile by CFP-A of PET over H$_3$PO$_4$ modified Al$_2$O$_3$ catalysts at 500 °C [39]. Terephthalic acid and related esters were the key intermediates for the production of terephthalonitrile from PET. Herein, the possible reaction pathways from PET to benzonitrile at 650 °C were also investigated by a series of experiments. Firstly, the pyrolysis experiment of PET without catalyst under ammonia was carried out. The detailed product distributions were detected by GC-MS (Agilent 7890-5977B, Santa Clara, CA; Figure S1 and Table S1 in Supplementary Materials). The pyrolytic products were benzoic acid and its derived esters, benzamide, terephthalic acid and its derived esters. Therefore, benzoic acid, methyl benzoate, benzamide, terephthalic acid, dimethyl terephthalate and benzonitrile were employed as the feedstocks for producing nitriles (Table 5). In entries 1–3, benzoic acid, benzamide and methyl benzoate could be easily converted to benzonitrile with high selectivity (> 98%) through the acid-catalyzed ammoniation and dehydration reactions over Al$_2$O$_3$-based catalysts. It indicated that benzoic acid, benzamide, methyl benzoate were the key intermediates from PET to benzonitrile. In addition, as terephthalic acid and dimethyl terephthalate served as feedstocks (entries 4 and 5), the main product was terephthalonitrile and benzonitrile. The selectivity of benzonitrile in nitriles was more than 45%. The results showed that benzonitrile could be produced from terephthalic acid and dimethyl terephthalate at higher pyrolysis temperature via selective decarboxylation reaction over Ca(OH)$_2$/γ-Al$_2$O$_3$ catalyst.

Besides the nitriles, a certain amount of aromatic hydrocarbons could also be detected. Under the optimal conditions for producing benzonitrile, the carbon yield of aromatics was around 13 C%. Under the same conditions, the carbon yields of model compounds were less than 6%, and more aromatics were produced from benzamide than those of aromatic acids and their derived esters. Meanwhile, as benzonitrile served as feedstocks, less than 1% of aromatics were detected in the products.

It indicated that aromatics could not be produced by the further cracking of benzonitrile, and benzonitrile was stable during the CFP-A process. The carbon-carbon bond dissociation energy of model compounds (Table S2) was calculated based on the DFT method with much weaker value for benzamide than those of nitriles, and aromatic acids and their derived esters. Aromatics (e.g., toluene, styrene, biphenyl) were detected during the thermal decomposition of PET without catalyst. Therefore, aromatics could be produced from PET by the direct thermal decomposition process, they could also be produced from benzamide, aromatic acids and derived esters via catalytic decarboxylation and cracking reactions over Ca(OH)$_2$/γ-Al$_2$O$_3$ catalyst during the CFP-A process.

Based on all the above findings, the reaction pathways from PET to terephthalonitrile and benzonitrile were proposed and summarized in Figure 4. At lower temperature (500 °C), terephthalic acid and its derived esters were the main pyrolytic products, and the key intermediates from PET to terephthalonitrile. At a higher temperature (650 °C), benzoic acid, methyl benzoate, benzamide were the main intermediates from PET to benzonitrile. Aromatics were produced as by-products from PET, benzamide, aromatic acids and derived esters via thermal decomposition, decarbonylation, cracking reaction during the CFP-A process.

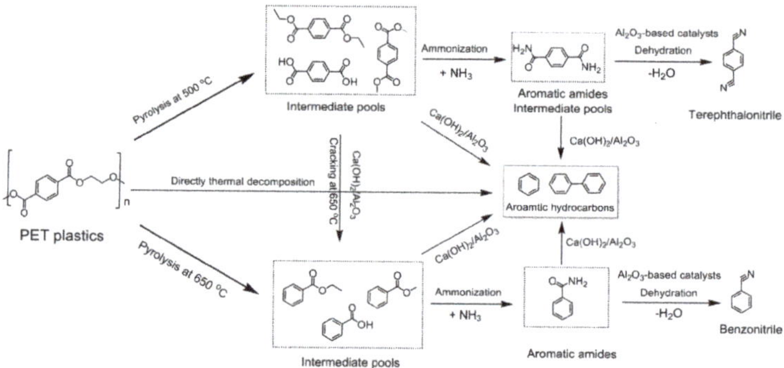

Figure 4. Possible reaction pathways from PET to terephthalonitrile and benzonitrile.

2.5. Catalysts Stability

The stability of catalysts was studied by conducting 4 reaction/regeneration cycles at the optimal conditions for terephthalonitrile and benzonitrile. BET (Barrett–Emmet–Taller) surface area, pore volume, acidity and basicity were also investigated to illustrate the deactivation of catalysts. For each cycle, the spent catalyst was calcined with air (100 mL/min) at 650 °C for 3 h to remove the coke formed on the surface of the catalyst. In Figure 5a, as compared to the fresh catalyst, the carbon yield of nitriles and terephthalonitrile did not decrease after 2 cycles, indicating that the catalyst (2% Ca(OH)$_2$/Al$_2$O$_3$) kept stable for the first and second cycles. After 4 cycles, the carbon yield of nitriles and terephthalonitrile decreased to 60.62 C% and 54.76 C%, slightly decreased by 6% as compared to that at the first cycle (63.17 C% and 58.30 C%). In Figure 5b, 4% Ca(OH)$_2$/Al$_2$O$_3$ deactivated little at the first and second cycles for producing benzonitrile. However, after 4 cycles, the carbon yield of nitriles and benzonitrile decreased to 34.49% and 25.18%, respectively. Therefore, the above results suggested that 2% and 4% Ca(OH)$_2$/Al$_2$O$_3$ catalysts were stable for the production of terephthalonitrile and benzonitrile.

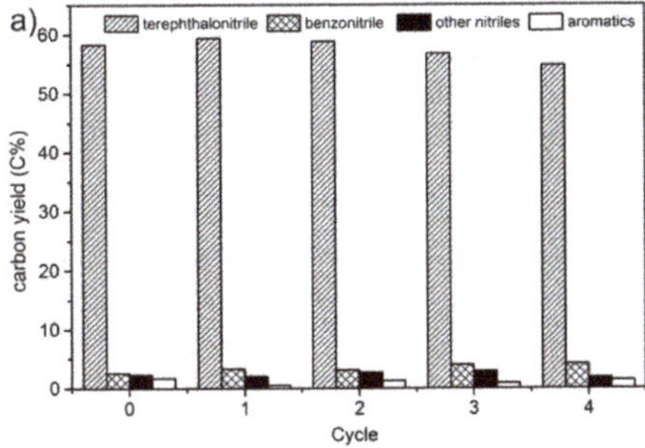
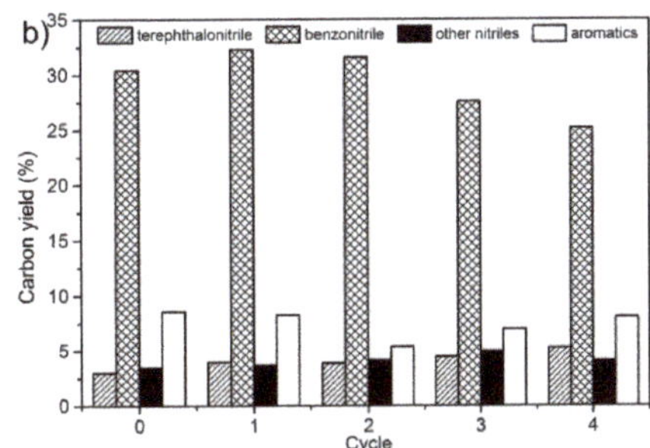

Figure 5. Yield of products vs. catalyst cycle for (**a**) terephthalonitrile at 500 °C, and (**b**) benzonitrile at 650 °C. ((**a**) 1g 2% Ca(OH)$_2$/Al$_2$O$_3$; N$_2$ and NH$_3$ flow rate, 120 mL/min and 40 ml; PET feeding, 0.5 g/batch; (**b**) 1.2g 4% Ca(OH)$_2$/Al$_2$O$_3$; ammonia, 160 mL/min; PET feeding, 0.5 g/batch).

2.6. Catalysts Characterzation

The fresh and used catalysts after 4 cycles were characterized by XRD, N$_2$ adsorption/desorption, NH$_3$-TPD and CO$_2$-TPD. In Figure 6 for XRD patterns, the peaks at 18.5°, 20.4°, 36.8°, 27.8° and 66.7° (2θ) of fresh 2% and 4% Ca(OH)$_2$/Al$_2$O$_3$ were the characteristic ones for Al$_2$O$_3$·3H$_2$O, the peaks at 37°, 46° and 66.7° of recycled catalysts were the characteristic ones for Al$_2$O$_3$. The crystallinity of catalysts changed a lot during the reactions. In addition, no characteristic peaks of Ca(OH)$_2$ and its derives were detected in Figure 6, indicating that the crystallinity of Ca(OH)$_2$ was not very high and Ca(OH)$_2$ was dispersed well on the surface of γ-Al$_2$O$_3$. BET surface area, pore volume, total acid amounts and total basic amounts were given in Table 6. The BET surface area and pore volume of fresh 2% and 4% Ca(OH)$_2$/Al$_2$O$_3$ were 148.14 m^2/g & 0.70 cm^3/g, and 147.13 m^2/g & 0.69 cm^3/g, but decreased to 116.76 m^2/g & 0.56 cm^3/g, and 122.63 m^2/g & 0.60 cm^3/g after 4 cycles, respectively. It also indicated the changes of catalyst micro structure during cycles. The total acid amounts and basic amounts of fresh and used catalysts measured by NH$_3$-TPD and CO$_2$-TPD were summarized in Table 6 with detailed spectra given in Figures S2 and S3. As compared with 2% Ca(OH)$_2$/Al$_2$O$_3$, 4% Ca(OH)$_2$/Al$_2$O$_3$ has higher total acid amounts (377.05 μmol NH$_3$/g) and basic amounts (241.24 μmol CO$_2$/g). After 4 cycles for producing terephthalonitrile at 500 °C with 2% Ca(OH)$_2$/Al$_2$O$_3$, the acid amounts of used catalyst decreased slightly to 295.74 μmol NH$_3$/g, while the basic amounts decreased significantly to 77.40 from 197.54 μmol CO$_2$/g. For 4% Ca(OH)$_2$/Al$_2$O$_3$, the total acid amounts and basic amounts decreased significantly after 4 cycles at 650 °C.

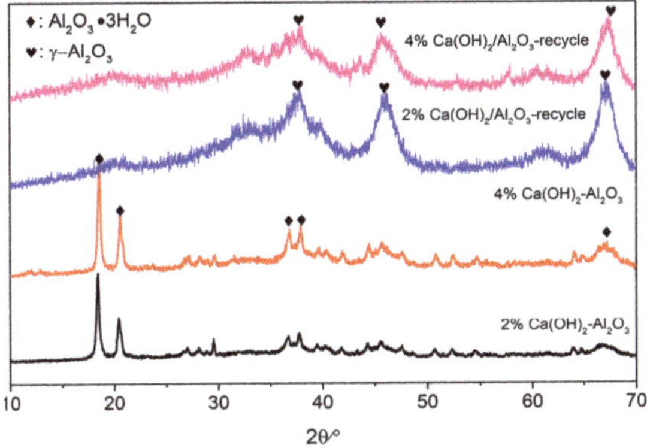

Figure 6. XRD patterns of fresh and used catalysts.

Table 6. Typical properties of catalysts.

Entry	Catalysts	BET Surface Area (m²/g)	Pore Volume (cm³/g)	Total Acid Amounts (μmol NH₃/g)	Total Basic Amounts (μmol CO₂/g)
1	2% Ca(OH)$_2$/γ-Al$_2$O$_3$	148.14	0.70	303.91	197.54
2	4% Ca(OH)$_2$/γ-Al$_2$O$_3$	147.13	0.69	377.05	241.24
3	Used 2% Ca(OH)$_2$/Al$_2$O$_3$ [1]	116.76	0.56	295.74	77.40
4	Used 4% Ca(OH)$_2$/Al$_2$O$_3$ [2]	122.63	0.60	160.58	99.16

[1] Used 2% Ca(OH)$_2$/Al$_2$O$_3$: After 4 cycles for producing terephthalonitrile at 500 °C and 160 mL of pure ammonia, and remove coke with air (100 mL/min) at 650 °C for 3 h. [2] Used 4% Ca(OH)$_2$/Al$_2$O$_3$: After 4 cycles for producing benzonitrile at 650 °C and 160 mL of pure ammonia, remove coke with air (100 mL/min) at 650 °C for 3 h.

3. Materials and Methods

3.1. Materials

Methanol (≥99.5%), benzene (≥99.5%), toluene (≥99.5%), xylene (≥99.5%), biphenyl (≥99.5%) and Ca(OH)$_2$ (≥99.5%) were purchased from Sinopharm Chemical Reagent Co. Ltd., Beijing, China. Bi-cyclohexane (≥99.5%), naphthalene (≥99.5%), terephthalonitrile (≥99.5%), benzonitrile (≥99.5%) and terephthalamide (>98%) as calibrants, benzoic acid (99%), p-phthalic acid (99%), benzamide (≥99.5%), methyl benzoate (≥99.5%), dimethyl terephthalate (≥99%) and γ-Al$_2$O$_3$ (20 nm) were purchased from Aladin Chemical Reagent Co. Ltd (Shanghai, China).

PET powder (AR, ~100 meshes) were purchased from Shanghai Youngling-Tech Co. Ltd., Shanghai, China. All these chemicals and materials were used without further purification. The elemental analysis of PET powders was shown in Table S3.

Air, NH$_3$ (≥99.995%, AR), N$_2$ (99.999%), Ar (99.999%), He (99.999%) were purchased from Nanjing Special Gases Factory, Jiangsu, China. The standard gas (C6+ 0.0920%, CH$_4$ 53.922%, C$_2$H$_2$ 0.561%, C$_3$H$_4$ 0.523%, trans -butane 0.501%, cis-butane 0.505%, 1,3-butadiene 0.523%, N$_2$ 10.40%, C$_2$H$_4$ 1.01%, C$_3$H$_8$ 1.01%, iso-butane 2.01%, n-butene 0.532%, isopentane 0.505%, CO$_2$ 2.02%, CO 0.986%, C$_2$H$_6$ 1.05%, C$_3$H$_6$ 1.01%, n-butane 1.98%, isobutene 0.505%, n-pentane 0.515%, O$_2$ 5.10%, H$_2$ 14.74%) was purchased from Dalian Special Gases Co., Ltd (Dalian, China) for gas calibration.

3.2. Catalysts Preparation and Regeneration

For Ca(OH)$_2$/γ-Al$_2$O$_3$: The modified γ-Al$_2$O$_3$ catalysts were prepared by wetness impregnation with aqueous Ca(OH)$_2$ solution [(mass ratio from 2% to 8% based on Ca(OH)$_2$ (the mass ratio of γ-Al$_2$O$_3$ to water was 1:10)]. After the impregnation, the catalysts were dried in an oven at 110 °C for 12 h, and calcined at 550 °C for 4 h in air. All the above catalysts were crushed and screened for about 40 meshes. For catalyst regeneration, after each cycle, catalyst was calcined with air (100 mL/min) at 550 °C for 3 h to remove coke.

3.3 Catalyst Characterization

Catalysts were analyzed on a theta rotating anode X-ray diffractometer (TTP-III, Rigaku, Tokyo, Japan) using CuKα radiation at 40 kV and 40 mA, with 2θ ranges of 10°–70° at scan rate of 10 °/min. The nitrogen adsorption/desorption isotherms of the catalysts were measured by Autosorb-iQ (Quantachrome, Boynton Beach, FL, USA). The surface area and total volume were determined through the BET method. The NH$_3$-TPD and CO$_2$-TPD tests of the catalyst were conducted with Chembet PULSAR temperature-programmed reduction/desorption (TPR/TPD) (Quantachrome, Boynton Beach, FL, USA). The detailed method of NH$_3$-TPD was referred to previous work [16]. The CO$_2$-TPD for acidity test of the catalyst was also conducted with the Chembet PULSAR TPR/TPD. Four different volumes (0.5, 1, 1.5, 2 mL) of a standard CO$_2$ gas were used to calibrate total basic density with $R^2 > 0.999$. About 200 mg of sample were put in a reactor and pre-treated in situ for 4 h at 550 °C in a flow of argon. After cooling to 100 °C, CO$_2$ adsorption was performed by feeding pulses of CO$_2$ to the reactor. After the catalyst surface became saturated, the sample was kept at 100 °C for 2 h to remove the base excess. CO_2 was thermally desorbed by rising the temperature with a linear heating rate of approximately 10 °C/min from 100 to 500 °C.

3.3. Pyrolysis Experiments

Catalytic pyrolysis of PET was carried out in a fixed bed reactor (Anhui Kemi Machinery Technology Co. LTD, Hefei, China) at 450 to 700 °C (height: 400 mm, internal diameter: 10 mm; Figure S4). All the experiments were isothermal and batch. For each run, the catalyst was fixed in the reactor as catalyst bed, and solid PET was fed manually into the reactor under a certain rate and purged with carrier gas. The volatile products were trapped in a cold trap, and diluted with hot methanol for GC and GC/MS analysis. The non-condensable gas products were collected with a gas bag, and washed with phosphoric acid solution to remove excess ammonia. The volume of gas products was measured via drainage method, and its compositions were determined by GC. Due to p-phthalic acid, benzoic acid, benzamide, methyl benzoate, dimethyl terephthalate were the solid feedstocks, they were fed the same as PET for catalytic pyrolysis. Benzonitrile was liquid at room temperature and injected to the reactor for pyrolysis experiments. The detailed pyrolysis system was shown in Figure S5.

3.4. Products Analyses

The liquid samples were analyzed by a GC-MS (Agilent 7890B-5977B, Agilent Technologies Inc. Santa Clara, CA, USA) equipped with an HP-5 MS capillary column (30 m × 0.25 mm × 0.25 mm). Split injection was performed at a split ratio of 50 using helium (99.999%) as carrier gas. The oven temperature was held at 40 °C for 3 min, heated to 280 °C at 10 °C /min, and held at 280 °C for 5 min. The GC-MS mode was shown in Table S4.

The carbon yield and selectivity of coke, bio-oil and gases products were quantitatively determined by elemental analysis (Elementar, Langenselbold, Germany), GC. The liquid products (such as aromatic nitriles, aromatic hydrocarbons) were quantitatively determined by GC (GC-2010 plus, Shimadzu, Kyoto) employing a 30m × 0.25mm × 0.25μm fused-silica capillary column (DB-Wax, Shimadzu). The products in the cold trap were mixed with bi-cyclohexane as the internal standard with calibration factor of 1.084 for benzonitrile, 1.261 for terephthalonitrile, 0.941 for benzene, 1.023 for toluene, 1.112 for naphthalene and diluted by hot methanol (25 mL) for GC analysis. The GC operating conditions were as follows: Carrier gas—nitrogen; injection port—250 °C in a split mode; detector (FID)—250 °C; column temperature—40 °C; oven temperature program—heating up to 250 °C at a rate of 10 °C/min, and holding at a final temperature for 10.0 min.

For gas product analysis, the entire gas of each run was collected with an air bag, weighed, and analyzed using GC (GC-2014C, Shimadzu, Kyoto, Japan) with two detectors and four columns (PC1: P-N 80/100 mesh, 3.2 × 2.1 mm × 1 m; MC-1: P-N 80/100 mesh, 3.2 × 2.1 mm × 1.0 m; MC-2: MS-13X, 80/100 mesh, 3.2 × 2.1 mm × 2 m; MC-3: HP-Al$_2$O$_3$, 30 m × 0.53 mm × 15 μm), a thermal conductivity detector (TCD), PC-1, MC-1 and MC2 columns for analysis of H_2, CO, CH_4 and CO_2, and a flame ionization detector (FID) and MC-3 column for gas hydrocarbons. The moles of gas products were externally calibrated with three different concentrations diluted with N_2 from the standard gas mixture. The GC operating conditions were as follows: Carrier gas—nitrogen and argon; detector (FID and TCD)—150 °C; column temperature—50 °C; oven temperature program—holding at 50 °C for 3 min, heating up to 130 °C at a rate of 10 °C/min, and holding for 3 min. In addition, due to the excess ammonia existing in this process, which is easy to react with CO_2 (formed in the pyrolysis process), thus CO_2 couldn't be detected by GC in this study.

The carbon yield of coke, gases, nitriles, aromatics and nitriles selectivity were calculated from Equations (1) to (5) as described in previous work [29]. The unidentified carbon yield was calculated from Equation (6) by mass balance closure. The residence time was calculated from Equation (7)

$$\text{Coke yield (C\%)} = \text{Carbon moles in coke/Carbon moles in PET feeding} \times 100\% \qquad (1)$$

$$\text{Gases yield (C\%) = Carbon moles in gases/Carbon moles in PET feeding} \times 100\% \qquad (2)$$

$$\text{Nitriles yield (C\%) = Carbon moles in nitriles/Carbon moles in PET feeding} \times 100\% \qquad (3)$$

$$\text{Aromatics yield(C\%) = Carbon moles in aromatics/Carbon moles in PET feeding} \times 100\% \qquad (4)$$

$$\text{Nitriles selectivity (\%) = Carbon moles in specific nitrile/Carbon moles in all nitriles} \times 100\% \qquad (5)$$

$$\text{Unidentifed carbon yield (C \%) = 100\%-identified carbon yield} \qquad (6)$$

$$\text{Residence time (s) = Catalyst volume/Carrier gas flow rate} \qquad (7)$$

4. Conclusions

$Ca(OH)_2/Al_2O_3$ catalysts were used to produce terephthalonitrile and benzonitrile from polyethylene terephthalate (PET) via catalytic fast pyrolysis with the ammonia process. The best conditions for the production of terephthalonitrile were selected as 0.8 g of 2% $Ca(OH)_2/\gamma$-Al_2O_3, 500 °C under pure ammonia with 58.30 C% terephthalonitrile yield and 92.28% selectivity in nitriles. In addition, 4% $Ca(OH)_2/$ Al_2O_3 was the suitable catalyst for producing benzonitrile. Under conditions with catalyst dosage of 1.2 g, residence time of 1.87 s at 650 °C and pure ammonia flowing of 160 mL/min, the yield and selectivity of benzonitrile was 30.42 C% and 82.60%, respectively. After 4 cycles, the catalysts deactivated slightly and kept stable. The fresh and used catalysts were further characterized with XRD, N_2 adsorption/desorption, NH_3-TPD and CO_2-TPD.

Supplementary Materials: The following are available online at http://www.mdpi.com/2073-4344/9/5/436/s1, Figure S1. The GC-MS spectra of pyrolysis PET with ammonia at 650 °C; Figure S2. The NH_3-TPD spectra of fresh and used catalysts; Figure S3. The CO_2-TPD spectra of fresh and used catalysts; Figure S4. The schematic diagram of the pyrolysis system; Figure S5. The schematic diagram of the liquid feeding pyrolysis system; Table S1. The detailed chemical compositions of pyrolysis PET with ammonia at 650 °C; Table S2. C-C Bond dissociation energy of some model compounds @25 °C and @650 °C; Table S3. Elemental analyses of PET plastic; Table S4. Integration parameters and their values set in mass spectrometry detector (MSD) ChemStation.

Author Contributions: Data curation, L.X.; Investigation, X.-w.N., L.-y.Z., Q.D. and G.-h.D.; Supervision, Z.F.; Conceptualization, L.X., Funding acquisition, L.X. and Z.F.; Project administration, Z.F.; Writing-original draft, L.X.; Writing-review and editing, Y.-t.W. and Z.F.

Funding: The work was financially supported by the Natural Science Foundation of Jiangsu Province (No. BK20180548), Nanjing Agricultural University (77J-0603, 68Q-0603), National Natural and Science Foundation of China (21878161), Open Funding of Key Laboratory of Energy Thermal Conversion and Control of the Ministry of Education of China at Southeast University, "Innovation & Entrepreneurship Talents" Introduction Plan of Jiangsu Province.

Conflicts of Interest: There are no conflicts to declare.

References

1. Halden, R.U. Plastics and Health Risks. *Ann. Rev. Public Health* **2010**, *31*, 179–194. [CrossRef] [PubMed]
2. Shin, J.; Lee, Y.; Park, S. Optimization of the pre-polymerization step of polyethylene terephthalate (PET) production in a semi-batch reactor. *Chem. Eng. J.* **1999**, *75*, 47–55. [CrossRef]
3. Al-Sabagh, A.M.; Yehia, F.Z.; Harding, D.R.K.; Eshaq, G.; ElMetwally, A.E. Fe_3O_4-boosted MWCNT as an efficient sustainable catalyst for PET glycolysis. *Green Chem.* **2016**, *18*, 3997–4003. [CrossRef]
4. Artetxe, M.; Lopez, G.; Amutio, M.; Elordi, G.; Olazar, M.; Bilbao, J. Operating conditions for the pyrolysis of poly-(ethylene terephthalate) in a conical spouted-bed reactor. *Ind. Eng. Chem. Res.* **2010**, *49*, 2064–2069. [CrossRef]
5. Blackmon, K.P.; Fox, D.W.; Shafer, S.J. Process for Converting Pet Scrap to Diamine Monomers. U.S. Patent 4,973,746,11, 1990.
6. Datta, J.; Kopczyńska, P. From polymer waste to potential main industrial products: Actual state of recycling and recovering. *Crit. Rev. Environ. Sci. Technol.* **2016**, *46*, 905–946. [CrossRef]
7. Li, X.; Li, J.; Zhou, G.; Feng, Y.; Wang, Y.; Yu, G. Enhancing the production of renewable petrochemicals by co-feeding of biomass with plastics in catalytic fast pyrolysis with ZSM-5 zeolites. *Appl. Catal. A Gen.* **2014**, *481*, 173–182. [CrossRef]

8.   Carlson, T.R.; Cheng, Y.T.; Jae, J.; Huber, G.W. Production of green aromatics and olefins by catalytic fast pyrolysis of wood sawdust. *Energy Environ. Sci.* **2010**, *4*, 145–161. [CrossRef]

9.   Wong, S.L.; Ngadi, N.; Abdullah, T.A.T.; Inuwa, I.M. Current state and future prospects of plastic waste as source of fuel: A review. *Renew. Sustain. Energy Rev.* **2015**, *50*, 1167–1180. [CrossRef]

10.   Kunwar, B.; Cheng, H.N.; Chandrashekaran, S.R.; Sharma, B.K. Plastics to fuel: a review. *Renew. Sustain. Energy Rev.* **2016**, *54*, 421–428. [CrossRef]

11.   Sharuddin, S.D.A.; Abnisa, F.; Daud, W.M.A.W.; Aroua, M.K. A review on pyrolysis of plastic wastes. *Energy Conver. Manag.* **2016**, *115*, 308–326. [CrossRef]

12.   Lopez, G.; Artetxe, M.; Amutio, M.; Bilbao, J.; Olazar, M. Thermochemical routes for the valorization of waste polyolefinic plastics to produce fuels and chemicals. A review. *Renew. Sustain. Energy Rev.* **2017**, *73*, 346–368. [CrossRef]

13.   Yoshioka, T.; Handa, T.; Grause, G.; Lei, Z.; Inomata, H.; Mizoguchi, T. Effects of metal oxides on the pyrolysis of poly (ethylene terephthalate). *J. Anal. Appl. Pyrolysis* **2005**, *73*, 139–144. [CrossRef]

14.   Kumagai, S.; Yamasaki, R.; Kameda, T.; Saito, Y.; Watanabe, A.; Watanabe, C.; Teramae, N.; Yoshiok, T. Tandem μ-reactor-GC/MS for online monitoring of aromatic hydrocarbon production via CaO-catalysed PET pyrolysis. *React. Chem. Eng.* **2018**, *2*, 776–784. [CrossRef]

15.   Kumagai, S.; Yamasaki, R.; Kameda, T.; Saito, Y.; Watanabe, A.; Watanabe, C.; Teramae, N.; Yoshiok, T. Aromatic hydrocarbon selectivity as a function of CaO basicity and aging during CaO-catalyzed PET pyrolysis using tandem μ-reactor-GC/MS. *Chem. Eng. J.* **2018**, *332*, 169–173. [CrossRef]

16.   Du, S.; Valla, J.A.; Parnas, R.S.; Bollas, G.M. Conversion of polyethylene terephthalate based waste carpet to benzene-rich oils through thermal, catalytic, and catalytic steam pyrolysis. *ACS Sustain. Chem. Eng.* **2016**, *4*, 2852–2860. [CrossRef]

17.   Xue, Y.; Johnston, P.; Bai, X. Effect of catalyst contact mode and gas atmosphere during catalytic pyrolysis of waste plastics. *Energy Conver. Manag.* **2017**, *142*, 441–451. [CrossRef]

18.   Diaz-Silvarrey, L.S.; McMahon, A.; Phan, A.N. Benzoic acid recovery via waste poly(ethylene terephthalate) (PET) catalytic pyrolysis using sulphated zirconia catalyst. *J. Anal. Appl. Pyrolysis* **2018**, *134*, 621–631. [CrossRef]

19.   Xu, L.; Han, Z.; Yao, Q.; Ding, J.; Zhang, Y.; Fu, Y.; Guo, Q. Towards the sustainable production of pyridines via thermo-catalytic conversion of glycerol with ammonia over zeolite catalysts. *Green Chem.* **2015**, *17*, 2426–2435. [CrossRef]

20.   Xu, L.; Yao, Q.; Han, Z.; Zhang, Y.; Fu, Y. Producing pyridines via thermocatalytic conversion and ammonization of waste polylactic acid over zeolites. *ACS Sustain. Chem. Eng.* **2016**, *4*, 1115–1122. [CrossRef]

21.   Xu, L.; Yao, Q.; Deng, J.; Han, Z.; Zhang, Y.; Fu, Y.; Huber, G.W.; Guo, Q. Renewable N-Heterocycles Production by Thermocatalytic Conversion and Ammonization of Biomass over ZSM-5. *ACS Sustain. Chem. Eng.* **2015**, *3*, 2890–2899. [CrossRef]

22.   Chen, W.; Chen, Y.; Yang, H.; Li, K.; Chen, X.; Chen, H. Investigation on biomass nitrogen-enriched pyrolysis: Influence of temperature. *Bioresour. Technol.* **2018**, *249*, 247–253. [CrossRef] [PubMed]

23.   Xu, L.; Yao, Q.; Zhang, Y.; Fu, Y. Integrated Production of Aromatic Amines and N-Doped Carbon from Lignin via ex Situ Catalytic Fast Pyrolysis in the Presence of Ammonia over Zeolites. *ACS Sustain. Chem. Eng.* **2017**, *5*, 2960–2969. [CrossRef]

24.   Zhang, Y.; Yuan, Z.; Hu, B.; Deng, J.; Yao, Q.; Zhang, X.; Liu, X.; Fu, Y.; Lu, Q. Direct conversion of cellulose and raw biomass to acetonitrile by catalytic fast pyrolysis in ammonia. *Green Chem.* **2019**, *21*, 812–820. [CrossRef]

25.   Bornschein, C.; Werkmeister, S.; Wendt, B.; Jiao, H.; Alberico, E.; Baumann, W.; Junge, H.; Junge, K.; Beller, M. Mild and selective hydrogenation of aromatic and aliphatic (di)nitriles with a well-defined iron pincer complex. *Nature Commun.* **2014**, *5*, 4111. [CrossRef] [PubMed]

26.   Kharasch, M.S.; Beck, T.M. The Chemistry of Organic Gold Compounds. V. Auration of Aromatic Nitriles. *J. Am. Chem. Soc.* **2002**, *56*, 2057–2060. [CrossRef]

27.   Kim, J.; Kim, H.J.; Chang, S. Synthesis of Aromatic Nitriles Using Nonmetallic Cyano-Group Sources. *Angew. Chem. Int. Ed.* **2012**, *51*, 11948–11959. [CrossRef]

28.   Kholkhoev, B.C.; Burdukovskii, V.F.; Mognonov, D.M. Synthesis of polyamidines based on 1, 4-dicyanobenzene and 4, 4′-diaminodiphenyl oxide in ionic liquids. *Russ. Chem. Bull.* **2010**, *59*, 2159–2160. [CrossRef]

29. Kobayashi, M.; Nagasawa, T.; Yamada, H. Regiospecific hydrolysis of dinitrile compounds by nitrilase from *Rhodococcus rhodochrous* J1. *Appl. Microbial. Biotechnol.* **1988**, *29*, 231–233. [CrossRef]

30. Hao, L.; Luo, B.; Li, X.; Jin, M.; Fang, Y.; Tang, Z.; Jia, Y.; Liang, M.; Thomas, A.; Yang, J.; Zhi, L. Terephthalonitrile-derived nitrogen-rich networks for high performance supercapacitors. *Energy Environ. Sci.* **2012**, *5*, 9747–9751. [CrossRef]

31. Abuorabi, S.T.; Jibril, I.; Obiedat, R.; Hatamleh, L. Cycloaddition reactions of 2,4,6-trimethoxybenzonitrile oxide with disubstituted acetylenes. 3. *J. Chem. Eng. Data* **1988**, *33*, 540–541. [CrossRef]

32. Zhao, X.; Xiao, H.; Yue, C.Y.; He, Y.; Chen, H. Preparation, characterization and femtosecond time-resolved optical kerr effect of plasma polybenzonitrile derivative films. *J. Mater. Sci. Mater. Electron.* **2001**, *12*, 557–560. [CrossRef]

33. Qiang, F.; Ding, X.; Wu, X.; Yi, Y.; Jiang, L. Synthesis and characterization of novel aromatic ether nitrile monomer containing propenylphenoxy groups and properties of its copolymer with 4,4′-bismaleimidodiphenylmethane. *J. Appl. Polym. Sci.* **2010**, *83*, 1465–1472.

34. Cui, D.; Nishiura, M.; Hou, Z. Lanthanide-imido complexes and their reactions with benzonitrile. *Angew. Chem. Int. Ed.* **2010**, *117*, 981–984. [CrossRef]

35. Jia, Q.; Wang, J. N-heterocyclic carbene-catalyzed convenient benzonitrile assembly. *Org. Lett.* **2016**, *18*, 2212–2215. [CrossRef]

36. Den Ridder, J.J.J.; Van Ingen, H.W.T.J.; van den Berg, P.J. Kinetics of the gas-phase ammoxidation of toluene to benzonitrile on a Bi-Mo-O catalyst. *Rec. J. R. Netherland Chem. Soc.* **2015**, *98*, 289–293. [CrossRef]

37. Narayana, K.V.; Martin, A.; Bentrup, U.; Lücke, B.; Sans, J. Catalytic gas phase ammoxidation of o-xylene. *Appl. Catal. A Gen.* **2004**, *270*, 57–64. [CrossRef]

38. Dumitriu, E.; Hulea, V.; Kaliaguine, S.; Huang, M.M. Transalkylation of the alkylaromatic hydrocarbons in the presence of ultrastable zeolites transalkylation of toluene with trimethylbenzenes. *Appl. Catal. A Gen.* **1996**, *135*, 57–81. [CrossRef]

39. Xu, L.; Zhang, L.; Song, H.; Dong, Q.; Dong, G.; Kong, X.; Fang, Z. Catalytic Fast Pyrolysis of Polyethylene Terephthalate Plastic for the Selective Production of Terephthalonitrile under Ammonia Atmosphere. *Waste Mang.* **2019**. [CrossRef]

© 2019 by the authors. Licensee MDPI, Basel, Switzerland. This article is an open access article distributed under the terms and conditions of the Creative Commons Attribution (CC BY) license (http://creativecommons.org/licenses/by/4.0/).

Article

Experimental Studies on Co-Combustion of Sludge and Wheat Straw

Zeyu Xue, Zhaoping Zhong * and Bo Zhang

Key Laboratory of Energy Thermal Conversion and Control of the Ministry of Education, School of Energy and Environment, Southeast University, Nanjing 210096, Jiangsu, China; jszyxue@163.com (Z.X.); zhangbo8848@yeah.net (B.Z.)
* Correspondence: zzhong@seu.edu.cn; Tel.: +86-025-83794700

Received: 17 January 2019; Accepted: 12 February 2019; Published: 15 February 2019

Abstract: This work presents studies on the co-combustion of sludge and wheat straw (30 wt % sludge + 70 wt % wheat straw). Prior to the combustion experiment, thermogravimetric analysis was performed to investigate the combustion characteristic of the blended fuel. Results indicated that the blended fuel could remedy the defect of each individual component and also promote the combustion. Co-combustion experiments were conducted in a lab-scale vertical tube furnace and the ash samples were analyzed by Inductively Coupled Plasma Optical Emission Spectrometer (ICP-OES), X-ray Diffraction (XRD), and Scanning Electron Microscope (SEM). Thermodynamic calculations were also made to study the interactions that occurred. Addition of sludge could raise the melting point of wheat straw ash and reduce the slagging tendency. Co-combustion also restrained the release of K and transferred it into aluminosilicate and phosphate. Transfer of Pb and Zn in the co-combustion was also studied. The release and leaching toxicity of the two heavy metals in the co-combustion were weakened effectively by wheat straw. $PbCl_2(g)$ and $ZnCl_2(g)$ could be captured by K_2SiO_3 in wheat straw ash particles and generate silicates. Interactions that possibly occurred between K, Zn, and Pb components were discussed at the end of the paper.

Keywords: sludge; wheat straw; co-combustion; thermodynamic calculation

1. Introduction

Biomass energy is attracting the world's attention due to its zero CO_2 emissions and considerable production worldwide. Of all the treatments, combustion has the advantages of a simple process, high heat utilization, and capacity reduction rate. However, its popularization was restrained by slagging, fouling, and ash deposit derived from the alkali contained [1]. Alkali metal, especially potassium, was necessary for plant growth and primarily existed in the form of ionic or soluble salts in straw biomass. Mass fraction of K in the raw dried straw accounted for 0.1~5 wt %; it was mobile and easily to transfer [2]. In the biomass combustion, potassium could: (1) release in the form of KCl(g); (2) react with Si and form low-melting silicates; and (3) combine with S and cause ash deposits [3–5]. Migration of K in biomass combustion requires monitoring and K-inducing problems need to be resolved.

On the other hand, sludge is a byproduct of sewage treatment, extremely complex, and composed of organic fragments, bacteria, colloids, microorganisms, organic, and inorganic matters. Unstable organic matters in the sludge decomposed and reeked odor, meanwhile the heavy metal contained could also leach out and pollute groundwater [6]. However, sludge was rich in P, which mainly comes from the suspended solids in the raw waste water and phosphorus-based flocculants added in the water treatment. The idea of co-combustion of biomass and sludge has been put forward in recent years and some researchers proved that the co-combustion did mitigate the problems related with alkali metals in the combustion of biomass [7–9]. In particular, Li revealed the mechanism of potassium phosphate formation during the co-combustion of sludge and straw [10].

Apart from P, Al-based flocculants were another component in the water treatment. Mass fraction of Al in the dried sludge accounted for 15~40% and Al_2O_3 was one of the main components of sludge ash [11]. Our previous studies reported that Al in the coal could react with alkali silicates in the biomass–coal co-combustion, which was the primary reaction between the two fuels and could relieve the agglomeration of bed materials in fluidized beds [12,13]. It could be inferred that similar reactions may also take place in the biomass–sludge co-combustion.

However, sludge from municipal sewage plants contained heavy metals including Zn, Pb, Cu, Cr, and Mn [14,15]. In the incineration of sludge, heavy metals could transfer into flue gas and pollute the ecosystem. According to the release ratio, the elements could be divided into easily volatile elements and semi-volatile elements. Their release was also affected by incineration temperature, excess air coefficient, type of reactor, fuel composition, and so on [16]. Considering the relative content and release characteristics, Pb and Zn in the sludge were paid more attention in sludge combustion. Yu and Zhang found that Cl in the fuel could enhance the release of Pb in that evaporation pressure of chlorides was usually larger than oxides [17,18]. However, some researchers also claimed that Cl did not exert a significant effect on Zn release in sludge combustion compared with other heavy metals [19,20]. Guo conducted the co-firing of sludge and biomass and found Zn and Pb appeared to have different migration characteristics and found Pb was inclined to be enriched in fly ash [21]. Several studies also assessed the risk of heavy metals in the sludge ash, confirming the decreasing effect on heavy metal leaching toxicity in co-combustion [22–24]. In recent years, the chemical equilibrium calculation was also used to investigate the existing form of alkali and heavy metals and it was believed that the theoretical conclusion corresponded basically with experimental results [25,26].

Despite the numerous studies concerning the combustion of biomass and sludge before, the mechanisms of co-combustion characteristics and slagging control in the co-combustion remain unclear and require further investigation. In addition, mineral transformation behaviors, especially heavy metal transfer pathways during co-combustion at different combustion stages, should be investigated.

In this paper, characteristics of sludge–wheat straw co-combustion was investigated firstly, and then interactions concerning K, Zn, and Pb were also studied by thermodynamic calculation and experiment in a tube furnace. Samples produced were analyzed by ICP-OES, XRD, and SEM, respectively. Moreover, thermodynamic calculations were applied to discuss the possible reactions of Pb and Zn in the co-firing. Mass fraction of sludge in the co-combustion was set as 30% considering its relatively lower heat value and higher amount of heavy metals, and could be applied to the existing biomass power plants [21,22].

2. Results and Discussion

2.1. TG and DTG Analysis

The TGA tests of wheat straw, sludge, and their mixture were carried out and the TG and DTG curves of the three fuels are given in Figure 1. For combustion of wheat straw, four distinct stages could be divided. In the initial preheating stage below 200 °C, the wheat straw experienced the loss of free/bound water. The second stage occurred around 310 °C, around which the pyrolysis took place, meanwhile the volatile matter separated out and burned rapidly. At about 500 °C the fixed carbon of the wheat straw began to combust. At about 680 °C the residue burned up and some minerals evaporated, causing a slight decrease of weight. After 700 °C the combustion was largely completed.

For combustion of sludge, three stages could be inferred from the DTG curve. At 100 °C the first peak was caused by the loss of water. The second peak at 300 °C and the third peak at 470 °C were the reaction of volatile matter and fixed carbon, respectively. The maximum weight loss rate was much smaller compared with wheat straw (WS). After 600 °C, the two curves both became flat.

In the condition of the blended fuels combustion, three stages would be concluded on the whole. In the dewater process, the two peaks at 100 °C and 200 °C were brought by the two individual components. In the second stage, the peak was also generated by the superposition of the two individual peaks and the value of the maximum weight loss rate was apparently higher than that of the linear superposition by the two individuals. Additionally, the value of the reaction temperature gained by the experiment was lower than the value calculated. In the third stage, the maximum weight loss rate of carbon residue was lower than the predicted value while the temperature was slightly higher. To conclude, co-combustion was not the simple superposition of the two individuals and interaction did take place, actually improving the combustion condition.

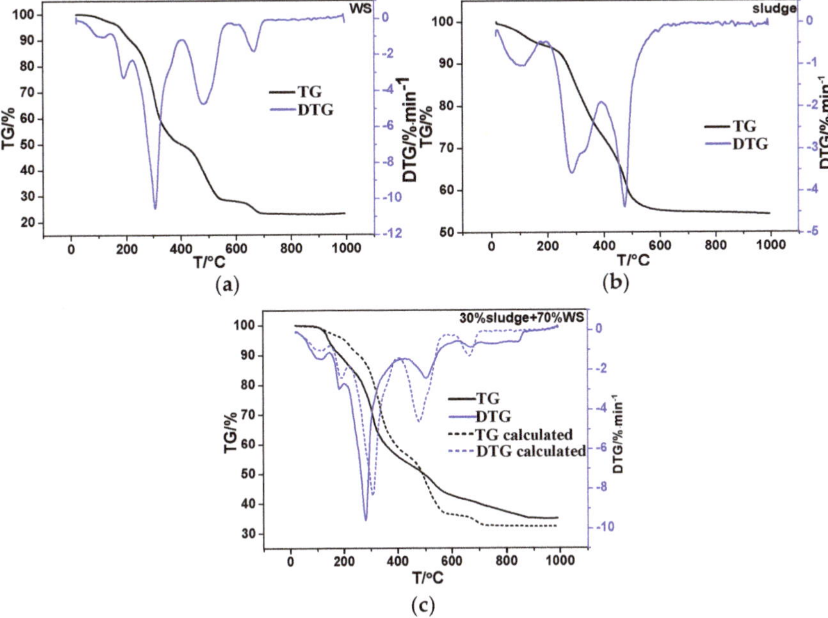

Figure 1. TG and DTG curves of: (**a**) wheat straw (WS); (**b**) sludge; (**c**) 70% WS + 30% sludge in air atmosphere.

2.2. Morphology of Bottom Ash

Figure 2a illustrates SEM micrographs of WS ash at 800 °C. It could be seen that a slight slagging and sintering phenomenon occurred in the surface of WS ash produced at this temperature. Every single submicron particle adhered with each other. However, the morphology of the ash was irregular, and the pore structure was relatively distinct.

WS bottom ash produced at 1000 °C in Figure 2b was found to have the character of experiencing severe liquid-phase melting, the surface was rather smooth and the boundaries between the ash particles became obscure. In addition, the pore structure of the ash almost disappeared and was occupied by eutectic compounds with a low melting point. This could increase the mass transfer resistance and affect the transfer of gaseous species.

In the case of co-combustion at 1000 °C, from Figure 2c it could be inferred that the sludge ash particles appeared in the shape of a group of bricks, of which the size was much larger than WS ash particles. WS ash particles scattered over the surface of the sludge ash. Addition of sludge diluted the ratio of WS ash, thus increasing the gaps between the particles and reducing the tendency of slagging. However, the chemical reaction that occurred also played a role.

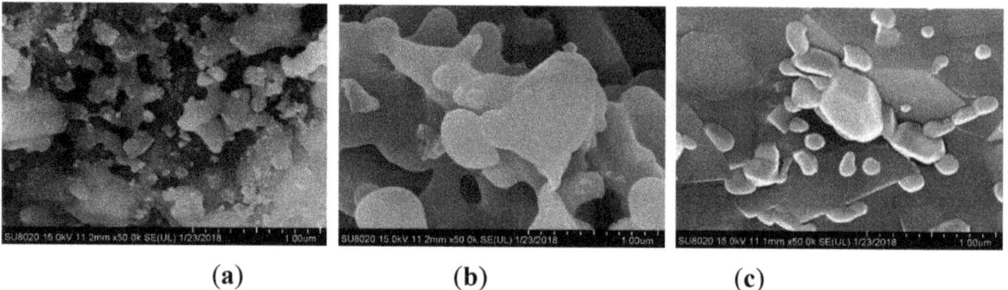

(a)   (b)   (c)

Figure 2. SEM images of ash samples: (**a**) WS ash produced at 800 °C; (**b**) WS ash produced at 1000 °C; (**c**) ash produced in co-combustion at 1000 °C at 50,000 magnification.

To quantify the slagging trend of the ashes, deformation temperature (DT), softening temperature (ST), hemispherical temperature (HT), and flowing temperature (FT) are listed in Table 1.

Table 1. Melting characteristic temperature of the fuel ashes produced at 800 °C. DT: deformation temperature; ST: softening temperature; HT: hemispherical temperature; FT: flowing temperature.

	DT/°C	ST/°C	HT/°C	FT/°C
WS	947	1136	1172	1201
sludge	1195	1289	1310	1324
70% WS + 30% sludge	1021	1194	1221	1243

2.3. Ash Component Analysis

Ash samples were collected and an XRD experiment was conducted. The crystalline phase patterns are shown in Figure 3. For WS ash produced at 600 °C, the crystalline phase detected was KCl, $CaSiO_3$, SiO_2, and Na_2MgCl_4. Alkali metals mainly existed in the form of chlorides and the peak of Si species was not clear. Sludge ash at 600 °C mainly consisted of SiO_2, $KAl_2Si_3AlO_{10}(OH)_2$, Fe_2O_3, Al_2SiO_5, and $Ca_3(PO_4)_2$. In the co-combustion at 600 °C, principal constituents of the ash were SiO_2, $KAl_2Si_3AlO_{10}(OH)_2$, Fe_2O_3, $K_4Ca(PO_4)_2$, and $3Al_2O_3 \cdot 2SiO_2$. Potassium chlorides in WS ash could react with Al in the sludge ash and generate muscovite, transforming the soluble, active K into insoluble, stable parts. Additionally, potassium phosphate was also detected. The following reactions may take place in the co-combustion at this temperature.

$$2KCl + 3Al_2O_3 + 6SiO_2 + 3H_2O(g) \rightarrow 2KAl_2Si_3AlO_{10}(OH)_2 + 2HCl(g) \tag{1}$$

$$6KCl + Ca_3(PO_4)_2 \rightarrow 2K_3PO_4 + 3CaCl_2 \tag{2}$$

At 800 °C, WS ash mainly consisted of SiO_2, $CaSiO_3$, K_2SO_4, and $K_2Si_2O_5$. It should be noted that potassium chloride totally transferred into other forms. Reactions in Equations (3)−(4) took place with the rise of temperature.

$$6KCl + Ca_3(PO_4)_2 \rightarrow 2K_3PO_4 + 3CaCl_2 \tag{3}$$

$$2KCl + H_2O(g) + 2SiO_2 \rightarrow K_2Si_2O_5 + 2HCl(g) \tag{4}$$

Meanwhile, the chemical composition of sludge ash remained constant during the raising temperature. Blended fuel ash produced at 800 °C consisted of $KAlSiO_4$, $KAlSi_3O_8$, SiO_2, $Ca_3(PO_4)_2$, and $K_4Ca(PO_4)_2$. Reactions concerning K in the co-combustion at 800 °C are listed below.

$$K_2Si_2O_5 + Al_2O_3 \rightarrow 2KAlSiO_4 \tag{5}$$

$$3K_2Si_2O_5 + Ca_3(PO_4)_2 \rightarrow 3CaSiO_3 + 2K_3PO_4 + 3SiO_2 \tag{6}$$

When the temperature was enhanced to 1000 °C, the crystallinity became poor due to the melting/sintering that occurred in the WS ash, which was caused by amorphous glassy silicates [12]. The only species detected in the WS ash was SiO_2. Sludge ash at 1000 °C consisted of $3Al_2O_3 \cdot 2SiO_2$, SiO_2, $KAlSi_3O_8$, Fe_2O_3, $Ca_3(PO_4)_2$, and $Ca_9Fe(PO_4)_7$. It could be seen that kaolin was the dominant substance, of which the melting temperature was 1785 °C and was rather stable at high temperatures.

At 1000 °C, ash of the blended fuel was mainly composed of SiO_2, $KAlSi_2O_6$, $KFeSi_3O_8$, $CaAl_2SiO_6$, $Ca_7Mg_2P_6O_{24}$, and $Ca_3(PO_4)_2$. K mainly existed in the Al–Si structure and P in the sludge could also react with the K species. Similar results were also found in earlier research [13,27]. The transfer pathway related to K is listed below.

$$K_2Si_2O_5 + Fe_2O_3 + 4SiO_2 \rightarrow 2KFeSi_3O_8 \tag{7}$$

It could be seen that in the co-combustion, K-inducing slagging and melting could mainly be solved by sludge through two mechanisms: (1) Al could react with K and form aluminosilicates in the range of 600~1000 °C; (2) P could react with K and form phosphates at 800 °C. In addition, the reaction degree of K in the WS and sludge was enhanced by the rising temperature. A 30% mass fraction of sludge could also reduce the release of K, Pb and Zn species were not detected due to its relatively low content (<5% in the crystalline phase), which was under the detection limit.

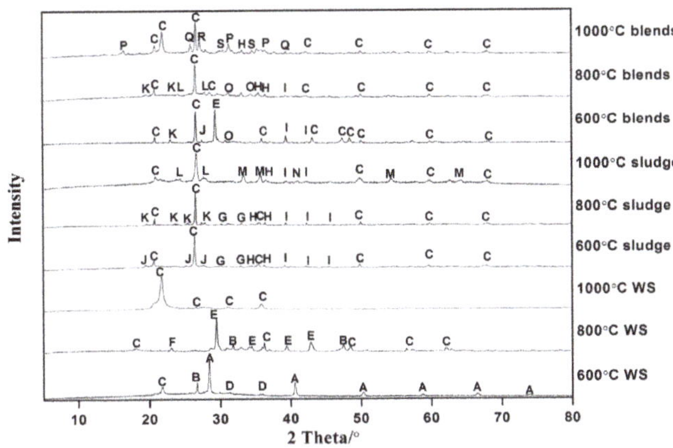

Figure 3. XRD patterns of the ash samples A-KCl, B-$CaSiO_3$, C-SiO_2, D-Na_2MgCl_4, E-K_2SO_4, F-$K_2Si_2O_5$, G-Al_2SiO_5, H-$Ca_3(PO_4)_2$, I-Fe_2O_3, J-$KAl_2Si_3AlO_{10}(OH)_2$, K-$KAlSiO_4$, L-$KAlSi_3O_8$, M-$3Al_2O_3 \cdot 2SiO_2$, N-$Ca_9Fe(PO_4)_7$, O-$K_4Ca(PO_4)_2$, P-$KAlSi_2O_6$, Q-$KFeSi_3O_8$, R-$CaAl_2SiO_6$, S-$Ca_7Mg_2P_6O_{24}$.

2.4. K, Pb, and Zn Distribution in the Combustion

2.4.1. Mass Balance Analysis

Figure 4 shows K distribution in the gas phase and bottom ash during the combustion of WS and 70% WS + 30% sludge. In the combustion of WS, release of K was enhanced as the temperature rose up to 900 °C, which could be owing to the improvement in the combustion. At 900 °C, the maximum evaporation rate of 56.6% was reached. In addition, the ratio of soluble K in the ash was reduced, indicating that potassium silicates and aluminosilicates were formed. At 1000 °C, soluble K in the ash only accounted for 5.3%. Compared with the release rate at 900 °C, a decrease at 1000 °C was observed. This phenomenon could be explained by the change in the gas–solid mass transfer coefficient [28]. In the co-combustion, transfer from the soluble K component to the insoluble parts also took place with the increase of temperature, meaning the mobility of K was weakened. For the release of K in the co-combustion, it was smaller than the combustion of WS alone from 600 to 1000 °C. As the temperature was rising, the release ratio initially increased from 600 to 700 °C and it then fluctuated from 700 to 1000 °C. This was the consequence of the interaction of WS and sludge. In the beginning, the high temperature evaporation dominated the transfer of K before 700 °C. Then the release was relatively restrained due to the intensification of the interaction of K and sludge.

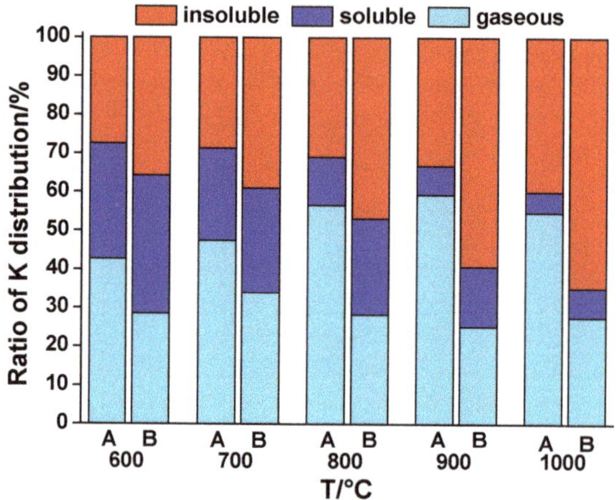

Figure 4. Distribution of K in the case of: (A) WS combustion; (B) co-combustion.

Detailed reactions have been discussed in the former part. To conclude, sludge substituted the element Cl from KCl(g) and released it in the form of HCl(g) instead. The reduction rate of K release by sludge was 33.3%, 28.6%, 50.0%, 57.7%, and 53.4% at 600 °C, 700 °C, 800 °C, 900 °C, and 1000 °C, respectively.

Distribution of Pb and Zn were studied by comparing the co-combustion with sludge combustion. As shown in Figure 5, in combustion of sludge alone, release of Pb was enhanced by the rising temperature in the range 600~800 °C. This was mainly caused by the evaporation of $PbCl_2(g)$, of which the melting point is as low as 501 °C. However, at 900 °C, a clear decrease was observed. This decrease was also substantial in the co-combustion process. This indicated that the released Pb species could be recaptured by other components in the ash while the effect of the evaporation of $PbCl_2(g)$ was relatively impaired.

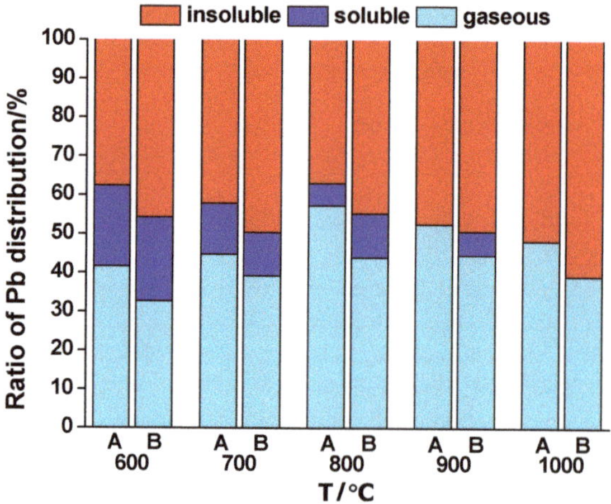

Figure 5. Distribution of Pb in the case of: (A) sludge combustion; (B) co-combustion.

With the presence of wheat straw, the release of Pb was reduced in the co-combustion. As the temperature was rising, the effect of biomass was also augmented. Wang believed that the molten WS ash could also capture released Pb aerosol particles via a physical adhesive force [28]. Meanwhile, share of the soluble Pb component decreased with the temperature; the soluble Pb species was not detected at 900 °C and 1000 °C for sludge combustion and co-combustion, respectively.

For Zn distribution in the sludge combustion, from Figure 6 it was obvious that release of Zn and transfer from the soluble part to the insoluble part were also enhanced by the rising temperature. At 600 °C, the release ratio of Zn was only 25.7% and its maximum value reached 37.1% at 900 °C. The release rate was obviously lower than Pb. The soluble part of Zn in the solid phase reduced from 14.7% to 0% meanwhile. Similar to Pb, co-combustion restrained the release of Zn compared with sludge combustion and the value was in the range of 30.5% to 44.3% and the reducing effect of co-combustion was the most notable at 800 °C. With the temperature rising, the dissimilarity tended to be insubstantial.

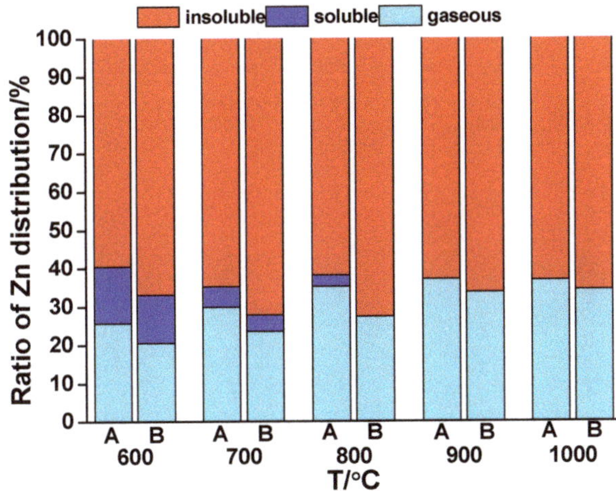

Figure 6. Distribution of Zn in the case of: (A) sludge combustion; (B) co-combustion.

It should be pointed out that co-combustion reduced the soluble part of Zn in the ash and the ratio was below the detection limit since 800 °C. To conclude, the effect of co-combustion on Zn was inferior to Pb due to their different binding capacity to Cl in this temperature range. In addition, high temperature facilitated the combination of potassium silicate and Al in the sludge, thus transferring the silicate into potassium aluminosilicate, increasing the melting point of the bottom ash.

2.4.2. Thermodynamic Calculations on Pb and Zn in the Combustion

Since the relative low content of Pb and Zn in the fuel and their transfer behaviors could not be observed via conventional characterization methods, thermomechanical analysis was applied by the chemical equilibrium calculating tool to further interpret the experiment results. The range of temperature was varied from 550 to 1050 °C and the air-to-fuel ratio was set as 1.2:1 in the calculations. Sludge combustion alone and the co-combustion were calculated respectively.

Figures 7 and 8 indicate the effect of temperature on the transformation of Pb in the sludge combustion and co-combustion, respectively. It could be seen in sludge combustion, Pb mainly existed in the form of $PbSiO_3$ and $Pb_3(PO_4)_2$ at low temperatures, meaning that P and Si in the raw sludge could bind with Pb and restrain its release. As the temperature rose, the phosphate and silicate transferred to $PbO(g)$, $Pb(g)$, $PbCl_2(g)$, and $PbCl(g)$. Release of the chlorides was enhanced by the rising temperature from 550 to 950 °C. After 1000 °C, $PbO(g)$ became the dominant Pb-containing species and almost all Pb existed in the gas phase by thermodynamic calculation.

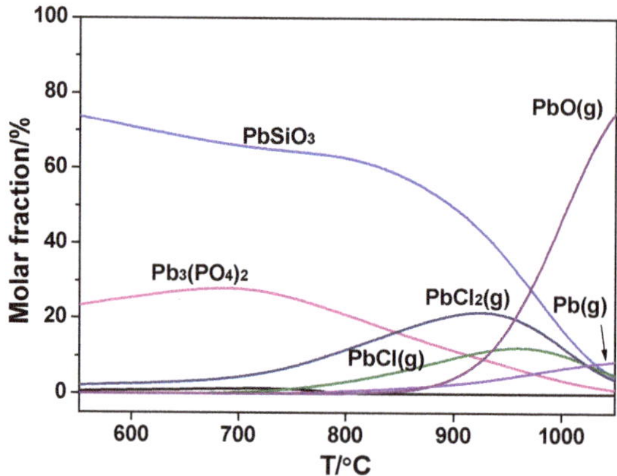

Figure 7. Effect of temperature on the distribution of Pb species in sludge combustion in thermodynamics.

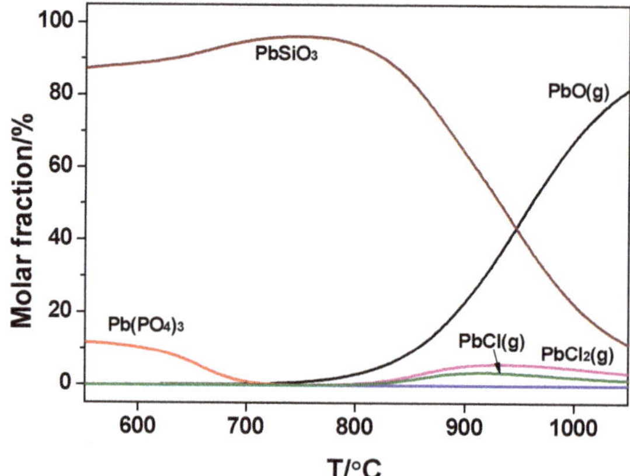

Figure 8. Effect of temperature on the distribution of Pb species in the co-combustion in thermodynamics.

To conclude, the release ratio of Pb in sludge combustion was promoted by the rising temperature, especially after 900 °C due to the decomposition of phosphate and silicate.

However, in the co-combustion, Pb mainly existed in the silicates initially while the content of phosphate was smaller due to the reaction between P and K in the WS via the reaction in Equation (8). This has been discussed in the previous section. Release of chloride was enhanced by temperature in the range 550~920 °C.

$$Pb_3(PO_4)_2 + 3K_2SiO_3 \rightarrow 3PbSiO_3 + 2K_3PO_4 \tag{8}$$

Comparing the two cases, it could be inferred that the binding capacity of Cl to Pb was impaired obviously in the co-combustion by calculation. Decrease of Pb release may occur via Equation (9).

$$PbCl_2(g) + K_2SiO_3 \rightarrow PbSiO_3 + 2KCl(g) \tag{9}$$

As shown in Figure 9, in the combustion of sludge alone, Zn mainly existed in the form of $ZnFe_2O_4$ and $Zn_3(PO_4)_2$ initially. However, the two metal composite oxides decomposed and released free ZnO into the solid phase gradually when the temperature went up. Meanwhile, the release of $ZnCl_2(g)$ increased slowly as the temperature went up in the range 550~870 °C. By calculation, the maximum release rate reached 23% at 870 °C and then decreased slowly. Similar results were also gained in previous reports [29].

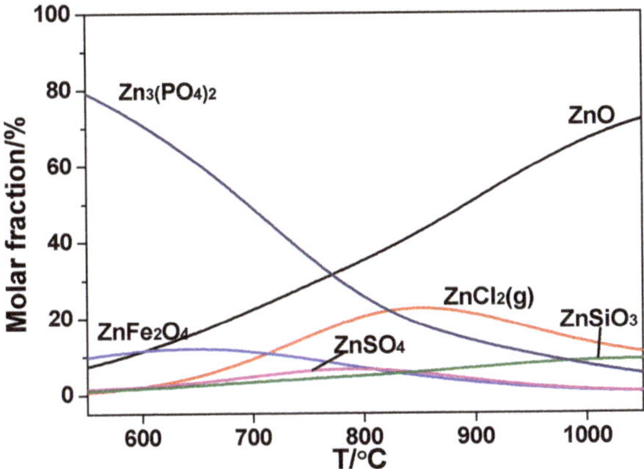

Figure 9. Effect of temperature on the distribution of Zn species in the combustion of sludge in thermodynamics.

In Figure 10, sorts of Zn species are essentially the same, but the relative amounts varied in the co-combustion. Initially the relative content of $Zn_3(PO_4)_2$ was smaller compared with sludge combustion alone, which could be owing to the combination of P and K too.

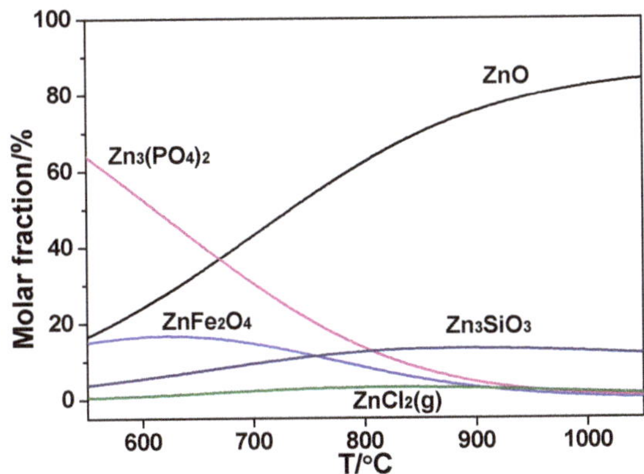

Figure 10. Effect of temperature on the distribution of Zn species in the co-combustion in thermodynamics.

In the range 600~1000 °C, the release rate of Zn was much smaller than sludge combustion due to the decrease in the portion of $ZnCl_2(g)$ by calculation and the relative content only accounted for 4%. Although the value varied widely from the actual release ratio gained by the experiment, it could reveal the decrease of combining capacity of Zn and Cl in the co-combustion. Reactions concerning K and Zn are listed below.

$$Zn_3(PO_4)_2 + 3K_2SiO_3 \rightarrow 3ZnSiO_3 + 2K_3PO_4 \tag{10}$$

$$ZnCl_2(g) + K_2SiO_3 \rightarrow ZnSiO_3 + KCl(g) \tag{11}$$

In the WS combustion, Cl primarily occurred in the form of KCl in the solid phase, and in the co-combustion, some share of KCl transferred into HCl(g). Release of the two heavy metals are also affected by Cl, thus there existed a competitive mechanism between K, Pb, and Zn to Cl in the blended fuels. According to the data obtained, possible transfer pathways inferred are shown in Figure 11. The red lines in the figure represent the interactions which took place possibly in the co-combustion.

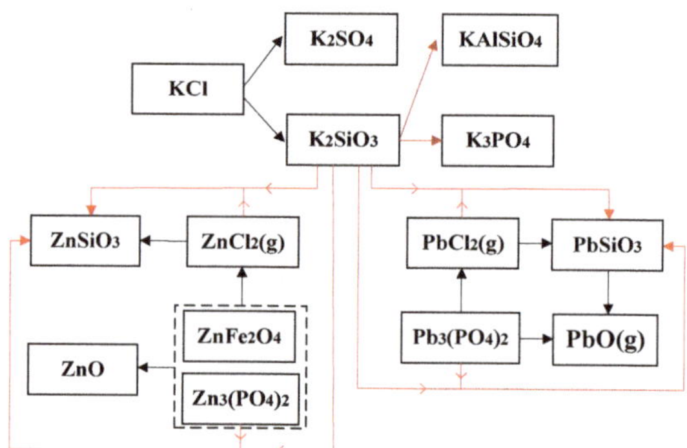

Figure 11. K, Pb, and Zn migration pathways in the combustion.

In the combustion of WS, K mainly existed in chlorides at low temperatures, and with the temperature rising, a portion of KCl evaporated into the gas phase, and the remaining part in the solid phase was inclined to transform into sulfates and silicates, causing the melting phenomena. In the presence of sludge, the incoming Cl was also likely to combine with K, which in the process of release could be recaptured by the Al–Si structure in the sludge ash particles. Wang once reported that the effect of Cl on Pb distribution depended on the content of alkali metals. In the combustion of sludge alone, content of Cl was larger than the sum of K and Na, which indicated that the excess Cl could bond with Pb and Zn at high temperatures [30]. While in the co-combustion, the excess Cl in the sludge, which should have facilitated the release of heavy metals, was inclined to combine with K preferentially. However, the released K then reacted with the Al–Si component in the process of transferring from the inside to the outside of sludge particles. Simultaneously, Zn and Pb were retained in the solid phase, transformed from chlorides to insoluble species. As for Pb, its oxides could combine with Si, however the silicate could decompose at higher temperatures. Likewise, ZnO could also combine with Fe and Al in the combustion. Specially, for Pb, its oxides could also evaporate in the combustion, thus the combination of PbO and SiO_2 determined the release of PbO. The overall reaction at high temperatures could be concluded in Equation (12).

$$PbCl_2 + K_2SiO_3 + H_2O(g) + Al_2O_3 + 2SiO_2 \rightarrow PbSiO_3 + 2KAlSiO_4 + 2HCl(g) \tag{12}$$

3. Materials and Methods

3.1. Materials

In this study, sludge samples were collected from a local municipal wastewater treatment plant in Nanjing city, Jiangsu province. The wheat straw was collected from Suzhou city, Jiangsu province. The sewage sludge and wheat straw were dried at 105 °C for 24 h and then crushed into particles with the size between 0.10 and 0.15 mm. The dried fuel particles were stored in sample bags for use. Properties of sludge and wheat straw are shown in Table 2.

Table 2. Properties of sludge and wheat straw.

Analyses	Sludge	Wheat Straw
Proximate analysis (wt %, dry basis)		
Ash	55.51	14.80
Volatiles	40.89	64.99
Fixed carbon	3.60	20.21
Ultimate analysis (wt %, dry and ash-free basis)		
C	59.90	51.67
H	7.43	6.21
O	23.28	40.45
N	8.76	1.44
S	0.13	0.53
Main ash forming element (wt %, dry basis)		
Na	0.13	0.05
K	0.07	1.13
Mg	0.23	0.12
Ca	1.96	3.4
Al	11.74	0.25
Si	11.68	2.29
P	1.04	0.21
Fe	1.42	0.17
Cl	0.39	0.35
Heavy metal content ($mg \cdot kg^{-1}$, dry basis)		
Zn	1240.00	45.00
Pb	105.00	13.00

3.2. Experimental Apparatus and Procedure

The thermogravimetric experiment of the fuels was carried out in TGA (SETSYS-1750 CS Evol, SETARAM, Caluire-et-Cuire, France). The fuel samples were heated from room temperature to 1000 °C with the heating rate of 20 °C/min in air atmosphere. The injected sample mass was approximately 6 mg every time.

A horizontal tube furnace was used to investigate the partitioning of potassium and heavy metals and the ash contents in solid residues of wheat straw and sludge during combustion. The quartz tube had an inner diameter of 45 mm and a length of 1000 mm. The combustion air was supplied by an air compressor and the flow rate was set as 0.03 Nm^3/h for 45 min. When the temperature reached the value set in advance, about 2 g of wheat straw (WS), sludge powder samples, or their mixture were weighted and mixed evenly in a porcelain combustion boat and pushed into the quartz tube set horizontally in the heating furnace rapidly.

When the combustion was accomplished, bottom ash in the porcelain boat was cooled, collected, and then weighted. Afterwards, the concentration of the digestion solution was analyzed by ICP-OES (Optima8000, Perkin-Elmer, Waltham, USA). The volatilization rate was calculated according to Equations (13)–(16).

$$R = \frac{C_a \times A}{C_{fuel}} \times 100\% \tag{13}$$

$$V = 100\% - R \tag{14}$$

$$R_s = \frac{C_{as} \times A}{C_{fuel}} \times 100\% \tag{15}$$

$$R_{ins} = R - R_s \tag{16}$$

where C_{fuel} is the initial element concentration in the fuel sample, Ca is the total element concentration in the bottom ash, A is the ash yield, and C_{as} is the soluble element concentration in the bottom ash. R means the total element retention rate in the bottom ash; R_s and R_{ins} is the water-soluble and insoluble element share in the ash, accordingly.

The crystalline phase of the bottom ash of fuel was determined by X-ray diffraction (SmartLab 3, Rigaku, Tokyo, Japan) over the range 5~90° and the data generated was analyzed by MDI Jade 6.5 (Materials Data Incorporated, Freemont, CA, USA). To analyze the microstructure of bottom ash samples, a scanning electron microscope (SU8010, HITACHI, Tokyo, Japan) was also applied. In addition, the melting point of the ash was analyzed by an ash fusibility tester (SDAF2000D, Sande, Changsha, China) according to GB/219–74.

3.3. Thermodynamic Equilibrium Calculations

To gain information on the transfer of heavy metals in the combustion, thermodynamic equilibrium calculations were also carried out with HSC-chemistry 6.0 software (Outokumpu, Pori, Finland), based on the method of Gibbs free energy minimization. In this paper, the elements C, H, O, N, S, Cl, Si, Al, P, K, Na, Ca, Mg, Fe, Pb, and Zn in the fuels were taken into consideration. The temperature range was between 550 and 1050 °C and the calculated point was set every 25 °C. Air-fuel ratio in the research was set as 1.2:1 and the pressure was 1 bar.

4. Conclusions

Co-combustion experiments at a proportion of 70% WS and 30% sludge were conducted in a TG instrument and a horizontal tube furnace, respectively, to study the combustion characteristics. Results indicated that co-combustion could lower the ignition temperature and increase the combustion quality of sludge. For the blended fuel, the first combustion stage was controlled by wheat straw while the second was mainly affected by sludge.

Sludge could restrain the release of K in the WS combustion effectively and also solve the melting and slagging. Mechanisms could be concluded in two aspects. First, sludge ash separated the WS ash away and reduced the adhesive force between the ash particles. Second, Al and P in the sludge could react with K species and generate high-melting point components. As the temperature rose, the reaction was enhanced.

Transfer of Pb and Zn in the co-combustion was also studied. Similar to K, their release was also restrained in the co-combustion compared with sludge combustion alone. This could be owing to the competition mechanism of Pb/Zn and K on Cl in the blended fuel. K species could substitute Cl from $PbCl_2$/$ZnCl_2$ and remained the latter in the solid phase. Due to the relative low content of heavy metals in the fuel, the transformation and migration of the trace elements were easily affected by other elements.

Author Contributions: Conceptualization, Z.Z. and Z.X.; methodology, Z.X.; software, Z.X., B.Z.; validation, Z.X., Z.Z. and B.Z.; formal analysis, Z.X.; investigation, Z.X.; resources, Z.Z.; data curation, Z.Z.; writing—original draft preparation, Z.X.; writing—review and editing, Z.Z., B.Z.; visualization, Z.X.; supervision, Z.Z.; project administration, Z.Z.; funding acquisition, Z.Z.

Funding: This research was funded by National Key Research and Development Plan, grant number 2018YFB0605102.

Acknowledgments: This authors wish to acknowledge the financial support provided by the National Key Research and Development Plan (2018YFB0605102).

Conflicts of Interest: The authors declare no conflict of interest.

References

1. Lin, W.; Dam-Johansen, K.; Frandsen, F. Agglomeration in bio-fuel fired fluidized bed combustors. *Chem. Eng. J.* **2003**, *96*, 171–185. [CrossRef]
2. Khan, A.A.; de Jong, W.; Jansens, P.J.; Spliethoff, H. Biomass combustion in fluidized bed boilers: Potential problems and remedies. *Fuel Process. Technol.* **2009**, *90*, 21–50. [CrossRef]

3. Nielsen, H.P.; Baxter, L.L.; Sclippab, G.; Morey, C.; Dam-Johansen, K. Deposition of potassium salts on heat transfer surfaces in straw-fired boilers: A pilot-scale study. *Fuel* **2000**, *79*, 131–139. [CrossRef]

4. Davidsson, K.O.; Åmand, L.; Leckner, B.; Kovacevik, B.; Svane, M.; Hagström, M.; Pettersson, J.B.C.; Pettersson, J.; Asteman, H.; Svensson, J.; et al. Potassium, Chlorine, and Sulfur in Ash, Particles, Deposits, and Corrosion during Wood Combustion in a Circulating Fluidized-Bed Boiler. *Energy Fuel* **2007**, *21*, 71–81. [CrossRef]

5. Davidsson, K.O.; Åmand, L.E.; Steenari, B.M.; Elled, A.L.; Eskilsson, D.; Leckner, B. Countermeasures against alkali-related problems during combustion of biomass in a circulating fluidized bed boiler. *Chem. Eng. Sci.* **2008**, *63*, 5314–5329. [CrossRef]

6. Zha, J.; Huang, Y.; Xia, W.; Xia, Z.; Liu, C.; Dong, L.; Liu, L. Effect of mineral reaction between calcium and aluminosilicate on heavy metal behavior during sludge incineration. *Fuel* **2018**, *229*, 241–247. [CrossRef]

7. Chen, W.; Chang, F.; Shen, Y.; Tsai, M. The characteristics of organic sludge/sawdust derived fuel. *Bioresour. Technol.* **2011**, *102*, 5406–5410. [CrossRef] [PubMed]

8. Degereji, M.U.; Gubba, S.R.; Ingham, D.B.; Ma, L.; Pourkashanian, M.; Williams, A.; Williamson, J. Predicting the slagging potential of co-fired coal with sewage sludge and wood biomass. *Fuel* **2013**, *108*, 550–556. [CrossRef]

9. Aho, M.; Yrjas, P.; Taipale, R.; Hupa, M.; Silvennoinen, J. Reduction of superheater corrosion by co-firing risky biomass with sewage sludge. *Fuel* **2010**, *89*, 2376–2386. [CrossRef]

10. Wang, X.; Ren, Q.; Li, L.; Li, S.; Lu, Q. TG–MS analysis of nitrogen transformation during combustion of biomass with municipal sewage sludge. *J. Therm. Anal. Calorim.* **2016**, *123*, 2061–2068. [CrossRef]

11. Parsons, S.A.; Daniels, S.J. The Use of Recovered Coagulants in Wastewater Treatment. *Environ. Technol.* **1999**, *20*, 979–986. [CrossRef]

12. Zhang, B.; Zhong, Z.; Xue, Z.; Xue, J.; Xu, Y. Release and transformation of potassium in co-combustion of coal and wheat straw in a BFB reactor. *Appl. Therm. Eng.* **2018**, *144*, 1010–1016. [CrossRef]

13. Xue, Z.; Zhong, Z.; Zhang, B.; Zhang, J.; Xie, X. Potassium transfer characteristics during co-combustion of rice straw and coal. *Appl. Therm. Eng.* **2017**, *124*, 1418–1424. [CrossRef]

14. Marani, D.; Braguglia, C.M.; Mininni, G.; Maccioni, F. Behaviour of Cd, Cr, Mn, Ni, Pb, and Zn in sewage sludge incineration by fluidised bed furnace. *Waste Manag.* **2003**, *23*, 117–124. [CrossRef]

15. Liu, J.; Fu, J.; Ning, X.; Sun, S.; Wang, Y.; Xie, W.; Huang, S.; Zhong, S. An experimental and thermodynamic equilibrium investigation of the Pb, Zn, Cr, Cu, Mn and Ni partitioning during sewage sludge incineration. *J. Environ. Sci.* **2015**, *35*, 43–54. [CrossRef] [PubMed]

16. Ogada, T.; Werther, J. Combustion characteristics of wet sludge in a fluidized bed: Release and combustion of the volatiles. *Fuel* **1995**, *75*, 617–626. [CrossRef]

17. Yu, S.; Zhang, B.; Wei, J.; Zhang, T.; Yu, Q.; Zhang, W. Effects of chlorine on the volatilization of heavy metals during the co-combustion of sewage sludge. *Waste Manag.* **2017**, *62*, 204–210. [CrossRef]

18. Zhang, B.; Bogush, A.; Wei, J.; Xu, W.; Zeng, Z.; Zhang, T.; Yu, Q.; Stegemann, J. Influence of Chlorine on the Fate of Pb and Cu during Clinkerization. *Energy Fuel* **2018**, *32*, 7718–7726. [CrossRef]

19. Van de Velden, M.; Dewil, R.; Baeyens, J.; Josson, L.; Lanssens, P. The distribution of heavy metals during fluidized bed combustion of sludge (FBSC). *J. Hazard. Mater.* **2008**, *151*, 96–102. [CrossRef]

20. Cenni, R.; Frandsen, F.; Gerhardt, T.; Spliethoff, H.; Hein, K.R.G. Study on trace metal partitioning in pulverized combustion of bituminous coal and dry sewage sludge. *Waste Manag.* **1998**, *18*, 433–444. [CrossRef]

21. Guo, F.; Zhong, Z.; Xue, H. Partition of Zn, Cd, and Pb during co-combustion of sedum plumbizincicola and sewage sludge. *Chemosphere* **2018**, *197*, 50–56. [CrossRef] [PubMed]

22. Wang, T.; Xue, Y.; Zhou, M.; Yuan, Y.; Zhao, S.; Tan, G.; Zhou, X.; Geng, J.; Wu, S.; Hou, H. Comparative study on the mobility and speciation of heavy metals in ashes from co-combustion of sewage sludge/dredged sludge and rice husk. *Chemosphere* **2017**, *169*, 162–170. [CrossRef] [PubMed]

23. Åmand, L.E.; Leckner, B. Metal emissions from co-combustion of sewage sludge and coal/wood in fluidized bed. *Fuel* **2004**, *83*, 1803–1821. [CrossRef]

24. Zhang, S.; Jiang, X.; Liu, B.; Lv, G.; Jin, Y.; Yan, J. Co-combustion of Bituminous Coal and Pickling Sludge in a Drop-Tube Furnace: Thermodynamic Study and Experimental Data on the Distribution of Cr, Ni, Mn, As, Cu, Sb, Pb, Cd, Zn, and Sn. *Energy Fuel* **2017**, *31*, 3019–3028. [CrossRef]

25. Sun, Y.; Chen, J.; Zhang, Z. Distributional and compositional insight into the polluting materials during sludge combustion: Roles of ash. *Fuel* **2018**, *220*, 318–329. [CrossRef]
26. Contreras, M.L.; Arostegui, J.M.; Armesto, L. Arsenic interactions during co-combustion processes based on thermodynamic equilibrium calculations. *Fuel* **2009**, *88*, 539–546. [CrossRef]
27. Fryda, L.; Sobrino, C.; Cieplik, M.; van de Kamp, W.L. Study on ash deposition under oxyfuel combustion of coal/biomass blends. *Fuel* **2010**, *89*, 1889–1902. [CrossRef]
28. Wang, X.; Huang, Y.; Zhong, Z.; Yan, Y.; Niu, M.; Wang, Y. Control of inhalable particulate lead emission from incinerator using kaolin in two addition modes. *Fuel Process. Technol.* **2014**, *119*, 228–235. [CrossRef]
29. Mason, P.E.; Darvell, L.I.; Jones, J.M.; Williams, A. Observations on the release of gas-phase potassium during the combustion of single particles of biomass. *Fuel* **2016**, *182*, 110–117. [CrossRef]
30. Wang, X.; Huang, Y.; Niu, M.; Wang, Y.; Liu, C. Effect of multi-factors interaction on trace lead equilibrium during municipal solid waste incineration. *J. Mater. Cycles Waste* **2016**, *18*, 287–295. [CrossRef]

© 2019 by the authors. Licensee MDPI, Basel, Switzerland. This article is an open access article distributed under the terms and conditions of the Creative Commons Attribution (CC BY) license (http://creativecommons.org/licenses/by/4.0/).

 *catalysts* 

Article

Carbonate-Catalyzed Room-Temperature Selective Reduction of Biomass-Derived 5-Hydroxymethylfurfural into 2,5-Bis(hydroxymethyl)furan

Jingxuan Long, Wenfeng Zhao, Yufei Xu, Hu Li * and Song Yang *

State Key Laboratory Breeding Base of Green Pesticide & Agricultural Bioengineering,
Key Laboratory of Green Pesticide & Agricultural Bioengineering, Ministry of Education,
State-Local Joint Laboratory for Comprehensive Utilization of Biomass, Center for R&D of Fine Chemicals,
Guizhou University, Guiyang 550025, China; gs.jxlong17@gzu.edu.cn (J.L.); fci.wfzhao18@gzu.edu.cn (W.Z.);
gs.yfxu17@gzu.edu.cn (Y.X.)
* Correspondence: hli13@gzu.edu.cn (H.L.); jhzx.msm@gmail.com (S.Y.); Tel.: +86-(851)-8829-2171 (H.L. & S.Y.)

Received: 7 November 2018; Accepted: 4 December 2018; Published: 7 December 2018

Abstract: Catalytic reduction of 5-hydroxymethylfurfural (HMF), deemed as one of the key bio-based platform compounds, is a very promising pathway for the upgrading of biomass to biofuels and value-added chemicals. Conventional hydrogenation of HMF is mainly conducted over precious metal catalysts with high-pressure hydrogen. Here, a highly active, sustainable, and facile catalytic system composed of K_2CO_3, Ph_2SiH_2, and bio-based solvent 2-methyltetrahydrofuran (MTHF) was developed to be efficient for the reduction of HMF. At a low temperature of 25 °C, HMF could be completely converted to 2,5-bis(hydroxymethyl)furan (BHMF) in a good yield of 94% after 2 h. Moreover, a plausible reaction mechanism was speculated, where siloxane in situ formed via hydrosilylation was found to be the key species responsible for the high reactivity.

Keywords: biomass conversion; biofuels; hydrogenation; green chemistry; sustainable catalysis

1. Introduction

As the depletion of traditional energy resources such as coal and petroleum continues, the efficient synthesis of fuels and value-added molecules from biomass is becoming both urgent and essential [1–6]. Among the biomass derivatives, 5-hydroxymethylfurfural (HMF), which can be obtained from sugars (e.g., fructose [7–10] and glucose [11–13]) via dehydration, is hailed as a key platform molecule for producing a variety of valuable products, see Figure 1 [14,15]. For example, catalytic oxidation of HMF can be converted to 2,5-diformylfuran (DFF), 2,5-furandicarboxylic acid (FDCA), 2-formyl-5-furancarboxylic acid (FFCA), and 5-hydroxymethyl-2-furancarboxylic acid (HMFCA) [14–17]. On the other hand, HMF can be hydrogenated to 2,5-bis(hydroxymethyl)- furan (BHMF), 1,6-hexanediol, 5-methylfurfural (MF), 2,5-dimethylfuran (DMF), 2,5-dimethoxytetrahydrofuran (DMTHF), and 2,5-dihydroxymethyl-tetrahydrofuran (DHMTHF) using various metals (e.g., Pt, Ru, Pd, Cu, Zr) on different supports (e.g., C, SiO_2, Co_3O_4) [17–22]. As one of promising bio-based downstream products, BHMF can be applied in the preparation of various functional materials (e.g., polymers, resins, crown ethers, and artificial fibers), and used as versatile scaffolds in the preparation of medicines and relevant bioactive compounds [23].

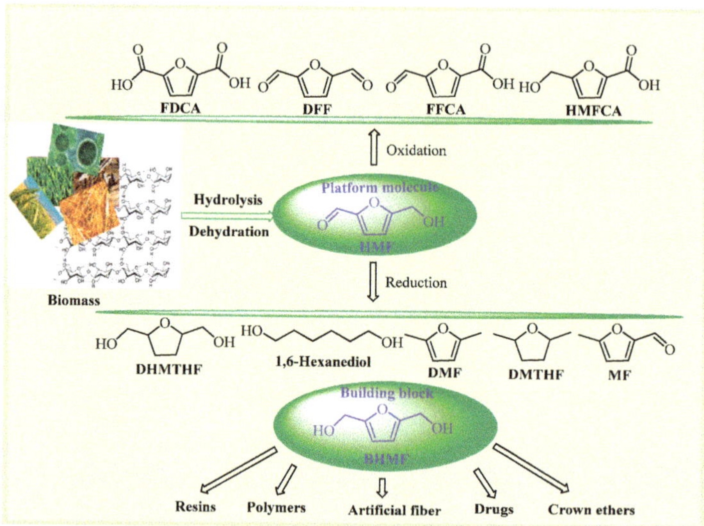

Figure 1. Schematic illustration for upgrading bio-based HMF to value-added products. (HMF: 5-hydroxymethylfurfural, BHMF: 2,5-bis(hydroxymethyl)-furan, FDCA: 2,5-furandicarboxylic acid, DFF: 2,5-diformylfuran, FFCA: 2-formyl-5-furancarboxylic acid, DMF: 2,5-dimethylfuran, HMFCA: 5-hydroxymethyl-2-furancarboxylic acid, DHMTHF: 2,5-dihydroxymethyl-tetrahydrofuran, DMTHF: 2,5-dimethoxytetrahydrofuran, MF: 5-methylfurfural).

In most cases, the efficient catalytic hydrogenation of HMF into BHMF mainly occurs over precious metals (e.g., Ru, Pd) [20,24], or transition metals (e.g., Cu) [25–28] under a relatively high pressure of H_2 (1-10 MPa), see Table 1. The employed metallic catalysts are typically expensive, with relatively harsh reaction conditions [20,25]. Likewise, transfer hydrogenation using alcohol (e.g., ethanol, 2-propanol) or formic acid as the liquid hydrogen source has also attracted much attention from many researchers [21,25,29–32]. However, the use of formic acid may lead to the corrosion of the equipment and metal leaching, while the alcohol can further react with BHMF by etherification to reduce its selectivity [33]. Thereby, it is still essential to explore cheap and green alternative catalytic systems for the reduction of HMF to BHMF in high efficiencies.

Table 1. Activity comparison with previously reported results of HMF-to-BHMF hydrogenation.

Entry	Catalyst	Temperature (°C)	Time (h)	Mole/Pressure of H-Donor	Yield (%)	Conversion (%)	Reference
1	Pd/C	80	20	10 MPa H_2	82	97	[19]
2	Ru/CeO$_x$	130	2	2.7 MPa H_2	81	100	[23]
3	Ru-Mg–Zr	130	2	2.72 MPa H_2	94		[23]
4	Cu/C	180	8	5 MPa H_2	53.3	70	[24]
5	Ru/CoNi-LDO	180	4	1 MPa H_2	49.9	100	[27]
6	Cu(50)-SiO$_2$	100	4	1.5 MPa H_2	93	95	[28]
7	ZrO(OH)$_2$	150	2.5	ethanol	83.7	94.1	[29]
8	ZrBa(3)-SBA	150	2.5	isopropanol	55.4	58.8	[30]
9	Cu/AlO$_x$	220	CF [a]	1,4-butanediol	93	94	[25]
10	Ru/Co$_3$O$_4$	190	6	isopropanol	82.8	100	[20]
11	Cm*Ru(HTsDPEN)	40	2	formic acid	99	100	[31]
12	Pd/C	70	4	formic acid	94	-	[32]
13	K$_2$CO$_3$	25	2	Ph$_2$SiH$_2$	94	100	This work

[a] CF: Continuous flow.

Hydrosilylation of carbonyl compounds to siloxanes shows great potential in the selective synthesis of alcohols [34,35]. This type of conversion process is very attractive because the used silanes are a water- and air-stable hydrogen source, which is able to be stimulated by metal-containing catalysts under mild conditions [22,36–40]. In addition, the strongly basic solvent used was proposed to facilitate its coordination with the catalytically active species (e.g., carbonate) [41], which is favorable for the reaction with silane to release hydride. In this regard, 2-methyltetrahydrofuran (MTHF) with relatively strong Lewis basicity [42], which is derivable from renewable lignocellulosic biomass [43], may have the superior capability to promote the hydrogenation of carboxides using hydrosilanes and is also preferable to protect the environment [44–46]. These features make MTHF a very attractive solvent for organic reactions, potentially in the efficient synthesis of BHMF from HMF.

In the present work, potassium carbonate, an abundant, readily available, and cost-effective salt, is developed to be highly efficient for chemoselective hydrogenation of HMF into BHMF at room temperature employing Ph_2SiH_2 and MTHF as the hydrogen source and solvent, respectively. The effects of different experimental parameters, including various catalysts, solvent type, reaction time, catalyst dosage, and hydrosilane amount on the reaction activity were studied. In addition, the reaction mechanism was also elucidated.

2. Results and Discussions

2.1. Effect of Different Catalysts

The reactivity of various catalysts on the synthesis of BHMF from HMF was first studied using Ph_2SiH_2 as an H-donor at room temperature in MTHF. Without a catalyst, no reaction took place, see Table 2, entry 1, indicating that the reaction was chemically stable at examined conditions (room temperature for 2 h). The presence of K_2CO_3 was found to be highly efficient for the hydrogenation of HMF (100% conversion) under identical conditions, giving BHMF in 70.1% yield, see Table 2, entry 2. Gratifyingly, after post-treatment with methanol, a much higher BHMF yield (94.2%) was obtained, see Table 2, entry 3. The relatively lower selectivity toward BHMF without methanol treatment was due to the formation of the siloxane intermediate, as identified by LC-MS, see Figure S1, while the addition of methanol could facilitate the breakage of the siloxane bond to give BHMF in high selectivity. By comparison, almost no reaction occurred when employing Li_2CO_3, Na_2CO_3, KCl, $CaCO_3$, or $SrCO_3$ as a catalyst (entries 6–10), while $KHCO_3$ and CH_3COOK gave no more than 40% BHMF yield at an HMF conversion of around 50% (entries 4 and 5). These results indicated that both cations and anions played a crucial role in the reaction. With the increase of the cation radius, the salt ionic property increased, thus showing enhanced solubility [47]. Among the alkali/alkaline earth metals, K^+ (with a relatively larger ionic radius) allows the paired carbonate to be more freely available to contact the hydrosilane and substrate. Apart from the cation, the influence of the anion on the reaction is not negligible, mainly with respect to its basicity that has a positive correlation with nucleophilicity. In this regard, carbonate with good accessibility and relatively high nucleophilicity is more prone to attack and activate the silane, Ph_2SiH_2, to release hydride [47]. In view of the superior BHMF yield, K_2CO_3 was chosen as the optimized catalyst for the reduction process.

Table 2. The results for reduction of HMF to BHMF over difference catalysts.

Entry	Catalyst	Yield (%)	Conversion (%)
1	none	0	0
2	K_2CO_3	70.1	100
3 [a]	K_2CO_3	94.2	100
4	$KHCO_3$	40.0	53.1
5	CH_3COOK	35.0	50.2
6	Li_2CO_3	0	<1
7	Na_2CO_3	0	<1
8	KCl	0	0
9	$CaCO_3$	0	0
10	$SrCO_3$	0	<1

Reaction conditions: HMF (0.5 mmol), catalyst (0.1 mmol), Ph_2SiH_2 (1.4 mmol H^-), MTHF (2 mL), 25 °C, and 2 h.
[a] After the reaction, 3 mL methanol was added into the resulting mixture, followed by stirring under ambient conditions for 30 min.

2.2. Influence of Reaction Time

In the assistance of the optimized catalyst, K_2CO_3, the influence of the reaction time on the hydrogenation of HMF to BHMF was studied. All experiments were conducted at 25 °C with the reaction time of 0.5–6 h and the results are displayed in Figure 2. Both the yield of BHMF and the conversion of HMF increased with prolonging the time-course from 0.5 h to 2 h, demonstrating that a comparatively longer reaction time could promote the formation of BHMF. However, the BHMF selectivity decreased as the time was extended to 4 h, possibly due to generating cross-linked siloxanes. These results obviously revealed that the hydrogenation of HMF to BHMF employing this facile and benign catalytic system was extremely selective and active, in comparison with other previous ones utilizing H_2, alcohol, or formic acid as the hydrogen source, see Table 1. Therefore, 2 h and 25 °C were chosen to be the optimized reaction time and temperature, respectively, with K_2CO_3 as the catalyst for the subsequent study.

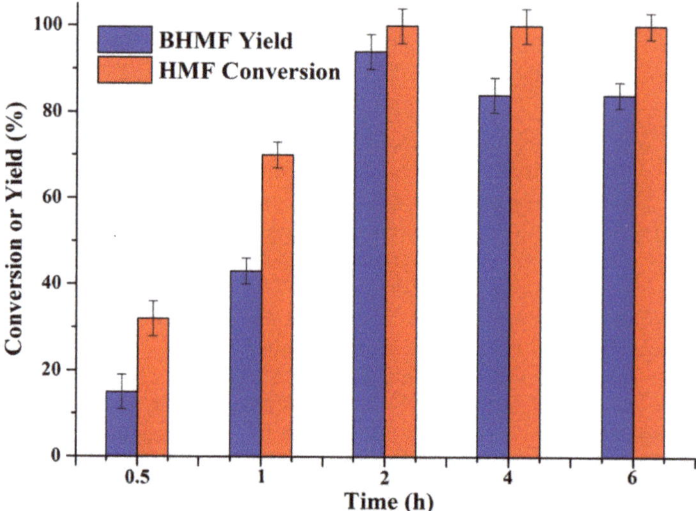

Figure 2. The effect of reaction time on the reduction of HMF into BHMF. (Reaction conditions: HMF (0.5 mmol), K_2CO_3 (0.1 mmol), Ph_2SiH_2 (1.4 mmol H^-), MTHF (2 mL), and 25 °C; After the reaction, 3 mL methanol was added into the resulting mixture, followed by stirring under ambient conditions for 30 min.)

2.3. Influence of Various Solvents

The effect of the used solvents in the reduction reaction was also investigated. MTHF, an aprotic solvent, with relatively high polarity, displayed high HMF conversion (100%) and satisfactory BHMF yield (94%), see Figure 3. In contrast, the solvents (e.g., DCM, 1,4-dioxane, CH_3CN, THF, and GVL) with relatively low polarity exhibited inferior performance, affording BHMF in yields of 20.1%, 25.2%, 60.0%, 67.1%, and 70%, respectively. A solvent with moderate polarity could affect the solvent–reactant interactions [48,49], where the increased interaction of HMF and solvent might promote the formation of BHMF. On the other hand, the solubility of Ph_2SiH_2 can also be enhanced by compatible polarity of the solvent, thereby facilitating the interaction of Ph_2SiH_2 with the catalyst to release H^-. Overall, the appropriate interaction between the solvent, hydrosilane, and catalyst can facilitate the release of hydride to attack HMF, thus giving BHMF in a good yield.

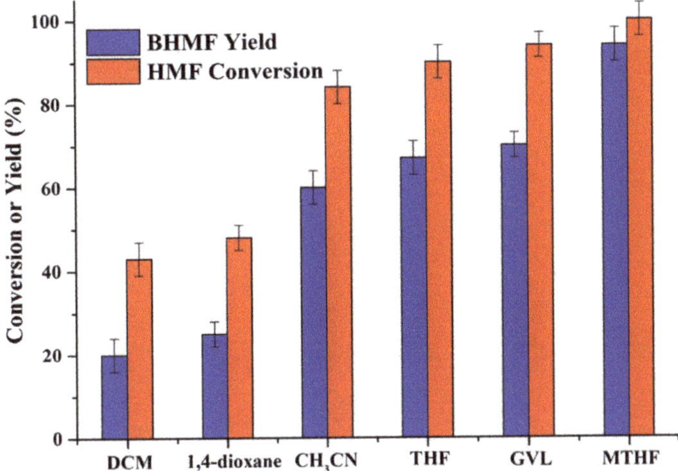

Figure 3. Effect of solvent type on reduction of HMF into BHMF. (Reaction conditions: HMF (0.5 mmol), K_2CO_3 (0.1 mmol), Ph_2SiH_2 (1.4 mmol H^-), solvent (2 mL), 25 °C, and 2 h; After the reaction, 3 mL methanol was added into the resulting mixture, followed by stirring under ambient conditions for 30 min.)

2.4. Influence of Different Hydrosilanes

The influence of various hydrosilanes on the catalytic reduction of HMF into BHMF was studied by using K_2CO_3 as the catalyst, see Table S1. Among the employed simple silanes, phenylsilane ($PhSiH_3$) and diphenylsilane (Ph_2SiH_2), see Table S1, entries 5 and 6, revealed relatively higher yields of BHMF, which might be due to their superior hydride-donating capability [50]. Although the conversion of HMF reached 100% with PMHS as an H-donor, the obtained BMHF yield was relatively low (66.0%). The possible reason is that PMHS might react with products and substrates, forming insoluble PMHS-derived resin in a relatively large molecular weight under alkaline conditions, see Figure S2B. Even after adding methanol, the formed solid was not completely dissolved, see Figure S2C, indicating the relatively stable resin structure. It can be speculated that some of the formed silicon–oxygen bonds could not be cleaved during the post-treatment with methanol, thus impeding the formation of BHMF with a high selectivity. In view of the superior BHMF yield, Ph_2SiH_2 as a comparatively cheap hydrogen source was screened out for further optimization of HMF hydrogenation.

2.5. Influence of Ph_2SiH_2 Dosage

The Ph_2SiH_2 dosage was found to have a significant effect on BHMF synthesis, as shown in Figure 4. As the dosage of Ph_2SiH_2 rose from 0.05 to 0.7 mmol, the yield of BHMF was observed to increase from 20.0 to 94.2%, accordingly. Nevertheless, when the PMHS dosage was more than 0.7 mmol, the BHMF yield was decreased despite constant HMF conversion. The decline in the yield of BHMF was most likely caused by the reaction of excessive Ph_2SiH_2 with potassium carbonate and BHMF to form cross-linked siloxanes that are too stable to hydrolyze even proceeding post-treatment with methanol. In a word, the use of 0.7 mmol Ph_2SiH_2 was favorable to attain the highest BMHF yield at 25 °C for 2 h in MTHF.

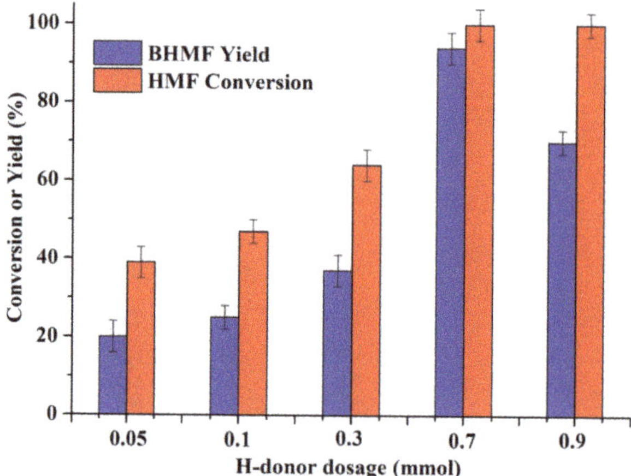

Figure 4. Effect of H^- dosage of Ph_2SiH_2 on the reduction of HMF into BHMF. (Reaction conditions: HMF (0.5 mmol), K_2CO_3 (0.1 mmol), MTHF (2 mL), 25 °C, and 2 h; After the reaction, 3 mL methanol was added into the resulting mixture, followed by stirring under ambient conditions for 30 min.)

2.6. Influence of Catalyst Dosage

The effect of the K_2CO_3 dosage on the production of BHMF from HMF under the optimal conditions was further investigated at 25 °C over a period of 2 h, see Figure 5. When the K_2CO_3 dosage increased from 0 to 0.1 mmol, the conversion of the HMF and BHMF yield rose to 100% and 94.2%, respectively. However, as the catalyst dosage was beyond 0.1 mmol, the yield of BHMF displayed a remarkable reduction, which could be attributed to BHMF adsorption onto the surface of solid K_2CO_3, thus inhibiting the alcoholysis of the siloxane intermediate to release the product BHMF [32]. In addition, the fast depletion of Ph_2SiH_2 was most likely to occur by forming pentavalent silicate species ((Ph_2SiH_2)-CO_3) with carbonate [51]. Although the pentavalent silicate species were active for hydrogenating carboxides [52], serious polymerization was prone to happen due to utilizing superfluous K_2CO_3 (e.g., 0.25 mmol), hence possibly decreasing the catalyst activity.

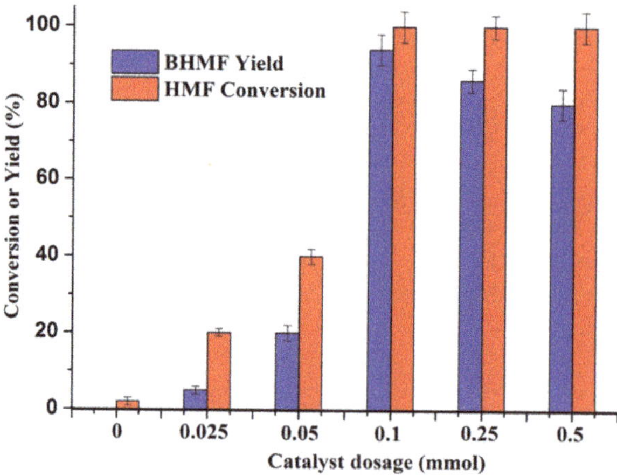

Figure 5. Influence of K_2CO_3 dosage on synthesis of BHMF from HMF. (Reaction conditions: HMF (0.5 mmol), Ph_2SiH_2 (1.4 mmol H^-), MTHF (2 mL), 25 °C, and 2 h; After the reaction, 3 mL methanol was added into the resulting mixture, followed by stirring under ambient conditions for 30 min.)

2.7. Reaction Mechanism for HMF-to-BHMF Hydrogenation

To explicitly interpret the reaction mechanism for hydrogenating HMF to BHMF, ex situ ^{1}H NMR spectra were recorded at 25 °C in the time course of 0.5–2 h, see Figure 6. The aldehydic proton **1a** of HMF gradually disappeared, while the protons of hydroxyl **2c** and methine of the furan ring **2a** of BHMF were also detected and significantly enlarged by further extending the reaction duration. The mechanism for the hydrogenation of aldehydes or ketones catalyzed by K_2CO_3 was thus proposed to be very similar to the reduction process of amides [53], where Ph_2SiH_2 might act as an H-donor to promote the reduction of HMF to BHMF, as demonstrated by in situ NMR in our previous study [39]. The possible reaction mechanism of the hydrogenation of HMF to BHMF is shown in Scheme 1. Initially, the interaction of a Lewis base (carbonate) and Ph_2SiH_2 forms a high-valency silicate species, which is a nucleophile in nature [54] and can activate the aldehyde group by hydride transfer to afford the siloxane intermediate, followed by hydrolysis to give BHMF. This catalytic system is simple to operate and does not produce observable byproducts during the reduction process, compared with other previous reaction systems [26,27].

To make our process more promising for real application, relatively stable sugar like fructose has been used as the starting material instead of unstable HMF. It was found that a moderate yield of BHMF (66%) could be obtained via a one-pot two-step reaction process separately catalyzed by Amberlyst-15 and K_2CO_3. This sequential process makes the biomass derivative HMF a promising platform molecule for the production of value-added chemicals.

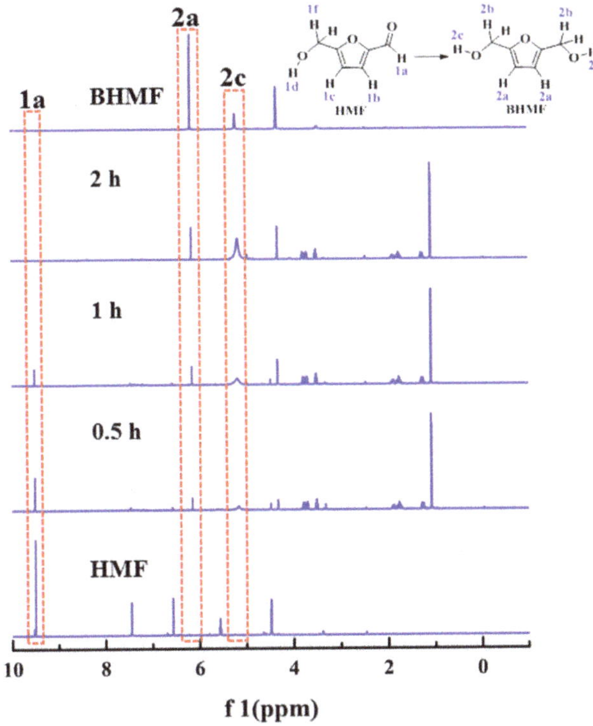

Figure 6. Ex situ ^{1}H NMR spectra of hydrogenating HMF to BHMF at 25 °C in 0.5–2 h. (Reaction conditions: HMF (0.5 mmol), Ph_2SiH_2 (1.4 mmol H$^-$), K_2CO_3 (0.1 mmol), and MTHF (2.0 mL); After the reaction proceeded for 2 h, 3 mL CD_3OD was added into the resulting mixture, followed by stirring under ambient conditions for 30 min.)

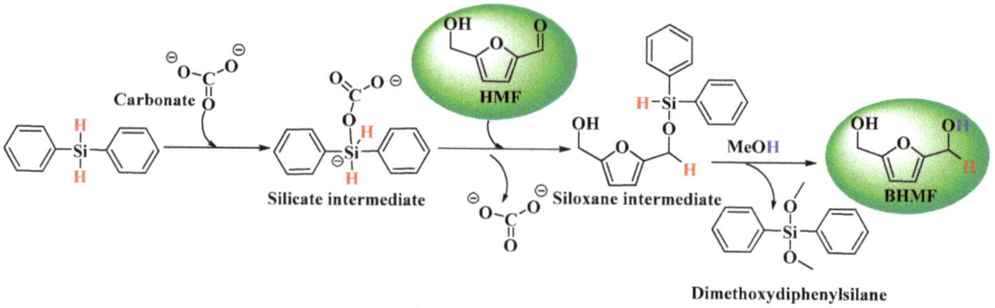

Scheme 1. Plausible reaction pathway for the HMF-to-BHMF reduction.

3. Materials and Experiments

3.1. Materials

5-hydroxymethylfurfural (HMF, 99%), 2,5-bis(hydroxymethyl)furan (BHMF, 99%), phenylsilane (97%), polymethylhydrosiloxane (PMHS, 99%), triethoxysilane (97%), diphenylsilane (97%), deuterated dimethyl sulfoxide (DMSO-d_6, 99.8 at.% D), deuterated methanol (CD_3OD, 99.8 at.% D), 1,4-dioxane (99%), potassium carbonate (K_2CO_3, 99%), gamma-valerolactone (GVL, 99%), naphthalene (99%), methanol (99%), triethylsilane (99%), trimethoxysilane (99%), Amberlyst-15 (>99%), fructose (>99%), and 1,1,3,3-tetramethyldisiloxane (97%) were bought from Innochem Technology Co., Ltd. (Beijing, China). Tetrahydrofuran (THF, 99%) was bought from Shanghai TCI Develop. Co., Ltd. (Shanghai, China). Lithium carbonate (Li_2CO_3, 99%) and sodium carbonate (Na_2CO_3, 99%) were purchased from Aladdin Ind. Inc. (Shanghai, China). Potassium bicarbonate ($KHCO_3$, 99%), potassium chloride (KCl, 98%), acetonitrile (MeCN, 99%), 2-methyltetrahydrofuran (MTHF, 99%), and dichloromethane (DCM, 99%) were bought from Sigma-Aldrich Co. LLC. (Shanghai, China). Other used chemicals were of analytical grade without further purification, unless stated otherwise.

3.2. Procedures for Catalytic Reduction of HMF into BHMF

The tested reactions were all conducted in Ace pressure tubes (15 mL). In a general reaction process, 0.5 mmol HMF, 0.1 mmol K_2CO_3, 2.0 mL MTHF, and 0.7 mmol Ph_2SiH_2 were added into the reactor and sealed. Afterwards, the tube was moved into an oil-bath under magnetic stirring at 400 rpm for a certain period. The zero time was established when the reactor was placed in the bath (25 °C). Upon completion, methanol (3 mL) was added into the resulting mixture, followed by stirring under ambient conditions for 30 min to proceed alcoholysis of the siloxane intermediate. Then, the resulting reaction solution was filtered to remove solid particles, while the yield of BHMF was quantified by GC and the conversion of HMF was analyzed by HPLC. When fructose (0.5 mmol) instead of HMF was used as the substrate, an additional reaction process catalyzed by Amberlyst-15 (15 mg) at 120 °C for 45 min was conducted prior to the reduction process under otherwise identical reaction conditions. All the reactions were repeated three times, and the obtained results were the average values of three individual experiments, with a standard deviation (σ) of 0.5–4.2%.

3.3. Analytical Methods 1100 (Agilent Technologies) system

HMF conversion was analyzed by HPLC (1100 series, Agilent Technologies Inc., CA, USA) equipped with a refractive index detector, an ultraviolet detector ($\lambda = 284$ nm), and an Aminex HPX-87H column (Bio-Rad Laboratories, Hercules, CA, USA). The column temperature was set at 25 °C, and the mobile phase used was acetonitrile/0.1 wt.% acetic acid aqueous solution (30:70, v/v) at a flow rate of 1.0 mL min^{-1} (HMF retention time: 3.7 min). BHMF was quantitatively analyzed by an Agilent 7890B GC (Agilent Technologies Inc., Santa Clara, CA, USA) fitted with a flame ionization detector and an HP-5 column. Analysis conditions were optimized as 1 μL injection volume, 0.7 mL/min nitrogen carrier gas, 75:1 split ratio, with injector and detector temperatures of 240 and 280 °C, respectively; oven temperature programmed from 60 °C (keeping for 1 min) to 100 °C at a rate of 5 °C/min, then raised to 270 °C (keeping for 2 min) at a rate of 10 °C/min (BHMF retention time: 14.7 min). Naphthalene (10 mg) was used as an internal standard. The products were quantified based on the standard curves of the corresponding commercial reagents ($R^2 \geq 0.9997$). In addition, liquid products were identified by GC-MS (6890N GC/5973 MS, Agilent Technologies Inc., CA, USA), while a siloxane intermediate was identified by liquid chromatography/mass spectrometry (LC-MS) (Ultimate 3000 LC/Q Exactive MS, Thermo Fisher Scientific, San Jose, CA, USA).

3.4. Reaction Time-Course Study

For the time-course study of the reaction with NMR, after the reaction had proceeded for a specific time (0.5–2 h), 3 mL deuterated methanol was added into the reaction solution, followed by stirring under ambient conditions for 30 min, and then 0.5 mL of the reaction solution was taken. ^{1}H NMR spectra of the reaction mixtures were performed using an NMR spectrometer (JEOL-ECX 500, JEOL, Tokyo, Japan).

4. Conclusions

In this work, a benign and efficient catalytic protocol was developed for selective hydrogenation of HMF into BHMF in good yields (ca. 94%) at room temperature for 2 h, using abundant, readily available, and cost-effective K_2CO_3 as the catalyst, eco-friendly Ph_2SiH_2 as the hydrogen source, and green and renewable MTHF as the solvent. The developed silane-mediated catalytic system was extremely active and selective, in comparison with previous reports that use noble or transition metals as a catalyst while H_2, formic acid, or alcohol as the hydrogen source. The mechanistic study illustrated the reaction proceeded via the hydrosilylation process with a high-valency silicate species as the key hydride species formed by the activation of Ph_2SiH_2 with carbonate, which can greatly facilitate the hydride transfer toward the aldehyde group to afford the siloxane intermediate. Subsequently, BHMF is obtained by the in situ hydrolysis of the siloxane intermediate, which can be significantly promoted by post-treatment with methanol. Moreover, this catalytic system is simple to operate and produces fewer byproducts during the process, exhibiting great promise for biomass upgrading.

Supplementary Materials: The Supplementary Materials are available online at http://www.mdpi.com/2073-4344/8/12/633/s1. Figure S1. LC-MS spectrum of the siloxane intermediate obtained in the hydrogenation of HMF to BHMF; Figure S2. Images of the K_2CO_3-catalyzed hydrogenation of HMF to BHMF using PMHS as H-donor; Figure S3. GC-MS spectrum of BMHF obtained from hydrogenation of HMF; Figure S4. GC-MS spectrum of dimethoxydiphenylsilane obtained from the interaction of Ph_2SiH_2 and MeOH; Table S1. Effect of different hydrosilanes on the hydrogenation of HMF to BHMF.

Author Contributions: Data curation, J.L.; Investigation, J.L., W.Z. and Y.X.; Supervision, S.Y.; Conceptualization, H.L.; Funding acquisition, S.Y.; Project administration, S.Y.; Writing-original draft, J.L.; Writing-review and editing, H.L. and S.Y.

Funding: The work was financially supported by the Key Technologies R&D Program of China (2014BAD23B01), National Natural Science Foundation of China (21576059, 21666008), Fok Ying-Tong Education Foundation (161030), and Guizhou Science & Technology Foundation ([2018]1037, [2017]5788).

Conflicts of Interest: There are no conflicts to declare.

References

1. Li, H.; Fang, Z.; Luo, J.; Yang, S. Direct conversion of biomass components to the biofuel methyl levulinate catalyzed by acid-based bifunctional zirconia-zeolites. *Appl. Catal. B. Environ.* **2017**, *200*, 182–191. [CrossRef]
2. Hu, L.; Xu, J.; Zhou, S.; He, A.; Tang, X.; Lin, L.; Zhao, Y. Catalytic advances in the production and application of biomass-derived 2,5-dihydroxymethylfuran. *ACS Catal.* **2018**, *8*, 2959–2980. [CrossRef]
3. Patel, M.; Kumar, A. Production of renewable diesel through the hydroprocessing of lignocellulosic biomass-derived bio-oil: A review. *Renew. Sustain. Energy Rev.* **2016**, *58*, 1293–1307. [CrossRef]

4. Qin, Y.; Yu, L.; Wu, R.; Yang, D.; Qiu, X.; Zhu, J.Y. Biorefinery lignosulfonates from sulfite-pretreated softwoods as dispersant for graphite. *ACS Sustain. Chem. Eng.* **2016**, *4*, 2200–2205. [CrossRef]

5. Li, H.; Fang, Z.; Smith Jr, R.L.; Yang, S. Efficient valorization of biomass to biofuels with bifunctional solid catalytic materials. *Proc. Energy Combust.* **2016**, *55*, 98–194. [CrossRef]

6. Li, H.; Riisager, A.; Saravanamurugan, S.; Pandey, A.; Sangwan, R.S.; Yang, S.; Luque, R. Carbon-increasing catalytic strategies for upgrading biomass into energy-intensive fuels and chemicals. *ACS Catal.* **2017**, *8*, 148–187. [CrossRef]

7. Román-Leshkov, Y.; Chheda, J.N.; Dumesic, J.A. Phase modifiers promote efficient production of hydroxymethylfurfural from fructose. *Science* **2006**, *312*, 1933–1937. [CrossRef] [PubMed]

8. Qi, X.; Watanabe, M.; Aida, T.M.; Smith, R.L., Jr. Efficient process for conversion of fructose to 5-hydroxymethylfurfural with ionic liquids. *Green Chem.* **2009**, *11*, 1327–1331. [CrossRef]

9. Chen, J.; Li, K.; Chen, L.; Liu, R.; Huang, X.; Ye, D. Conversion of fructose into 5-hydroxymethylfurfural catalyzed by recyclable sulfonic acid-functionalized metal–organic frameworks. *Green Chem.* **2014**, *16*, 2490–2499. [CrossRef]

10. Zhao, H.; Holladay, J.E.; Brown, H.; Zhang, Z.C. Metal chlorides in ionic liquid solvents convert sugars to 5-hydroxymethylfurfural. *Science* **2007**, *316*, 1597–1600. [CrossRef] [PubMed]

11. Hu, S.; Zhang, Z.; Song, J.; Zhou, Y.; Han, B. Efficient conversion of glucose into 5-hydroxymethylfurfural catalyzed by a common Lewis acid $SnCl_4$ in an ionic liquid. *Green Chem.* **2009**, *11*, 1746–1749. [CrossRef]

12. Choudhary, V.; Mushrif, S.H.; Ho, C.; Anderko, A.; Nikolakis, V.; Marinkovic, N.S.; Vlachos, D.G. Insights into the interplay of Lewis and Brønsted acid catalysts in glucose and fructose conversion to 5-(hydroxymethyl) furfural and levulinic acid in aqueous media. *J. Am. Chem. Soc.* **2013**, *135*, 3997–4006. [CrossRef] [PubMed]

13. Teong, S.P.; Yi, G.; Zhang, Y. Hydroxymethylfurfural production from bioresources: Past, present and future. *Green Chem.* **2014**, *16*, 2015–2026. [CrossRef]

14. Rout, P.K.; Nannaware, A.D.; Prakash, O.; Kalra, A.; Rajasekharan, R. Synthesis of hydroxymethylfurfural from cellulose using green processes: A promising biochemical and biofuel feedstock. *Chem. Eng. Sci.* **2016**, *142*, 318–346. [CrossRef]

15. Zhang, J.; Li, J.; Tang, Y.; Lin, L.; Long, M. Advances in catalytic production of bio-based polyester monomer 2,5-furandicarboxylic acid derived from lignocellulosic biomass. *Carbohyd. Polym.* **2015**, *130*, 420–428. [CrossRef] [PubMed]

16. van Putten, R.J.; Van Der Waal, J.C.; De Jong, E.D.; Rasrendra, C.B.; Heeres, H.J.; de Vries, J.G. Hydroxymethylfurfural, a versatile platform chemical made from renewable resources. *Chem. Rev.* **2013**, *113*, 1499–1597. [CrossRef] [PubMed]

17. Hu, L.; Lin, L.; Liu, S. Chemoselective hydrogenation of biomass-derived 5-hydroxymethylfurfural into the liquid biofuel 2,5-dimethylfuran. *Ind. Eng. Chem. Res.* **2014**, *53*, 9969–9978. [CrossRef]

18. Román-Leshkov, Y.; Barrett, C.J.; Liu, Z.Y.; Dumesic, J.A. Production of dimethylfuran for liquid fuels from biomass-derived carbohydrates. *Nature* **2007**, *447*, 982–985. [CrossRef] [PubMed]

19. Audemar, M.; Vigier, K.D.O.; Clacens, J.M.; De Campo, F.; Jérôme, F. Combination of Pd/C and Amberlyst-15 in a single reactor for the acid/hydrogenating catalytic conversion of carbohydrates to 5-hydroxy-2,5-hexanedione. *Green Chem.* **2014**, *16*, 4110–4114. [CrossRef]

20. Wang, T.; Zhang, J.; Xie, W.; Tang, Y.; Guo, D.; Ni, Y. Catalytic transfer hydrogenation of biobased HMF to 2,5-bis-(hydroxymethyl) furan over Ru/Co_3O_4. *Catalysts* **2017**, *7*, 92. [CrossRef]

21. Revunova, K.; Nikonov, G.I. Main group catalysed reduction of unsaturated bonds. *Dalton Trans.* **2015**, *44*, 840–866. [CrossRef] [PubMed]

22. Kwon, Y.; de Jong, E.; Raoufmoghaddam, S.; Koper, M.T. Electrocatalytic hydrogenation of 5-hydroxymethylfurfural in the absence and presence of glucose. *ChemSusChem* **2013**, *6*, 1659–1667. [CrossRef] [PubMed]

23. Alamillo, R.; Tucker, M.; Chia, M.; Pagán-Torres, Y.; Dumesic, J. The selective hydrogenation of biomass-derived 5-hydroxymethylfurfural using heterogeneous catalysts. *Green Chem.* **2012**, *14*, 1413–1419. [CrossRef]

24. Chen, B.; Li, F.; Huang, Z.; Yuan, G. Carbon-coated Cu-Co bimetallic nanoparticles as selective and recyclable catalysts for production of biofuel 2,5-dimethylfuran. *Appl. Catal. B-Environ.* **2017**, *200*, 192–199. [CrossRef]

25. Aellig, C.; Jenny, F.; Scholz, D.; Wolf, P.; Giovinazzo, I.; Kollhoff, F.; Hermans, I. Combined 1,4-butanediol lactonization and transfer hydrogenation/hydrogenolysis of furfural-derivatives under continuous flow conditions. *Catal. Sci. Technol.* **2014**, *4*, 2326–2331. [CrossRef]

26. Upare, P.P.; Hwang, Y.K.; Hwang, D.W. An integrated process for the production of 2,5-dihydroxymethylfuran and its polymer from fructose. *Green Chem.* **2018**, *20*, 879–885. [CrossRef]

27. Li, Q.; Man, P.; Yuan, L.; Zhang, P.; Li, Y.; Ai, S. Ruthenium supported on CoFe layered double oxide for selective hydrogenation of 5-hydroxymethylfurfural. *Mol. Catal.* **2017**, *431*, 32–38. [CrossRef]

28. Shen, R.; Zhang, M.; Xiao, J.; Dong, C.; Han, L.B. Ph$_3$P-Mediated Highly Selective C(α)-P Couplings of Quinone Monoacetals with R$_2$P(O)H: Convenient and Practical Synthesis of ortho-Phosphinyl Phenols. *Green Chem.* **2018**, *20*, 5111–5116. [CrossRef]

29. Hao, W.; Li, W.; Tang, X.; Zeng, X.; Sun, Y.; Liu, S.; Lin, L. Catalytic transfer hydrogenation of biomass-derived 5-hydroxymethyl furfural to the building block 2,5-bishydroxymethyl furan. *Green Chem.* **2016**, *18*, 1080–1088. [CrossRef]

30. Wei, J.; Cao, X.; Wang, T.; Liu, H.; Tang, X.; Zeng, X.; Lin, L. Catalytic transfer hydrogenation of biomass-derived 5-hydroxymethylfurfural into 2,5-bis (hydroxymethyl) furan over tunable Zr-based bimetallic catalysts. *Catal. Sci. Technol.* **2018**, *8*, 4474–4484. [CrossRef]

31. Thananatthanachon, T.; Rauchfuss, T.B. Efficient route to hydroxymethylfurans from sugars via transfer hydrogenation. *ChemSusChem* **2010**, *3*, 1139–1141. [CrossRef] [PubMed]

32. Thananatthanachon, T.; Rauchfuss, T.B. Efficient production of the liquid fuel 2,5-dimethylfuran from fructose using formic acid as a reagent. *Angew. Chem.* **2010**, *122*, 6766–6768. [CrossRef]

33. Wu, W.; Zhao, W.; Fang, C.; Wang, Z.; Yang, T.; Li, H.; Yang, S. Quantitative hydrogenation of furfural to furfuryl alcohol with recyclable KF and hydrosilane at room temperature in minutes. *Catal. Commun.* **2018**, *105*, 6–10. [CrossRef]

34. Roy, A.K. A review of recent progress in catalyzed homogeneous hydrosilation (hydrosilylation). *Adv. Organomet. Chem.* **2007**, *55*. [CrossRef]

35. Oestreich, M.; Hermeke, J.; Mohr, J. A unified survey of Si-H and H-H bond activation catalysed by electron-deficient boranes. *Chem. Soc. Rev.* **2015**, *44*, 2202–2220. [CrossRef] [PubMed]

36. Muthukumar, A.; Mamillapalli, N.C.; Sekar, G. Potassium phosphate-catalyzed chemoselective reduction of α-keto amides: Route to synthesize passerini adducts and 3-phenyloxindoles. *Adv. Synth. Catal.* **2016**, *358*, 643–652. [CrossRef]

37. Xie, W.; Zhao, M.; Cui, C. Cesium carbonate-catalyzed reduction of amides with hydrosilanes. *Organometallics* **2013**, *32*, 7440–7444. [CrossRef]

38. Li, H.; Zhao, W.; Dai, W.; Long, J.; Watanabe, M.; Meier, S.; Saravanamurugan, S.; Yang, S.; Riisager, A. Noble metal-free upgrading of multi-unsaturated biomass derivatives at room temperature: Silyl species enable reactivity. *Green Chem.* **2018**. [CrossRef]

39. Li, H.; Zhao, W.; Saravanamurugan, S.; Dai, W.; He, J.; Meier, S.; Yang, S.; Riisager, A. Control of selectivity in hydrosilane-promoted heterogeneous palladium-catalysed reduction of furfural and aromatic carboxides. *Commun. Chem.* **2018**, *1*, 32. [CrossRef]

40. Zhao, W.; Yang, T.; Li, H.; Wu, W.; Wang, Z.; Fang, C.; Saravanamurugan, S.; Yang, S. Highly recyclable fluoride for enhanced cascade hydrosilylation-cyclization of levulinates to γ-valerolactone at low temperatures. *ACS Sustain. Chem. Eng.* **2017**, *5*, 9640–9644. [CrossRef]

41. Ding, K.; Zannat, F.; Morris, J.C.; Brennessel, W.W.; Holland, P.L. Coordination of *N*-methylpyrrolidone to iron(II). *J. Organomet. Chem.* **2009**, *694*, 4204–4208. [CrossRef]

42. Bisz, E.; Szostak, M. 2-Methyltetrahydrofuran: A green solvent for iron-catalyzed cross-coupling reactions. *ChemSusChem* **2018**, *11*, 1290–1294. [CrossRef] [PubMed]

43. Yan, K.; Wu, G.; Lafleur, T.; Jarvis, C. Production, properties and catalytic hydrogenation of furfural to fuel additives and value-added chemicals. *Renew. Sustain. Energy Rev.* **2014**, *38*, 663–676. [CrossRef]

44. Sheldon, R.A. Fundamentals of green chemistry: Efficiency in reaction design. *Chem. Soc. Rev.* **2012**, *41*, 1437–1451. [CrossRef] [PubMed]

45. Byrne, F.P.; Jin, S.; Paggiola, G.; Petchey, T.H.M.; Clark, J.H.; Farmer, T.; Hunt, A.J.; McElroy, C.R.; Sherwood, J. Tools and techniques for solvent selection: Green solvent selection guides. *Sustain. Chem. Process* **2016**, *4*, 7. [CrossRef]

46. Alder, C.M.; Hayler, J.D.; Henderson, R.K.; Redman, A.M.; Shukla, L.; Shuster, L.E.; Sneddon, H.F. Updating and further expanding GSK's solvent sustainability guide. *Green Chem.* **2016**, *18*, 3879–3890. [CrossRef]

47. Kumar, G.; Muthukumar, M.A.; Sekar, G. A Mild and chemoselective hydrosilylation of α-keto amides using Cs$_2$CO$_3$/PMHS/2-MeTHF System. *J. Org. Chem.* **2017**, *33*, 4883–4890. [CrossRef]

48. Fulajtárova, K.; Soták, T.; Hronec, M.; Vávra, I.; Dobročka, E.; Omastová, M. Aqueous phase hydrogenation of furfural to furfuryl alcohol over Pd-Cu catalysts. *Appl. Catal. A* **2015**, *502*, 78–85. [CrossRef]

49. Li, H.; Smith, R.L. Solvents take control. *Nat. Catal.* **2018**, *1*, 176–177. [CrossRef]

50. Bornschein, C.; Werkmeister, S.; Junge, K.; Beller, M. TBAF-catalyzed hydrosilylation for the reduction of aromatic nitriles. *New J. Chem.* **2013**, *37*, 2061–2065. [CrossRef]

51. Zhao, W.; Wu, W.; Li, H.; Fang, C.; Yang, T.; Wang, Z.; Yang, S. Quantitative synthesis of 2,5-bis (hydroxymethyl)furan from biomass-derived 5-hydroxymethylfurfural and sugars over reusable solid catalysts at low temperatures. *Fuel* **2018**, *217*, 365–369. [CrossRef]

52. Kira, M.; Sato, K.; Sakurai, H. Reduction of carbonyl compounds with pentacoordinate hydridosilicates. *J. Org. Chem.* **1987**, *52*, 948–949. [CrossRef]

53. Ryoichi, K.; Masatoshi, T.; Yoshihiko, I. Reduction of amides to amines via catalytic hydrosilylation by a rhodium complex. *Tetrahedron Lett.* **1998**, *39*, 1017–1020. [CrossRef]

54. Rendler, S.; Oestreich, M. Hypervalent Silicon as a reactive site in selective bond-forming processes. *Synthesis* **2005**, *17*, 1727–1747. [CrossRef]

© 2018 by the authors. Licensee MDPI, Basel, Switzerland. This article is an open access article distributed under the terms and conditions of the Creative Commons Attribution (CC BY) license (http://creativecommons.org/licenses/by/4.0/).

Review

Clostridium sp. as Bio-Catalyst for Fuels and Chemicals Production in a Biorefinery Context

Vanessa Liberato [1], Carolina Benevenuti [1], Fabiana Coelho [1], Alanna Botelho [1], Priscilla Amaral [1], Nei Pereira, Jr. [1] and Tatiana Ferreira [2,*]

[1] Department of Biochemical Engineering, School of Chemistry, Universidade Federal do Rio de Janeiro, Rio de Janeiro 21941-909, RJ, Brazil; nessa.saab@gmail.com (V.L.); carolbenevenuti@hotmail.com (C.B.); fabicoelhoquimica@gmail.com (F.C.); lanna.mbotelho@gmail.com (A.B.); pamaral@eq.ufrj.br (P.A.); nei@eq.ufrj.br (N.P.J.)

[2] Department of Organic Processes, School of Chemistry, Universidade Federal do Rio de Janeiro, Rio de Janeiro 21941-909, RJ, Brazil

* Correspondence: tatiana@eq.ufrj.br; Tel.: +55-21-3938-7623

Received: 24 October 2019; Accepted: 12 November 2019; Published: 15 November 2019

Abstract: *Clostridium* sp. is a genus of anaerobic bacteria capable of metabolizing several substrates (monoglycerides, diglycerides, glycerol, carbon monoxide, cellulose, and more), into valuable products. Biofuels, such as ethanol and butanol, and several chemicals, such as acetone, 1,3-propanediol, and butyric acid, can be produced by these organisms through fermentation processes. Among the most well-known species, *Clostridium carboxidivorans*, *C. ragsdalei*, and *C. ljungdahlii* can be highlighted for their ability to use gaseous feedstocks (as syngas), obtained from the gasification or pyrolysis of waste material, to produce ethanol and butanol. *C. beijerinckii* is an important species for the production of isopropanol and butanol, with the advantage of using hydrolysate lignocellulosic material, which is produced in large amounts by first-generation ethanol industries. High yields of 1,3 propanediol by *C. butyricum* are reported with the use of another by-product from fuel industries, glycerol. In this context, several *Clostridium* wild species are good candidates to be used as biocatalysts in biochemical or hybrid processes. In this review, literature data showing the technical viability of these processes are presented, evidencing the opportunity to investigate them in a biorefinery context.

Keywords: *Clostridium* sp.; fuels; ethanol; butanol; acetone; 1,3-propanediol; biorefinery; biomass

---

1. Introduction

In 2018, Global Footprint Network stated that natural resources were being used 1.7 times faster than the Earth's capacity for regeneration [1]. Water, soil, and air pollution generated by anthropogenic actions have caused climate change, resource depletion, global warming, water crises, and the increase of natural disasters in recent years [2]. Many researchers and initiatives are committed to search and develop new techniques, feedstocks, and processes towards a sustainable economy that will be less dependent on oil as a source of energy and chemical production [3]. One approach is to progressively change our primary source of fuels and chemicals, decreasing the dependency on oil and oil products due to oil's environmental impact, depletion possibility, and market price oscillations caused by oil and financial crises. The Sustainable Development Goals adopted by global leaders at the 2015 United Nations Summit set the path for 17 goals to be achieved by 2030, such as: affordable and clean energy (SDG7); industries, innovation and infrastructure (SDG9); responsible production and consumption (SDG12); and climate action (SDG13) [4].

Many feedstocks can be used to produce renewable and sustainable fuels and chemicals, especially residual organic materials, industrial waste-streams, and by-products. The use of glycerol, molasses, corn steep liquor, energy crops, grass silage, cattle slurry, harvest residues, and food processing waste can reduce operational costs (associated with feedstock acquisition) and environmental impact due to improper disposal of residue [3,5–10]. These are renewable feedstocks that can be transformed into valuable biofuels (e.g., biobutanol and bioethanol) and biochemicals (lactic acid, butyric acid, and succinic acid), achieving a balance between environmental sustainability and economic growth in a process named biorefining [10]. Biorefining is defined as "the sustainable processing of biomass into a spectrum of marketable food and feed ingredients, bio-based products (chemicals, materials) and bioenergy (biofuels, power and/or heat)" [11]. In a biorefinery, components of biomass are organized as building blocks (e.g., cellulose, hemicellulose, lignin, and proteins) and the biorefinery concept encompasses the whole processing of biomass, considering upstream, midstream, and downstream processing [3,12]. The main challenge is the feasible production of biofuels and biochemicals in a cost effective, efficient, and sustainable way [3]. Although using renewable feedstock, biorefinery sustainability should also consider others aspects, such as land use for biomass production, soil, and air quality during processing, and environmental and economic impacts [13].

Considering biochemical conversion processes, microbial cells are a central aspect of biomass conversion, functioning as biocatalysts that convert liquid, solid, or gaseous substrates into alcohols, acids, enzymes, and other products through a series of metabolic reactions. In particular, *Clostridium* species have been involved in many important fermentative processes, such as the production of butanol, acetone, ethanol, acetic acid, lactic acid, succinic acid, 1,3-propanediol, and more [14–17]. Therefore, the current study reviews fundamental aspects of important *Clostridium* species and their biochemical pathways to highlight major theoretical concepts that are present in many processes currently discussed in literature. This work also highlights important biomolecules and feedstocks that have a great potential in the biorefinery concept and would play an important role towards a more sustainable bioeconomy.

2. Clostridium sp.

Prazmowski proposed the genus *Clostridium* through the type species *Clostridium butyricum* in 1880. This genus then became the general category for anaerobic, spore forming and Gram-positive microorganisms. There are about 228 species and subspecies currently known, some of which highly heterogeneous, that present several phenotypes [18]. Among the species of the genus there are those that synthesize quinones and cytochromes, acidophiles, thermophiles, and psychrophiles [19]. According to Lawson and Rainey there are phylogenetic and phenotypic incoherencies in the genus *Clostridium*. The researchers defend the restriction of the genus *Clostridium* to *Clostridium* cluster I as *Clostridium* sensu stricto [18].

2.1. General Aspects

Generally, bacteria of the *Clostridium* genus are rod-shaped with Gram-positive staining. Single cell diameters may vary from 0.3 to 2.0 µm and lengths from 1.5 to 20 µm; they are usually arranged in pairs or in short chains with rounded or pointed ends. With the presence or absence of flagella, a single cell may be motile, usually with peritrichous flagella, or immotile. They do not reduce sulfates and are, generally, catalase negative [18]. Most species have sporulation capacity, which can be directly affected by presence of oxygen, because they are anaerobic. However, species of *Clostridium* genus have different oxygen tolerance capacities [19]. Most species are chemotrophic; some are chemoautotrophic or chemolithotrophic. They may be saccharolytic, proteolytic, none, or both. Metabolically they are very diverse, with optimum temperature between 10 and 65 °C. They are distributed in the environment. Many species produce potent exotoxins, and some species are pathogenic to animals, both by infection of wounds and by absorption of toxins [20].

When exposed to nutrient-depleting conditions, alternative metabolic pathways of clostridia are activated. Sporulation starts at the beginning of the stationary phase as an adaptation response and can affect end products related to energy metabolism, since it involves several metabolic pathways associated to biochemical, morphological, and physiological changes. Although the lack of nutrients commonly triggers sporulation, a source of energy is still needed to fuel macromolecular synthesis. Complete depletion of all or even a single nutrient can lead to cell death [21].

2.2. Workhorse Species

Many *Clostridium* species produce a wide variety of organic acids and solvents of industrial interest, such as ethanol, butanol, 1,3-propanediol, and acetone, among others. The most important species for a biorefinery context will be discussed in this review. Table 1 shows the most important species and their main products.

Table 1. Main products and production data for *Clostridium* species of interest to biorefinery.

Species	Main Products	P_{Max} [1] (g/L)	Q_{Max} [1] (g/L.h)	Substrate or Feedstock	V [1] (L)	Time (h) [1]	Ref. [1]
C. carboxidivorans	Ethanol Butyric acid Butanol	5.55 1.43 2.66	NA [2]	100% CO	1.2	500	[22]
	Ethanol	23.93	0.143	20% CO, 5% H_2, 15% CO_2, 60% N_2	8	360	[23]
C. ragsdalei	Ethanol	13.2 [3]	NA	40% CO, 30% H_2, 30% CO_2	2.4	290	[24]
		19	NA	38% CO, 28.5% H_2, 28.5% CO_2, 5% N_2	2.4	470	[25]
C. ljungdahlii	Ethanol	20.7 [4]	0.374	60% CO, 35% H_2, 5% CO_2	4	2014	[26]
		48	NA	55% CO, 20% H_2, 10% CO_2, 15% N_2	1	560	[27]
C. kluyveri	n-caproic acid	8.42	0.07	Ethanol and acetate (10:1)	0.1	120	[28]
C. beijerinckii	Butanol	34.77 [4]	0.48	Corn stover hydrolysate	1.25	72	[29]
		11.92	0.17	Corn cob hydrolysate	50	72	[30]
		11.77	0.10	Sugarcane juice	1.5	100	[31]
	Isopropanol	3.41	NA	Glucose	0.06	48	[32]
C. saccharobutylicum	Butanol ABE [5]	12.8 19.9	0.257 -	Corn stover hydrolysate	5	48	[33]
C. saccharo- perbutylacetonicum	Butanol Acetone ABE	64.6 [6] 30.6 97.3	0.673 0.319 1.01	Glucose	0.5	96	[34]
C. thermocellum	Ethanol	13.66	NA	Cellulose	0.02	120	[35]
C. phytofermentans	Ethanol	7.0	0.03	AFEX™ -treated corn stover [7]	0.3	250	[36]
C. acetobutylicum	Butanol	44.6 [8]	2.13	Glucose	4	88	[37]
	Ethanol	33 [9]	0.47	Glucose	0.2	185	[38]
	Acetone	7.59 [10]	0.13	Glucose	5	54	[39]
C. butyricum	1,3-propanediol	93.7	3.30	Pure glycerol	1	32	[40]
C. cellulolyticum	Ethanol	2.5 [10]	0.007	Crystalline AVICEL [11] Cellulose	0.05	336	[41]

Table 1. *Cont.*

Species	Main Products	P_{Max} [1] (g/L)	Q_{Max} [1] (g/L.h)	Substrate or Feedstock	V [1] (L)	Time (h) [1]	Ref. [1]
C. cellulovorans	Butanol	3.37	0.046	Corn cob	2	72	[42]
C. pasteurianum	Butanol	17.8	7.8 [12]	Pure glycerol	0.09	48	[43]
	1,3-propanediol	9.5	0.2		0.09	48	
C. perfringens	1,3-propanediol	40	2.0	Pure glycerol	3	48	[44]

[1] P_{Max}: maximum production reported; Q_{Max}: maximum productivity for the maximum production reported; V: production volume; time: fermentation time; Ref: reference; [2] NA: not available data in reference; [3] using PLBC (Poultry litter biochar) medium with MES (4-morpholineethanesulfonic acid); [4] this production was achieved with integrated product recovery method; [5] ABE: sum of acetone, butanol, and ethanol concentrations; [6] butanol, acetone, and ABE production with immobilized cells; [7] ammonia fiber expansion (AFEX™) is an alkaline pretreatment using ammonia as a catalyst; [8] this production was achieved with a continuous process with ex situ butanol recovery and a mutant strain obtained by chemical mutagenesis; [9] mutant strain; [10] genetic engineered strain; [11] AVICEL: is a purified, partially depolymerized alphacellulose made by acid hydrolysis of specialty wood pulp; [12] productivity obtained in 1 L fermenter with 0.4 L of medium for 710 h, with cell recycle.

2.2.1. *Clostridium carboxidivorans*

C. carboxidivorans was first discovered in the sediment of an agricultural decantation pond at the State University of Oklahoma, USA, after enriching the environment with carbon monoxide (CO), as a strategy to identify bacteria species capable of assimilating CO as a substrate [14].

It is an aerotolerant Gram-positive anaerobic bacterium with rod-shape cells that may occur in pairs or isolated. It has motility and a rare sporulation mechanism. The optimum temperature for growth is 38 °C. This species presents autotrophic growth when in the presence of carbon dioxide (CO_2) and CO, but is also able to grow with several organic sources, such as fructose, galactose, glucose, and mannose [45]. It produces acetic acid, ethanol, butyrate, and butanol as the final metabolic products [14,45]. From syngas (mixture of mainly CO, CO_2, and H_2), almost 24 g/L of ethanol production have been reported [23] (Table 1). The surface of *C. carboxidivorans* cells are hydrophilic but also capable of being attracted by hydrophobic molecules, liquids, and surfaces, and capable of interacting with them when immersed in water [46].

In the genomic analysis of *C. carboxidivorans*, to understand biofuel production pathways genetic determinants were observed for CO use and production of acetate, ethanol, and butanol. One example is the presence of a gene encoding a dehydrogenase characteristic of the ABE (acetone-butanol-ethanol) fermentation, previously described for *C. acetobutylicum* and *C. beijerincki*, responsible for converting acetyl-CoA to acetaldehyde and ethanol, and acetyl-CoA to butyryl-CoA and butanol via butyraldehyde. However, genes present in other *Clostridium* strains that encode proteins for acetone production were not found in *C. carboxidivorans* [47].

2.2.2. *Clostridium ragsdalei*

C. ragsdalei (ATCC PTA-7826) "P11" was first isolated from duck pond sediment, in a laboratory in Norman, OK, USA. This species transforms waste gases (e.g., syngas and refinery wastes) into useful products [48]. It became of the major species capable of fermenting synthesis gas. *C. ragsdalei* uses the Wood–Ljungdahl pathway to assimilate CO, CO_2, and H_2 to produce acetate [49–51]. It is poorly inhibited by acetic acid compared to other syngas-fermenting strains, such as *Clostridium coskatii* [52]. This microorganism has been studied for acid and solvent production from synthesis gas. Some residues, such as corn steep liquor (CSL), have been used as a source of nutrients during fermentation in order to reduce production costs. The growth of *C. ragsdalei* in CSL depends on trace metals, NH_4, and reducing agents, due to its low resistance to oxygen [51].

Among the main products of its metabolism is ethanol, with the highest concentration of 19 g/L reported after 470 h of syngas fermentation [25] (Table 1). The conversion of the acetate produced in the acidogenic phase into ethanol in the solventogenic phase depends on the pH of the medium, because solventogenesis mainly begins when the PH value is reduced [53].

2.2.3. *Clostridium ljungdhali*

C. ljungdahlii ATCC 49,587 was isolated from chicken yard waste [54]. Rod-shaped cells with motility and rare formation of spores are some of the physical characteristics of this species. The pH and temperature range for cell growth are 4.0–7.0 and 30–40 °C, respectively. However, optimum growth conditions are 37 °C and pH 6.0.

Autotrophic or chemoorganotrophic growth are observed for this species, with H_2-CO_2 or CO and formate, ethanol, pyruvate, fumarate, erythrose, threose, arabinose, xylose, glucose, and fructose as carbon sources. It is incapable of assimilating methanol, ferulate, trimethoxyvenzoate, lactate, glycerol, citrate, succinate, galactose, mannose, sorbitol, sucrose, lactose, maltose, or starch. The major metabolic end product is acetic acid in the acidogenic phase, with trace amounts of ethanol in the solventogenic phase [19,54]. However, some studies have reported 20 g/L [26] and 48 g/L [27] of ethanol during syngas fermentation using this species (Table 1).

2.2.4. *Clostridium kluyveri*

C. kluyveri is distinguished by both its nutritional requirements and its ability to ferment ethanol to caproate [28]. Cells are motile, with peritrichous flagella, and generally occur singly, and occasionally in pairs or chains. They are weakly Gram-positive, but quickly become colored with Gram-negative staining. The endospores are oval, terminal, or subterminal and swell the cell. Ultra-structural studies demonstrated a five-layer cell wall and a three-layer plasma membrane. Growth is slow and occurs between 19 and 37 °C, with optimum temperature of 35 °C.

The strains require ethanol, CO_2, or sodium carbonate, and a high concentration of yeast autolysate or acetate, propionate, or butyrate for growth. In the presence of CO_2 or carbonate and acetate or propionate, ethanol is converted into butyrate, caproate, and H_2. H_2 is formed and small amounts of butyrate can be detected [55].

2.2.5. *Clostridium beijerinckii*

C. beijerinckii cells present a straight rod morphology and rounded ends. They are motile with peritrichous flagella, occurring singly, in pairs, or in short chains. They are Gram-positive, becoming Gram-negative in the older cultures. The endospores are oval, eccentric to the subterminal, and swell to the cell, without exosporium or appendages. The ideal growth temperature is 37 °C, but they also grow well at 30 °C. The growth is stimulated by a fermentable carbohydrate and is inhibited by 6.5% NaCl or 20% bile. Yeast extract supplies nutritional requirements of the cells [19,56].

C. beijerinckii is one of the major isopropanol producing bacteria with several known strains (ATCC 25752, CIP 104308, DSM 791, JCM 1390, LMG 5716, NCTC 13035) [38].It is also known as a butanol producer [30,31]. Aiming at biorefinery interest, this species uses lignocellulosic hydrolysate materials and produces butanol and acetone or isopropanol. High butanol production from sugarcane juice [30] or corn cob hydrolysate [31] has been reported (Table 1).This species produces alcohol dehydrogenase but only some strains specifically produce isopropanol dehydrogenase, responsible for the conversion of acetone to isopropanol via isopropanol-butanol-ethanol (IBE) fermentation [19,32]. It has been reported that Zn^{2+}, Ca^{2+}, and Fe^{2+} improve butanol and isopropanol production [32]. They also increase the specific activity of hydrogenase, stimulating H_2 production [57].

RNA transcription analysis has shown that the metabolic pathway used by *C. beijerinckii* to produce 1,2-propanediol, proprionate, and propanol is similar to the pathway used by *C. phytofermentans*. When L-rhamnose is used as a carbon source, a bacterial microcompartment (BMC) is formed, which is directly involved in propanol and proprionate formation. However, in the presence of glucose there is no formation of BMC and, consequently, propanol, and proprionate are not produced [56,58].

2.2.6. *Clostridium saccharobutylicum*

Morphologically *C. saccharobutylicum* cells are characterized by a rod shape with rounded ends of 3.8–10 µm in length and occur individually, in pairs, or short chains. Their cells are motile with peritrichous flagella. A Gram-positive cell wall is characteristic of this species, but older cultures stain with Gram-negative. The spores have an oval shape, with a terminal location inside the cell. The ideal temperature and pH ranges for growth and solvent production are 30–34 °C and pH 6.2–7.0, respectively [19].

C. saccharobutirycum produces acetone, butanol, ethanol, CO_2, H_2, and acetic and butyric acids as fermentation products [19]. It is mainly used to ferment lignocellulosic hydrolyzed materials, such as corn stover, to produce butanol in ABE (acetone-butanol-ethanol) fermentation (Table 1) [33,59].

2.2.7. *Clostridium saccharoperbutylacetonicum*

Morphologically, *C. saccharoperbutylacetonicum* cells are straight rods, which are found individually or in pairs. Cells are mobile with peritrichous flagella. They stain Gram-positive as young cells, whereas Gram-negative cells are observed in older cultures. They form oval spores and grow well in temperatures between 25 and 35 °C and a pH range of 5.6–6.7 [19].

C. saccharoperbutylacetonicum produces acetone, butanol, ethanol, CO_2, H_2, and acetic and butyric acid as final fermentation products [19,56]. It is considered outstanding as a model species for industrial application of acetone-butanol-ethanol (ABE) fermentation, producing 73%–85% of butanol and sporulates with low frequency [56]. Extractive fermentation with *C. saccharoperbutylacetonicum* immobilized cells resulted in high total butanol concentration (64.6 g/L) (Table 1) [34].

The use of genetic engineering techniques contributes to a better understanding of genetic information. Therefore, the strain *C. saccharoperbutylacetonicum* N1-4C demonstrated better plasmid stability and a slight increase in ABE production, probably due to the improvement in the efficiency of acids produced assimilation during the acidogenic phase [60]. The silencing of the *pta* and the *buk* genes, responsible for the expression of phosphotransacetylase and butyrate kinase, respectively, did not eliminate the formation of acetyl or butyrate. This indicated that *C. saccharoperbutylacetonicum* has additional routes for acid production [61].

2.2.8. *Clostridium thermocellum*

C. thermocellum is a thermophilic bacterium that grows between 60 and 64 °C. The culture medium influences the motility of this species. They are Gram-negative but possess a Gram-positive type cell wall without lipopolysaccharide. They are morphologically characterized by being straight or slightly curved cells, which can be found individually or in pairs. The spores are oval and terminal, and swell the cell [62].

C. thermocellum is capable of converting cellulosic waste into ethanol and H_2 [62]. It can also produce lactate and acetate as main products [63]. Growth in most media requires the presence of cellulose, cellobiose, or hemicelluloses [62]. A genetically modified strain produced 5 g/L of ethanol from microcrystalline cellulose [63] and, using a statistical approach, Balusu et al. [35] reported production of more than 13 g/L of ethanol from cellulosic biomass with a wild-type strain (Table 1).

A thermodynamic analysis of metabolomic data of *C. thermocellun* showed that bottlenecks of ethanol production from cellobiose are distributed across five reactions under high ethanol levels and identified best intervention strategies for ethanol production [64].

2.2.9. *Clostridium phytofermentans*

C. phytofermentans is morphologically characterized as straight sticks that measure 0.5–0.8 × 3–15 µm, able to occur alone or in pairs. They are stained Gram-negative despite having a Gram-positive wall and its motility is provided by one or two flagella. Sporulation for this species is terminal with round spores. It is an obligate anaerobic bacterium with wide temperature and pH ranges for growth: from 15 to 42 °C and pH of 6.0 to 9.0. Optimum temperature and pH are 37 °C and 8.0, respectively [65].

C. phytofermentans uses a wide variety of plant polysaccharides as a substrate, such as cellulose. It produces ethanol, acetate, CO_2, and H_2 as the main products [65]. Due to its capacity to convert cellulosic biomass to ethanol, *C. phytofermentans* seems promising in the biorefinery context, using lignocellulosic material as a source of substrate, with references reporting 7 g/L of ethanol from corn stover (Table 1) [36]. The use of algae biomass from sewage without pre-treatment by *C. phytofermentans* yielded 0.19 g of ethanol per g of biomass using 2% *w/v* of algae [66].

2.2.10. *Clostridium acetobutylicum*

A strictly anaerobic, Gram-positive and spore-forming bacterium, *C. acetobutylicum* was first isolated by Chaim Weizmann in 1916. Weizmann discovered that this species produces ethanol and butanol from starch. This isolated strain was used in the production of acetone during World War I as a precursor to cordite production [67]. The *C. acetobutylicum* strain responsible for ABE fermentation in the ratio 7:2:1 was only isolated in 1980 [68].

C. acetobutylicum has a high amylase activity, being able to metabolize a variety of carbohydrates, such as starch, glucose, and sucrose [69,70]. Recent studies point out the potential use of *C. acetobutylicum* for the conversion of algal hydrolysates, algae biomass, and galactose into value-added bioproducts [71]. However, *C. acetobutylicum* is less effective for complex carbon sources, i.e., lignocellulosic hydrolysates, which may contain toxic substances derived from the dehydration of sugars and aromatics from lignin that are generated during biomass pretreatment [70].

From the carbohydrate sources available in the culture medium, *C. acetobutylicum* synthesizes acetic acid and butyric acid, which are subsequently converted into organic solvents, such as acetone, butanol, and ethanol [21]. Among the known microorganisms capable of producing butanol, *C. acetobutylicum* is the one with the highest production and, therefore, one of the most commonly used species for butanol production [68]. In a continuous process with ex situ butanol recovery and a mutant strain obtained by chemical mutagenesis, 44.6 g/L of butanol was achieved (Table 1) [37].

2.2.11. *Clostridium butyricum*

An anaerobic and saccharolytic bacterium, fermentation of *C. butyricum* is similar to that of several species of *Clostridium*. *C. butyricum* is found in the intestinal tract of humans and animals, and this bacterium can be used as an animal feed additive because of its probiotic properties. Butyric acid, one of the short chain fatty acids produced in its fermentation, has anti-inflammatory effects and aids in the proliferation of intestinal mucosal cells [72]. *C. butyricum* is also frequently found in soil converting organic matter and helping soil mineralization, having a relevant role in ecology [73].

The metabolic products of *C. butyricum* are short chain fatty acids, acetic acid, butyric acid, and propionic acid. H_2, butanol, and 1,3-propanediol can also be obtained through fermentation of sugar and glycerol [72]. *C. butyricum* is considered one of the most important 1,3-propanediol producers. This is due to the high conversion efficiency, the soft cultivation system, and the non-sterile process that contribute to competitive and cheaper production [74]. More than 90 g/L of 1,3-propanediol was produced by this species with pure glycerol as the carbon source [40].

2.2.12. *Clostridium cellulolyticum*

C. cellulolyticum cells are mesophilic anaerobic bacteria capable of growing on crystalline cellulose [75,76]. Its name originates from the Greek words cellulosum (cellulose) and lyticus (dissolution). They are rod-shaped with peritrichous flagella [76]. *C. cellulolyticum* is referred to as mesophilic cellulolytic clostridium, which secretes cellulosome, a multi-enzyme complex capable of converting cellulose into metabolites of interest, such as ethanol, acetate, lactate, and H_2. The cellulosome converts cellulose to cellobiose and cellodextrins, being then incorporated and metabolized, generating other metabolites [75,76]. A genetically engineered strain was able to produce 8.5 times more ethanol from crystalline cellulose than wild-type cells (Table 1) [41].

2.2.13. *Clostridium cellulovorans*

C. cellulovorans is a cellulolytic mesophilic anaerobic bacterium capable of degrading pectin and hemicellulose, in addition to cellulose [77]. It had its genome sequenced in 2010 after being isolated from wood chips [78]. Its main product is butyric acid and among the by-products are acetic acid, lactic acid, formic acid, ethanol, H_2, and CO_2. It degrades cellulose efficiently due to the expressive number of cellulosomal genes. It is capable of using a wide range of carbon sources besides cellulose, such as lactose, glucose, galactose, maltose, sucrose, cellobiose, pectin, and xylan [78].

Due to the variety of carbon sources and the fact that it produces butyric acid, *C. cellulovorans* is a cellulolytic organism with great potential to produce butanol, requiring only one more stage of enzymatic activity [78]. The fermentation process engineering of *C. cellulovorans* enabled a high butanol production directly from corn cob (Table 1) [42].

2.2.14. *Clostridium pasteurianum*

An acidogenic saccharolytic, *Clostridium* is capable of fixing molecular nitrogen, dispensing other nitrogen sources [67]. It presents Gram-positive cells only in very young cultures. It has robust growth even in simple media and in non-sterile conditions.

It is capable of producing chemicals, such as 1,3-propanediol and n-butanol, from renewable raw materials, such as glycerol derived from biodiesel and glucose from hydrolysis of biomass, respectively [71,79]. Ethanol, acetic acid, butyric acid, lactic acid, H_2, and CO_2 are also among the products of its fermentation. Although necessary to ensure intracellular redox balance, the organic acid production pathway competes with that of the products of interest. There is also selectivity in the proportion of the products generated, between 1,3-propanediol and n-butanol, for example, which is strongly influenced by factors such as nutrient composition and growing conditions. Other factors that may influence this ratio are culture pH, inoculum condition, and medium supplementation with yeast extract, ammonia, organic acids, phosphate, and iron [79]. An increase in the production of butanol by *C. pasteurianum* DSM 525 was observed when lactic acid was used with glycerol. Lactic acid can buffer the medium, stabilizing the pH between 5.7 and 6.1 [80].

2.2.15. *Clostridium perfringens*

C. perfringens was first isolated by Welch and Nuttall in 1892 from a human corpse in an advanced state of decomposition [20]. It is widely distributed in soils and the food tracts of almost all warm-blooded animals. Generally, it is found to be an invader of the food tract after death. Caution is required in the isolation of this *Clostridium* after death. This agent is found in the gaseous gangrene of man and animals, or associated with other anaerobes, in this process [20]. *C. perfringens* is a straight and thick rod, usually individual, in pairs, or, rarely, in strings. It produces oval and small spores, which are not formed in very acidic medium. It is classified as a Gram-positive bacteria, but becomes Gram variable with subcultures [20].

It produces acid and gas from glucose, levulose, galactose, mannose, maltose, lactose, sucrose, xylose, trehalose, raffinose, starch, glycogen, and inositol. Fermentation products include acetic acid, butyric acid, and 1,3-propanediol [20]. In fed-batch fermentation, an isolated new strain (*C. perfringens* GYL) showed a maximum productivity of 2.0 g/(L·h), and produced 40.0 g/L 1,3-propanediol, with a high yield from glycerol (Table 1) [44].

2.2.16. *Clostridium tyrobutyricum*

Anaerobic and spore-forming, *C. tyrobutyricum* [81] is widely used in the production of butyric acid due to its high production in simple media [82]. It is also the biggest cause of problems in the dairy industry thanks to this ability of converting lactate to butyrate in the presence of acetate. During fermentation, lactate is the electron donor and the acetate is the electron acceptor, generating butyrate. Growth of saccharolytic clostridia is observed in most environments where the oxygen tension is suitably low [81,82].

2.2.17. Other Species

Clostridium drakei was initially described as *C. scalatogenes*, but 16S RNA analysis determined that it was a new species which was named after a researcher, Drake, in recognition of his contributions to the physiology and ecology of acetogens. Spores are located in the terminal portion of the cell. They are mandatory anaerobes, with an optimal growth temperature of 30 to 37 °C and optimal pH of 5.5–7.5. *C. drakei* grows autotrophically with H_2/CO_2 or CO, and also chemiotrophically. The main end products are acetic acid, ethanol, butyrate, and butanol [14].

Clostridium difficile was initially isolated from newborn feces and referred to as *Bacillus difficilis*. The name refers to the difficulty found in isolating and studying this species. *C. difficile* produces isocaproic and valeric acid. The cells are shaped like rods that join in aligned groups by the tips of two to six cells. They have motility and the presence of peritrichous. The sporulation capacity is observed when induced by culturing in specific medium, and the growth temperature ranges from 25 to 45 °C. Acetic, butyric, lactic acid, and butyrate are the main products obtained by this species [19].

Clostridium autoethanogenum was isolated from rabbit feces. It is strictly anaerobic, with motility and forms sporulation. It is able to produce ethanol, acetate, and CO_2 from CO as a carbon source. The optimal pH and temperature for growth is 5.8–6.0 and 37 °C, respectively [83]. When cultured with H_2, the assimilation of CO by *C. autoethanogenum* increases and the metabolic pathway to ethanol production is stimulated, reducing CO_2 generation [84].

2.3. Fermentation and Biochemical Pathways

Organic acids and alcohols can be obtained by anaerobic fermentation performed by *Clostridium* from organic or inorganic carbon sources. The fermentation processes of the different species are divided into three main fermentations: ABE (acetone-butanol-ethanol), IBE (isopropanol-butanol-ethanol), and HBE (hexanol-butanol-ethanol). The latter was recently proposed by a research group that has been studying the *C. carboxidivorans* P7 strain [85,86].

Independently of the end products, each strain can use different metabolic pathways based on the available carbon source and its genetic information. Different carbohydrates, for example, are fermented through different pathways by clostridia. In the case of hexoses, their metabolism follows the Embden–Meyerhof–Parnas (EMP) pathway, whereas the metabolism of pentoses takes place through the pentose phosphate (PP) pathway [86]. On the other hand, to assimilate CO, CO_2, and H_2 for HBE fermentation, *C. carboxidivorans* uses the Wood–Ljungdahl pathway [85,86]. These fermentation processes and biochemical pathways will be discussed in this section.

2.3.1. Acetone-Butanol-Ethanol (ABE) Fermentation

ABE fermentation was used on a large scale during the First World War. Acetone was produced by *C. acetobutylicum* to be used by England for warlike applications. However, the development of the petrochemical industry after the Second World War, decreased the interest in ABE fermentation to obtain those products. Recently, environmental problems related to the use of fossil fuels and economic reasons have again increased the interest for ABE fermentation [87,88].

ABE fermentation is composed of two phases: acidogenic and solventogenic. The first is characterized by organic acid production (butyric and acetic), and the second by assimilation of these acids as a carbon source for subsequent transformation into acetone, butanol, and ethanol [56,87]. Generally, molar the acetone:butanol:ethanol ratio is 6:3:1, but it can vary from species to species [56,87]. The most common and best known solventogenic clostridial strains for commercial butanol fermentation are *C. acetobutylicum*, *C. beijerinckii*, *C. saccharobutylicum*, and *C. saccharoperbutylacetonicum* [86].

The use of lignocellulosic biomass in ABE fermentation for butanol production has been studied by several authors as a solution for the environmental issues mentioned. However, carbon catabolic repression (CCR) reduces the efficiency of the process since glucose, usually produced after a pre-treatment of the biomass, suppresses the use of other sugars, such as xylose and arabinose, which are also produced [89,90]. For butanol production by *C. acetoperbutylacetonicum*, Noguchi et al. [89] used a different pre-treatment and saccharification of lignocellulosic material that generates cellobiose rather than glucose, and xylose, avoiding CCR. Butanol productivity increased with this strategy. Recently, it was also demonstrated that biochar (a by-product obtained from thermochemical conversion of biomass under oxygen limited conditions) can serve as a cost-effective substitute for mineral and buffer solutions in ABE fermentation [91].

2.3.2. Isopropanol-Butanol-Ethanol (IBE) Fermentation

Butanol has been considered an advanced biofuel and its production at a scale of billions of liters by ABE fermentation may result in oversupply of acetone [92]. IBE fermentation is a solution to the acetone problem, because in this process naturally-occurring strains can convert acetone in isopropanol, resulting in the production of isopropanol, butanol, and ethanol. Additional advantages are the possibility to use this mixture directly as a fuel and the removal of a high corrosive substance with low fuel properties [93].

C. beijerinckii is one of the main species that produce IBE. It has been identified that the BGS1 strain has 16 coding sequences related to alcohol dehydrogenase, including the *sAdhE* gene that encodes isopropanol dehydrogenase, responsible for acetone to isopropanol conversion. However, not all *C. beijerinckii* strains produce isopropanol dehydrogenese [32]. Another inconvenience, as depicted by Vieira et al. [92], is that natural IBE producers are more sensitive to product inhibition (caused mainly by butanol) and, consequently, are less efficient than ABE producers.

2.3.3. Hexanol-Butanol-Ethanol (HBE) Fermentation

C. carboxidivorans produces hexanol and hexanoic acid from synthesis gas (mainly CO, H_2, and CO_2) [94]. As this bacterium also produces butanol andethanol, the term HBE (hexanol-butanol-ethanol) fermentation was introduced by Fernandez-Naveira et al. [86]. HBE fermentation is a term derived from ABE fermentation and known for solventogenic bacteria [94], as already mentioned.

C. carboxidivorans is able to start the solventogenic phase without a drastic pH reduction, which is the case of ABE fermentation. Thus, buffering the culture medium for HBE fermentation by *C. carboxidivorans* may induce higher acid production and promote greater conversion in the solventogenic phase. However, further investigation and applications of genetic engineering are necessary to reduce the inhibitory effect of high concentrations of solvents and acids produced [86].

2.3.4. Embden-Meyerhof-Parnas (EMP) Pathway and Pentose Phosphate (PP) Pathway

EMP and PP pathways for *Clostridium* species to metabolize hexoses and pentoses, respectively, have been shown by several authors [86,95,96]. Agro-industrial wastes, commonly used as feedstocks for biofuel production, generate mainly glucose, arabinose, mannose, xylose, fructose, sucrose, and lactose as substrates after a pre-treatment process [86].

The nature of the sugars is a key parameter that will affect the efficiency of the fermentation process and the production of metabolites [86]. Monosaccharides are transported by a faster system than disaccharides. In the case of glucose and fructose, EMP is directly used. For galactose, for example, after being phosphorylated, galactose-6-P is metabolized via the tagatose 6-P pathway and subsequently enters the glycolytic pathway. These differences result in different consumption rates between sucrose and lactose [86,97]. In the case of pentoses, which are usually mixed with hexoses, there is the carbon catabolic repression (CCR) issue, which suppresses pentose consumption because of the preferable use of glucose [89,90].

2.3.5. Wood–Ljungdahl Pathway

The Wood–Ljungdahl (WL) pathway provides acetate from inorganic sources, such as H_2, CO, and CO_2. Acetogenic bacteria follow the WL pathway to produce biofuels from CO, CO_2, and H_2, or syngas/waste gas [86].

The WL pathway comprises two branches: methyl and carbonyl branches both contribute to acetyl-CoA formation and start with the assimilation of CO or CO_2 [98]. In the methyl branch, formate is obtained through the reduction of CO_2. In the carbonyl branch, CO or CO_2 can be taken directly and converted into acetyl-CoA [86]. Growth of acetogenic bacteria is possible because of the formation of acetate from 2 mols of CO_2, with H_2 as a reducing agent and, therefore, this pathway must be coupled with ATP formation. This is believed to be the oldest biochemical pathway, responsible for producing biomass and ATP in the ancient world [85]. In the first step of the methyl group, the possibility of CO assimilation followed by the action of CO desidrogenase generates two protons and two electrons that can be used as inputs to format desidrogenase action on the following reaction. In other words, the WL pathway is capable of sustaining the autotrophic growth of the microorganism only with CO, even without H_2 for electron donation [68,71]. In heterotrophic conditions, alcohols, organic acids, and simple sugars are able to donate electrons to the WL pathway [72].

In the Wood–Ljungdahl pathway, acetyl-CoA is a common intermediate of both branches. The acetyl-CoA formed by the WL pathway can be used for direct biomass formation or to undergo action of the phosphotransacetylase (PTA) followed by the acetate kinase (ACK), generating 1 ATP and forming 1 acetate. Alternatively, acetyl-CoA can also be converted to acetaldehyde, then to ethanol or butyryl-CoA and, subsequently, into butyrate and/or butanol, or into hexanoyl-CoA and then hexanoate and/or hexanol [86].

Oxygen inhibits various enzymes from the WL pathway and its presence may increase the redox potential. Therefore, it is considered one of the most toxic gases during the syngas fermentation [72]. However, some acetogenic microorganisms are tolerant of O_2 concentrations ranging from 0.5% to 6% of saturation [73].

3. Feedstock for Anaerobic Processes

3.1. Glycerol

Known for thousands of years through soap production, the glycerol molecule is considered the oldest isolated by man [99]. Its name comes from the Greek word "glykys", which means sweet [98]. It is an organic molecule of alcohol function that presents three hydroxyl groups in its structure, which gives it the characteristics of solubility in water [99]. It has no known toxic effect on either man or the environment [100,101]. It is a viscous liquid, with no smell, no color, a sweet taste and a high boiling point [100,101]. It is stable under several conditions [99]. There are about 1500 known applications for glycerol, which include those of the pharmaceutical, food, and cosmetics industries [102]. It is commercialized in the form of glycerin, a mixture of glycerol and impurities from the production process with purity varying between 95% and 99% [100,101]. It is also found in the form of "crude glycerol", originating mainly from the biodiesel industry, with 60%–80% glycerol in its composition [100–102].

Equivalent to 10% (*w/w*) of biodiesel production [103], glycerol supply no longer depends on demand. Glycerol is in excess supply in the market due to the large growth of the biodiesel industry since 2006 [101], making it a low-cost raw material. The price reduction of glycerol enables its use in new productive processes of value-added chemicals and its entry into countries where it was not previously accessible [99].

In addition to being used in chemical processes for the generation of several products, glycerol can also be employed as a substrate in biotechnological processes. Some examples of products generated by glycerol metabolization by *Clostridium* species are: 1,3-propanediol (*C. butyricum, C. butylicum, C. pasteurianum, C. perfringens, C. beijerinckii, C. kainantoi*), acetic acid (*C. butyricum, C. butylicum, C. pasteurianum, C. perfringens, C. beijerinckii, C. carboxidivorans*), butyric acid (*C. butyricum, C. pasteurianum,*

C. carboxidivorans), lactic acid (*C. pasteurianum*), butanol (*C. pasteurianum, C. acetobutylicum, C. beijerinckii, C. butylicum, C. carboxidivorans*), ethanol (*C. butyricum, C. pasteurianum, C. perfringens, C. acetobutylicum, C. beijerinckii, C. butylicum, C. carboxidivorans*), acetone (*C. acetobutylicum, C. beijerinckii, C. butylicum*), propionic acid (*C. butylicum*), propanol (*C. butylicum*), 1,2-propanediol (*C. butylicum*), and H_2 (*C. butyricum, C. pasteurianum, C. beijerinckii*) [14,44,57,104–106]. The elemental composition of crude glycerol from biodiesel production is around 53% carbon, 36% oxygen, and 11% nitrogen, which makes it a good source of carbon and energy [101,105].

Only glycerol can be converted into 1,3-propanediol by anaerobic fermentation due to its great specificity [105]. During the fermentation of *Clostridium* sp., the formation of acetic and butyric acids occurs, which interfere in the pH of the medium pH, making it more acidic. Without the proper pH control during the fermentative process, the pH of the medium reaches values below the pKas of the acids generated, leading to their dissociated forms. With this, the dissociated acids permeate the cell membrane and remain in the associated form within the intracellular fluid (more alkaline region), without being able to cross the membrane, causing an imbalance by the energy expended for the maintenance of the intracellular pH; that is, causing inhibition of cell growth and production of 1,3-propanediol [107]. However, acetic acid has an important role as a reduction equivalent in the formation of 1,3-propanediol, and the complete conversion of glycerol to 1,3-propanediol is impossible due to this need. Meanwhile, the conversion of glycerol into butanol or ethanol by *C. pasteurianum* does not suffer from the influence of by-products, because all hydrogen carriers are regenerated in the metabolic pathway [107].

3.2. Molasses

Molasses is a syrupy material obtained during the sugar production process. Industrial sugar production starts by extracting sugarcane or beet juice, which is then clarified and concentrated. The concentrated juice is subjected to successive crystallizations, followed by separation of sucrose crystals by centrifugation. The remaining viscous liquid is molasses [108,109].

Molasses composition varies a great deal among batches. Usually this by-product of sugar production is composed of water, approximately45–50% (*w/w*) total sugars (sucrose, glucose, and fructose), suspended colloids, heavy metals, vitamins, and nitrogenous compounds [110]. Due to its composition, animal feed and fertilizer are possible destinations for this viscous liquid. However, ethanol production by fermentation is one of its main applications [109].

As it contains a high quantity of sugars, molasses is an inexpensive renewable carbon source for several bioprocesses. Since a carbon source is necessary in relatively higher concentrations compared to other medium components, molasses is a good option to reduce raw material cost [109,111]. Molasses can be quite inhibitory to several organisms due to its salinity and osmolarity, and the presence of toxic elements, such as aluminium and sulphites, and thermal sugar degradation compounds [109].

In spite of molasses being mainly used for fuel ethanol production by fermentation, other potential molecules can be obtained by fermentation of this by-product. The most promising bioprocesses using molasses as substrate for *Clostridium* fermentations is acetone-butanol-ethanol (ABE) production. Considering the importance of butanol, studies for optimization of butanol production have been undertaken. Syed et al. [112] achieved high butanol production using blackstrap molasses as substrate. The authors compared butanol production by *C. acetobutylicum* PTCC-23 to other mutant strains obtained by UV exposure, N-methyl-N-nitro-N-nitrosoguanidine and ethyl methane sulphonate treatments. Li et al. [110] achieved a similar butanol concentration from cane molasses using *C. beijerinckii* mutant obtained by combined low-energy ion beam implantation and N-methyl-N-nitro-N-nitrosoguanidine induction.

While *C. acetobutylicum* and *C. beijerinckii* are the most encouraging species due to their ability to produce butanol from molasses [113], the literature reports high butyric acid production from molasses using *C. tyrobutyricum* species [114]. Li et al. [82] improved butanol production by coculture of *C.*

beijerinckii and *C. tyrobutyricum*, since *C. tyrobutyricum* produces butyric acid from molasses increasing the butyrate availability to butanol production by *C. beijerinckii*.

In addition, several studies evidenced that molasses is a rich source of antioxidants, such as phenolic compounds. Among the phenolic compounds present in molasses it is possible to find the flavonoids, which show high antioxidant activity [115,116]. The availability and cost of sugar molasses make it an attractive feedstock to be used in many countries as raw material for fermentations, and also as a potential extra source of antioxidants, which are considered value added chemicals.

3.3. Corn Steep Liquor

Corn steep liquor (CSL) is a multicomponent and multiphase matrix, obtained as a major by-product of the corn wet-milling industry. It is a viscous concentrate of corn solubles containing amino acids, vitamins, minerals, and proteins. It has been used as a growth medium for *Lactobacillus intermedius* in mannitol production and for *Saccharomyces cerevisiae* in solvent production [117].

The standard medium for *Clostridium* sp. growth is basically composed of yeast extract, vitamins, minerals, trace metals, reducing agents, and the carbon source. The more expensive component is the reducing agent, followed by yeast extract. Yeast extract is a water-soluble portion produced by extracting the cells' contents from autolyzed yeast, which may cost $1000 ton^{-1}. An alternative to replace complex nutrient sources such as yeast extract is corn steep liquor, which costs around $200 ton^{-1} [118].

Maddipatti et al. [118] produced 32% more ethanol by *Clostridium* strain P11 using CSL instead of yeast extract. They also produced seven times more butanol using this by-product in relation to yeast extract. CSL also diversified the product range produced by *C. carboxidivorans* and *C. ragsdalei*, with maximum concentrations of ethanol, n-butanol, n-hexanol, acetic acid, butyric acid, and hexanoic acid of 2.78 g/L, 0.70 g/L, 0.52 g/L, 4.06 g/L, 0.13 g/L, and 0.42 g/L, respectively [119].

3.4. Lignocellulosic Biomass

Lignocellulosic biomass is a rigid and fibrous material present in herbaceous energy and food crops; forestry, agricultural, and agro-industrial residues; municipal solid waste; and waste streams from the pulp and paper industry [120]. This renewable and abundant feedstock represents a key opportunity towards a more sustainable economy, especially due to its capacity of sequestering carbon to the soil, which is not released to the atmosphere upon harvesting [121,122]. Biofuels, biomolecules, and biomaterials produced from lignocellulosic materials present a promising alternative to crude oil, oil products, and fossil fuels due to their renewability, biocompatibility, biodegradability, and reduction in greenhouse gas emissions to the atmosphere [123,124].

This biomass is mainly composed of three different biopolymers: cellulose (35%–50%), hemicellulose (20%–35%) and lignin (10%–25%), for which proportions depend on the feedstock origin, species, harvesting time, and growth stage [120,121,125]. Acetyl groups, minerals, phenolic substituents, and ash are other constituents that can be found in this biomass [120,121]. Cellulose is a linear microfibril formed by D-glucose subunits linked through β-1,4 glyosidic bonds and has cellobiose as a repetitive unit [121,126]. Its cohesive structure is assembled by intramolecular and intermolecular hydrogen bonds and is 85% crystalline and 15% amorphous [121,127]. Cellulose is embedded in hemicellulose and lignin. Hemicellulose is a branched and amorphous structure containing different heteropolysaccharides (xylan, glucomannan, arabinoxylan, etc.), which are composed of pentose (D-xylose and L-arabinose) and hexose (D-glucose, D-galactose, and D-mannose) sugar units [121,124–126]. In addition, hemicellulose has organic acids (acetic and glucuronic acids) in its structure, whose predominance is dependent on its origin (angiosperms vs. gymnosperms). The hemicellulose of angiosperms is highly acetylated. Upon pretreatment, these acetyl groups partially catalyze the hydrolysis of hemicellulose. Lignin is an amorphous and hydrophobic polymeric structure composed of the alcohols p-coumaryl, coniferyl, and sinapyl, and is responsible for the biomass's tough structure, impermeability, and resistance to microbial and oxidative attacks [7,121,125]. This phenolic macromolecule displays variations in both chemical composition and structure due to a low degree of order and a high level of heterogeneity.

Different routes can be considered to transform lignocellulosic biomass into important biofuels and biochemicals, which elucidate the importance of this feedstock in biorefineries. Table 2 summarizes processes and end-products commonly obtained using lignocellulosic biomass as a feedstock in three conversion routes. The thermo-chemical, biochemical, and hybrid routes use lignocellulosic materials by chemical and biochemical processes, generating important products and building blocks for the chemical and pharmaceutical industries. However, one of the main concerns regarding lignocellulosic biomass is the food crop competition between energy and food production, which is a reality in the production of first-generation biofuels from sugar and vegetable oil [3,124]. Therefore, the production of second-generation biofuels is an alternative to produce ethanol, biodiesel, and other products from non-food crops, such as agricultural and agro-industrial residues, and forestry and energy crops [3]. This leads to valuable biochemical and biofuel production from less valuable feedstock, mainly in the form of residue. The use of residues as feedstock improves the sustainability of the process by reducing residue accumulation in the environment and, therefore, reducing many problematic effects, such as greenhouse gas emissions in landfills and dumps, and contamination of groundwater [3,128].

Table 2. Conversion routes, processes, and end products using lignocellulosic biomass as feedstock.

Conversion Route	Process	End Products
Thermo-chemical	Combustion Gasification Liquefaction Pyrolysis	Heat and power Hydrogen, alcohol, olefins, gasoline, diesel Hydrogen, methane, oils Hydrogen, olefins, oils, specialty chemicals
Biochemical	Fermentation Anaerobic digestion	Second-generation fuels (ethanol, butanol), chemicals and biochemicals Methane, fertilizer (digestate)
Hybrid [7]	Pyrolysis + fermentation	Ethanol, organic acids (butyric acid, acetic acid), 2,3 butanediol

Adapted from [124].

In order to use all components of the lignocellulosic biomass, as in the case for second-generation biofuels, a chemical and/or enzymatic pre-treatment of the feedstock is necessary in order to access the cellulose embedded in lignin and hemicellulose. Physical pretreatment, such as milling, irradiation, and extrusion, are commonly used to increase the surface area of the biomass and reduce its crystallinity to improve the enzymatic hydrolysis of its components [124]. Catalyzed and non-catalyzed steam explosion, ammonia fiber explosion (AFEX), liquid hot water, and microwave-chemical are examples of physico-chemical pretreatments [124]. Hydrothermal processing has also been assessed in literature to separate lignocellulosic biomass into cellulose, hemicellulose, and lignin in a biorefinery concept [129]. Common to the paper industry, chemical pretreatments encompass the use of acid and alkaline compounds, ionic liquids, and hot water at a controlled pH to remove lignin and hemicellulose in order to decrease the lignocellulosic material polymerization degree and crystallinity, and enhance biomass porosity [124].

Cellulases and hemicellulases can be used to convert cellulose and hemicellulose to fermentable sugars [121,124]. Bagasse, straws, and fibers have an interesting quantity of C5 and C6 sugars that can be used to produce second-generation biofuels and other chemicals in the context of a biorefinery. Although the pretreated hydrolyzed lignocellulosic material is commonly used to produce ethanol by yeasts such as *Saccharomyces*, *Kluveryomyces*, *Debaryomyces*, *Pichia*, and *Zymomonas*, these materials can also be used in ABE fermentation by *Clostridium* species [121,124]. Moreover, it is possible to produce not only ethanol but also butanol, which is a promising liquid fuel. Butanol-producing *Clostridium* sp. can uptake a wide range of hexoses, pentoses, and oligomers obtained from the hydrolysis of cellulose and hemicellulose [130]. Corn fiber hydrolysate was used for *C. beijerinckii* fermentation resulting in the production of 9.3 g/L ABE [131]. The same species was used to convert the wheat straw steam-exploded hydrolysates, yielding 128 g ABE/kg wheat straw [132]. Seven grams per liter of butanol were produced from concentrated sugar maple hemicellulosic hydrolysate by *C. acetobutylicum* ATCC824 [133].

Even though lignin is recalcitrant to micro-organisms, it can play an important role through chemical conversions leading to the production of methanol and BTX (benzene-toluene-xylenes) phenols and phenolics [121,134].

The thermochemical conversion of lignocellulosic biomass involves the use of heat and catalysts, while in the biochemical conversion organic matter is converted by micro-organisms through a series of metabolic reactions. The thermochemical–biochemical conversion or hybrid route merges these two processes, converting biomass to a gaseous substrate using thermal energy, and the resulting gas is converted to bioproducts using bacterial cultures. The hybrid route consists of the fast pyrolysis or gasification of lignocellulosic biomass producing synthesis gas (syngas), which is later converted to biofuels such as ethanol, butanol, butyrate, and 2,3-butanediol, among others [8,14,16,135–137]. In this way, all components of the lignocellulosic material are converted to synthesis gas and later fermented by *Clostridium* bacteria, overcoming the recalcitrant characteristic of this biomass and eliminating high costs related to the enzymatic pretreatment [7]. Additionally, the adoption of this hybrid strategy for processing lignocellulosic biomass allows a more efficient use of it since the whole biomass content (including lignin) can be converted to fuels and chemicals, not only the polysaccharide fractions as in the biochemical route. Nonetheless, process integration and optimization need to be done to make these processes attractive from an industrial standpoint.

3.5. Syngas

Synthesis gas, or syngas, is a gaseous mixture mainly composed of CO, H_2, and CO_2 that can present smaller quantities of methane (CH_4), hydrogen sulfide (H_2S), and nitrogen (N_2) depending on the biomass composition and conversion conditions [7]. Different biomass sources, such as municipal solid waste, animal slurry, agricultural residue, energy crops, and coal can be used to produce syngas by pyrolysis or gasification [7]. Fast pyrolysis of biomass occurs in an oxygen-free environment with moderate temperatures, resulting in syngas, biochar, and bio-oil, an energy rich liquid composed of carboxylic acids, sugars, alcohols, aldehydes, esters, ketones, aromatics, and furans [7,138]. Bio-oil and synthesis gas composition depends on the pyrolysis technology implemented and biomass composition [7]. Reactors used in fast pyrolysis, commonly fluidized bed reactors, must have a controlled temperature, high heat transfer rates, and rapid cooling of vapors [138].

Biomass can be also converted to synthesis gas through gasification, which is conducted at higher temperatures than pyrolysis (800–1000 °C), promoted by heat and/or electricity, and in the presence of a gasifying agent [7,138,139]. Synthesis gas is the main product of this thermochemical process, and biomass composition has an important impact on gas proportion and content, which can have trace amounts of sulfur (H_2S), hydrogen chlorine (HCl), alkali metals (potassium and sodium), tars, and ammonia (NH_3) [7,139]. These contaminants can interfere in the microbial conversion of synthesis gas and, therefore, purification technologies should be implemented, such as wet scrubbing or hot/cold/warm gas clean up [139]. Biomass properties (ash, moisture, particle size), gasifying agent (air, steam, pure O_2), reactor type (updraft gasifier, downdraft gasifier, fluid-bed gasifier) and operational conditions (temperature, fuel:gasifying agent ratio) also influence the synthesis gas composition [136,139].

Synthesis gas conversion can produce important chemicals and fuels through the Fisher–Tropsch (chemical) process or fermentation (biochemical process) [7]. The Fischer–Tropsch process involves catalyzation by iron, cobalt, or ruthenium, producing alcohols and liquid hydrocarbons at 200–350 °C, a fixed H_2:CO:CO_2 ratio, and a limited amount of impurities [140]. The chemical conversion process has several drawbacks due to its catalytic and operational nature, including low catalyst specificity, high energetic demand, sensitivity to toxic gases and high pressure, and temperature conditions [141–143]. The fermentation of synthesis gas obtained by the pyrolysis/gasification of biomass is considered a hybrid route since it is a combination of two conversion processes: thermochemical and biochemical [144]. Microbial conversion of syngas occurs at a determined temperature and pH, and can be performed by different acetogenic microorganisms (*C. ljungdahlii*, *C. autoethanogenum*, *Acetobacterium woodii*, *C. carboxidivorans* P7, *C. ragsdalei* P11, *Archaeoglobus fulgidus*, etc.) and hydrogenic microorganisms (*Rhodospirillum rubum*, *Desulfotomaculum carboxydivorans*, *Themococcus onnurineus* NA1, etc.) producing acetate, butyrate, ethanol, 2,3-butanediol, lactate, and others [7,14,17,136,145].

Synthesis gas fermentation has many advantages when compared to well-established processes in the chemical industry. Considering the Fischer–Tropsch process, syngas fermentation presents a higher tolerance to sulfur compounds; a variety of CO, H_2, and CO_2 ratios in synthesis gas that can be used as a substrate and not a fixed ratio; lower operational pressure and temperature, which decreases operational costs; and high productivity and product uniformity [8]. Regarding other biotechnological conversions of lignocellulosic biomass, synthesis gas fermentation eliminates a complex pre-treatment stage, which would enhance operational cost due to enzyme acquisition [7]. Moreover, all components of lignocellulosic biomass are converted, biomass composition does not interfere in the gas composition obtained through pyrolysis or gasification, and the resulting H_2:CO:CO_2 ratio does not affect fermentation [146].

However, one of the major bottlenecks in this process, especially concerning commercialization, is the mass transfer between gas and liquid phases due to the low solubility of synthesis gas [147–150]. The increase of gas solubility in culture media may enhance the availability of gaseous substrate to cells, improving both the cell's autotrophic growth and product conversion [147]. For micro-organisms to convert the gaseous substrate into fuels and chemicals, it is important that the nutrient is internalized by the cell, which occurs in a certain pathway from the interior of the gas bubble to the cell cytoplasm [151]. Although mass transfer can present many resistances during this pathway, most may be neglected in most bioreactors except for the resistance near the gas–liquid interface, which is a function of gas diffusivity in the liquid phase, as well as the film thickness [152]. The gas adsorption in the liquid phase is a limiting step for gas–liquid mass transfer and is accounted for in the overall volumetric mass transfer, k_La, which is increased when the gas mass transfer to the liquid phase increases [152]. Consequently, productivity would increase in gas–liquid systems.

Many approaches have been proposed in literature in order to increase mass transfer in gas–liquid systems, such as gas specificity, increasing operational pressure, increasing gas and liquid flow rates, larger specific gas–liquid interfacial areas, different reactor configurations, different liquid phases, innovative impeller designs, mathematical approaches, and more [146,153–158]. Table 3 summarizes some approaches reported in literature so far. Recently, the use of perfluorodecalin (PFC) and Tween® 80 improved carbon monoxide mass transfer to a liquid phase composed of distilled water, PFC, and Tween® 80 in a stirred tank reactor with a sparger and Rushton-type and Smith-type impellers [157].

Synthesis gas obtained from residues is a cheap feedstock that can be used to produce sustainable biofuels and chemicals [136]. An ethanol productivity of 7.3 g/L.day was achieved in 5 L bioreactors during a two-stage continuous fermentation of synthesis gas using *C. ljungdahlii* [26]. Gaseous waste streams from the steel industry, with different CO, CO_2, and H_2 ratios, could be fermented by *C. ljungdahlii* and *Clostridium autoethangenum* to produce ethanol, integrating a biorefinery and the steel industry [159]. Commercially, LanzaTech collaborated with Concord Blue Energy to produce ethanol and 2,3-butanediol by the fermentation of high quality synthesis gas obtained via gasified municipal solid wastes and agricultural residues [7]. INEOS New Planet Bioenergy also produces ethanol from syngas obtained through the gasification of vegetative waste and agricultural biomass, and in 2008 had a production rate of 100 gallons of ethanol per dry ton of feedstock using *C. ljungdahlii* [7,15]. Coskata Inc. developed bacteria for ethanol production, *Clostridium coskatii*, and have been investing in syngas fermentation for ethanol production from wood chips and waste and reformed natural gas [15].

Table 3. Mass transfer related to synthesis gas fermentation reported in literature using water as the liquid phase.

Reactor	Agitation (rpm)	Specific Gas Flow Rate (vvm)	Microorganism	Gas	k_La (h^{-1})	Ref
BCR	n/a	0.4	*n/a*	CO	72.0	[160]
CHF	n/a	n/a	*n/a*	CO	85.7–946.6	[146]
CSTR	300	n/a	*C. ljungdahlii*	CO	14.9	[161]
CSTR	400	n/a	*C. ljungdahlii*	CO	21.5	[161]
CSTR	400	0–0.32	*n/a*	Syngas	38.0	[162]
CSTR	500	n/a	*C. ljungdahlii*	CO	22.8	[161]
CSTR	600	n/a	*C. ljungdahlii*	CO	23.8	[161]
CSTR	700	n/a	*C. ljungdahlii*	CO	35.5	[161]
GLR	n/a	1.67	n/a	CO	129.6	[154]
HFMBR	n/a	0.625	n/a	CO	1096.2	[23]
HFMBR	n/a	0.029	n/a	CO	385.0	[163]
HFR	n/a	0.037	n/a	CO	420	[158]
MBR	n/a	0.00625–0.0625	n/a	Syngas	450.0	[154]
PBC with microbubble sparger	n/a	0–0.021	*R. rubrum*	Syngas	2.1 for CO	[148]
STR	300	0–0.032	*C. ljungdahlii*	Syngas	35.0 for CO	[148]
STR	300	0–0.032	SBR mixed culture	Syngas	31.0 for CO	[148]
STR	300	2.7	n/a	CO	166.09 ± 13.15	[157]
STR	400	0.14–0.86	n/a	CO	10.8–155.0	[164]
STR	400	0.70–2.14	n/a	CO	72.0–153.0	[155]
STR	400	0.36–1.07	n/a	CO	72.0–122.4	[165]
STR	400	0–0.032	*R. rubrum*	Syngas	101.0 for CO	[148]
STR	450	0–0.032	*R. rubrum*	Syngas	101.0 for CO	[148]
STR	500	0.36–1.07	*n/a*	CO	129.6–144.0	[165]
STR	500	2.0–2.7	*n/a*	CO	11.48 ± 57.79–399.06 ± 26.80	[157]
STR	600	0.36–1.07	*n/a*	CO	147.6–208.8	[165]
STR	650	0.36–1.07	*n/a*	CO	172.8–252.0	[165]
STR	700	0.36–1.07	*n/a*	CO	187.2–288.0	[165]
STR with microbubble sparger	300	n/a	SBR mixed culture	Syngas	104.0 for CO	[148]
TBR	n/a	0–0.021	*n/a*	Syngas	22.0	[162]
TBR	n/a	0–0.021	*C. ljungdahlii*	Syngas	137.0 for CO	[148]
TBR	n/a	0–0.021	*R. rubrum*	Syngas	55.5 for CO	[148]
TBR	n/a	0–0.021	SBR mixed culture	Syngas	121 for CO	[148]

Where: n/a—not applicable; SBR—sulfate reducing bacteria; N—impeller speed; QCO—specific gas flow rate; k_La—overall volumetric mass transfer coefficient; BCR—bubble column reactor; STR—stirred tank bioreactor; CSTR—continuous stirred tank bioreactor; HFMBR—hollow fiber membrane bioreactor; CHF—composite hollow fiber membrane; HFR—hollow fiber reactor; MBR—membrane bioreactor; PBC—packed bubble column; TBR—trickle bed reactor/GLR—gas-lift reactor; Ref—references.

4. Biomolecules Produced by *Clostridium* sp.

4.1. Ethanol

Ethanol, often referred to as ethyl alcohol or simply alcohol, is a volatile, biodegradable, low-toxicity, flammable, and colorless liquid at room temperature [166]. Global ethanol production has grown considerably over the past decade. In 2017, its global production was approximately 98 billion liters [167]. Ethanol is a very versatile building block for industry. It can be used to generate chemicals such as ethylene, propylene, 1,3-butadiene, and hydrocarbons, as well as in the production of oxygenated molecules, such as acetaldehyde, butanol, acetic acid, acetone, and dimethyl ether [168,169]. Ethylene, the largest-volume petrochemical produced worldwide, is used to produce ethylene glycol, propylene oxide, ethylene oxide, acrylonitrile, and polyethylene (PE) [170].

Only about 7% of ethanol produced in the world is made by a petrochemical process through the hydration of ethylene, and the main producers by this route are Germany, South Africa, and Saudi Arabia [171]. The vast majority of ethanol is produced by a fermentation process, using renewable sources as feedstock and microbial catalysts. Two main types of crops are used in the ethanol production industry by a well-established technology: sugar rich crops, such as sugar cane, and amylaceous crops, such as corn. Synthetic ethanol processing is economically less attractive than fermentation due to the high production cost of ethylene and the great availability of agricultural products and byproducts [171].

Clostridium species can produce ethanol from several renewable sources, such as sugary or starchy materials (sugar cane, sugar beet, corn, etc.), lignocellulosic materials (e.g., cane bagasse and corn stover), as well as from gaseous substrates, such as CO, CO_2, and H_2. Table 4 shows some examples.

Table 4. Recent studies reported about ethanol production by *Clostridium* species.

Feedstock	Microorganism	Reactor	P_{Et} [1] (g/L)	Q_{Et} [2] (g/L.h)	Ref. [3]
Syngas	*C. carboxidivorans* P7	HFM-BR	23.93	0.14	[23]
Syngas	*C. rasgdalei* P11	STR	25.26	0.001	[172]
Syngas	*C. ljunghdalli* ERI-2	CSTR/Bubble column	19.73	0.37	[26]
Syngas	*C. ljunghdalli* PETC	CSTR	19	0.30	[173]
Syngas	*C. carboxidivorans* P7	h-RPB	7	0.28	[174]
Syngas	*C. ljunghdalli*	Batch	0.49	0.03	[175]
Syngas	*C. carboxidivorans* P7	Batch	2	0.02	[176]
CO	*C. autoethanogenum* DSM 10061	Continuous gas-fed	7.14	-	[177]
Glucose	*C. saccharoperbutylacetonium* pSH2	Batch	7.9	0.11	[178]
Glucose	*C. acetobutylicum* hbd::int(69)	Fed-batch	33	0.5	[38]
Glucose	*C. carboxidivorans* P7	Batch	2.34	0.01	[179]
Lignocellulose	*C. thermocellum* ATCC31924	Batch	2.45	0.02	[180]
Lignocellulose	*C. cellulolyticum* H10	Batch	2.5	-	[41]
Lignocellulose and starch	*C. acetobutylicum* NBRC13948	SHF	1.7	0.018	[180]
Lignocellulose and starch	*C. acetobutylicum* NBRC13948	SSF	1.5	0.01	[180]
Cellobiose	*C. thermocellum* LL1275	Batch	5	0.07	[63]
Cellobiose	*C. cellulovorans* 83151-adhE2	CBP	2.03	0.03	[181]
Cellobiose	*C. phytofermentans* ATCC700394	CBP	7	0.03	[36]
Glycerol	*C. pasteurianum* MTCC6013	Immobilized cells	1.94	0.01	[182]

[1] P_{Et}: ethanol production; [2] Q_{Et}: ethanol productivity; [3] Ref.: reference.

Fernandez-Naveira et al. [180] reported an ethanol production of 2.34 g/L from a glucose (30 g/L) fermentation process by *C. carboxidivorans* DSM 15,243 in a continuous bioreactor. A mutant strain of *C. acetobutylicum* ATCC824, developed by the disruption of the butyrate/butanol pathway, produced 33 g/L of ethanol in a fed-batch fermentation process using glucose as a substrate with a productivity of 0.5 g/L.h [38]. In another study, *C. saccharoperbutylacetonicum* N1-4 had some genes overexpressed to develop a more robust strain, which resulted in a 400% increase in ethanol production. Glucose (80 g/L) was used in a batch bioreactor process and 7.9 g/L of ethanol was produced after 72 h of fermentation [178].

Khanna et al. [178] reported the bioconversion of crude and pure glycerol into ethanol, butanol, and 1,3-propanediol by immobilized *C. pasteurianum* MTCC 6013, using a silica gel chromatography column (80–120 mesh grade) as immobilization support. Pure glycerol (25 g/L) formed more products (19 g/L) than crude glycerol (12 g/L). In these conditions, 1.94 g/L of ethanol, 9.23 g/L of 1,3-propanediol, and 7.73 g/L of butanol were obtained [182].

Ethanol can also be produced from lignocellulosic materials, which is commonly called second-generation ethanol. Agricultural residues, such as sugar cane bagasse and corn stover, grasses, such as switchgrass, and forestry and wood residues have been studied in order to develop a cost-effective second-generation process using these plentiful feedstocks [169,183]. In a very recent study, Singh et al. [180] reported an ethanol production of 2.45 g/L during 120 h of cellulose batch fermentation (30 g/L) by *C. thermocellum* ATCC31924. *C. thermocellum* LL1275 produced 5 g/L of ethanol (productivity of 0.07 g/L.h) through batch fermentation with cellobiose as a substrate [63]. It was reported that 8.5 times more ethanol was produced by a genetic engineered strain of *C. cellulolyticum* in comparison to the wild strain, using switchgrass as feedstock in a batch fermentation process [41].

The production of ethanol with lignocellulosic biomass can be processed through four different configurations, depending on the feedstock, the microorganism chosen, and the product desired. These configurations are known as SHF (separate hydrolysis and fermentation), SSF (simultaneous saccharification and fermentation), SSCF (simultaneous saccharification and co-fermentation), and CBP (consolidated bioprocessing) [184]. In the SHF, the hydrolysis of the biomass and the hexose/pentose fermentations are carried out in separate reactors. The main advantage of this configuration is the possibility of performing each step in its optimal conditions of pH and temperature. However, high concentrations of glucose and/or cellobiose in the first step inhibit the cellulases and reduce their efficiencies. The SSF configuration occurs with cellulose hydrolysis and hexose fermentation in the same bioreactor, reducing the inhibition of cellulases by end products as observed in the SHF, but making it difficult to process both steps in their optimum conditions of pH and temperature. Furthermore, microorganisms used in the SSF are able to ferment only glucose, not being able to utilize xylose. The SSCF configuration is similar to the SSF, but glucose and xylose can be fermented in the same bioreactor, as some genetically engineered strains are developed to ferment both substrates [185,186]. CBP configuration is represented by one single microorganism that produces enzymes to hydrolyze cellulose and hemicellulose, and converts the resulting sugars into ethanol. Thus, the enzyme production, hydrolysis, and fermentation steps occur in the same bioreactor, reducing costs and increasing process efficiency. However, this type of biocatalyst is still in the early development stage [185,187]. In 2014, a study reported the production of ethanol and butanol through SHF and SSF processes with *C. acetobutylicum* NBRC13948. In this work, corn and wood chips of *Quercus acutissima* were used as feedstock and 1.7 g/L of ethanol was produced after 96 h of fermentation in the SHF process, while in the SSF process 1.5 g/L of ethanol was obtained after 144 h [188]. *C. phytofermentans* was used in a CBP configuration with corn stover as feedstock. An ammonia fiber expansion (AFEX) pretreatment process was conducted and 7 g/L of ethanol was produced after 264 h of fermentation [36].

Another way to transform the lignocellulosic biomass into ethanol is the hybrid technology that involves thermochemical and biochemical steps, known as syngas fermentation [182], as already mentioned in the feedstock section. In addition to the syngas produced from gasification of biomass, industrial waste gas streams containing CO, H_2, and CO_2 can also be converted to ethanol by *Clostridium* species, in the biochemical step [184]. The main advantage of the syngas fermentation process is that all components of lignocellulosic material, including lignin, are converted into syngas and later fermented, overcoming the recalcitrant characteristic of this biomass and eliminating the elevated costs related to the enzymatic pretreatment step [7]. In 2014, it was reported that 24 g/L of ethanol was obtained from syngas fermentation (50% CO, 30% H_2, and 20% CO_2) by *C. carboxidivorans* P7 in a hollow fiber membrane biofilm reactor (HFM-BR) [23]. Richter et al. [26] reported a two-stage syngas fermentation process by *C. ljunghdalli* ERI-2, operating one bioreactor for cell growth and a bubble column bioreactor equipped with a cell recycle module for ethanol production. Syngas containing 60% CO, 35% H_2, and 5% CO_2 was used, and 18 g/L of ethanol was obtained, with a productivity of 0.37 g/L.h. A study carried out recently with *C. carboxidivorans* P7 reported an ethanol productivity of 0.28 g/L.h with a syngas composition of 20% CO, 5% H_2, 15% CO_2, and 60% N_2, in a horizontal rotating packed bed biofilm reactor (h-RPB). Seven grams per liter of ethanol were obtained using the h-RPB reactor, which was 3.3 times higher than that obtained in a CSTR (continuous stirred tank reactor) under the same conditions [189].

4.2. Butanol

Butanol (butyl alcohol) is a four-carbon alcohol with the molecular formula C_4H_9OH. It is a volatile, biodegradable, low-toxicity, flammable, and colorless liquid at ambient temperature. There are four isomeric structures of butanol: n-butanol, sec-butanol, isobutanol, and tert-butanol. This alcohol is an important building block for the chemical industry, being used as an intermediate to the production of methacrylate esters, butyl acrylate, butyl glycol ether, butyl acetate, and plasticizers. It can also be used as a diluent for brake fluid formulation, and for the production of antibiotics, vitamins, and hormones [190,191]. One of the most important applications of butanol is as a direct replacement of gasoline or as a fuel additive, due to the low vapor pressure and corrosivity, which allows its transportation and storage in the same infrastructure existing for gasoline, as well as its blend with existing gasoline at much higher proportions than ethanol [191,192]. Furthermore, as butanol is a four-carbon alcohol, it has the double carbon content of ethanol and contains 25% more energy [191]. These advantages make butanol outstrip ethanol as an alternative biofuel [193].

Butanol has been traditionally produced by anaerobic fermentation of sugar-rich substrates using *Clostridium sp.* via ABE fermentation process. With the emergence of the petrochemical industry in the 1950s, most butanol in the world started to be produced from fossil oil, through the hydroformylation of propene, a process known as oxo synthesis [190]. In the 1970s, the global energy crisis rekindled the worldwide focus towards development of alternative fuels, reigniting interest in ABE fermentation [190,194]. The fermentation technology of butanol production by *Clostridium sp.* has some drawbacks, such as the relatively high substrate cost arising from the use of edible biomass, the low final butanol concentration obtained (less than 20 g/L), the low butanol selectivity, the low volumetric butanol productivity (less than 0.5 g/L.h), and the high cost of recovery. Thus, butanol fermentation is less competitive than that of other biofuels [193]. Aiming at overcoming these disadvantages related to the fermentation technology, some studies reported the development of genetically engineered strains to improve butanol yield, different modes of operation to increase the productivity, different recovery processes, and the use of non-edible feedstocks focusing on residual biomass [193].

Lignocellulosic biomass, syngas, molasses, and glycerol are promising feedstocks for ABE fermentation. Qureshi et al. [29] achieved 34.77 g/L of butanol using corn stover hydrolysate, and 30.86 g/L using barley straw hydrolysate, as substrates in the fermentation process by *C. beijerinckii* P260, with simultaneous product recovery. Using crude glycerol (50 g/L) as a substrate, butanol production of 8.95 g/L was reported in a batch fermentation with *C. pasteurianum* DSM 525, and 0.119 g/L.h of productivity was achieved [195]. A fibrous bed reactor with gas stripping recovery was used to produce butanol with cassava bagasse hydrolysate by a hyper-butanol-producing *C. acetobutylicum* strain (JB200). More than 76 g/L of butanol was obtained with a productivity of 0.32 g/L.h [196]. Only 1 g/L of butanol and 3 g/L of ethanol were produced after 600 h in batch fermentation (70% CO, 20% H_2, 10% CO_2) using *C. carboxidivorans* P7 [197].

Mutagenesis and metabolic engineering techniques have been used to develop new strains capable of producing butanol in adverse situations that the wild strain could not. Zhang et al. [198] achieved a butanol productivity of 0.53 g/L.h using a metabolically engineered strain of *C. tyrobutyricum* to ferment sugar cane juice, with corn steep liquor as the nitrogen source, using a fibrous bed bioreactor (FBB) in a repeated batch mode for 10 consecutive cycles in 10 days. A butanol concentration of 12.8 g/L was detected [196]. Lee et al. [37] reported the development of a metabolically-engineered strain of *C. acetobutylicum* that produced 44.6 g/L of butanol. Glucose was used as the substrate in a continuous fermentative process, and in situ adsorptive recovery of butanol. A butanol productivity of 2.64 g/L.h was reported, which represents a high value compared to similar studies.

Different fermentation techniques can be adopted in the ABE process, including: batch, fed-batch, semi-continuous, free cell continuous, immobilized cell continuous, cell recycle cell, biofilm reactor, fed-batch extractive fermentation, and fed-back with in situ product recovery. The batch process is the most reported due to the low contamination risk and simple operation, but a maximum ABE solvent concentration of 25–30 g/L has been obtained [197]. Zhang et al. [198] reported an immobilized cell fermentation system with co-culturing of *C. beijerinkii* and *C. tyrobutyricum*, using hydrolysate cassava bagasse as feedstock. More than 13 g/L of butanol was produced with a productivity of 0.44 g/L.h.

Obstacles, such as the low productivity, low yield, and low substrate consumption, are a result of solvent accumulation that causes severe inhibition. Efficient recovery processes are able to overcome these drawbacks, but usually lead to high processing cost and to large quantities of disposal water. The most-used recovery technique in the industry is conventional distillation, but other techniques have been investigated in order to reduce costs, enhance productivity, eliminate inhibition effects, and increase sugar consumption. Among these, liquid–liquid extraction, pervaporation, gas stripping, perstraction, reverse osmosis, ionic liquid extraction, adsorption, and aqueous two-phase separation can be highlighted [199]. A very recent study reported a two-stage fed-batch fermentation of glucose by

C. acetobutylicum MTC11274 with magnesium limitation and calcium supplementation, integrated to a gas stripping system. More than 54 g/L of butanol was detected, with a productivity of 0.58 g/L.h [200]. Dong et al. [201] reported a ceramic hollow fiber-supported polydimethylsiloxane (PDMS) composite membrane used for the pervaporation of butanol. The results showed the PDMS composite membrane exhibited high and stable performance for butanol recovery from ABE systems. Raganati et al. [202] showed an efficient recovery of butanol from ABE fermentation broth by adsorption on Amberlite XAD-7, as this adsorbent presents a high affinity for butanol and a poor affinity for glucose.

4.3. Acetone

Acetone, also known as propanone, is a colorless, volatile, flammable, and toxic liquid. Represented by the chemical formula $(CH_3)_2CO$, it is the simplest, smallest, and commercially most-important aliphatic ketone. Almost 7 million tons of acetone are produced globally per year [203]. The development of acetone industrial production by fermentation was promoted by the outbreak of the First World War. The Weismann process was patented in March 1915 [87,204]. The main objective to produce acetone by this time was the conversion into cordite, a smokeless ammunition powder [87,204]. However, in the 1950s, routes to produce solvents from oil were made cost-competitive with fermentation. Furthermore, the major feedstock for fermentation processes was molasses, which spiked in price because of animal feed demand [204,205]. Today, approximately 100 years after the publication of the Weizmann process, its modern approaches are gaining interest once more, with a different driver, namely, the search for more sustainable processes [204].

At present, the global acetone industry is driven by the solvents sector, representing approximately 34% of global demand in 2017. As a solvent, it is used in products, such as nail polish removers, cement, lacquers, cleaners, paint, coatings, films, and adhesives [206]. It is also used for MMA (methyl methacrylate) production, to form PMMA (polymethyl methacrylate), and as a fuel additive [206,207].

Acetone can be produced through different methods, such as the cumene process, from alkane nitriles, hydrolysis of geminal dihalides, dehydrogenation of isopropyl alcohol, ozonolysis of alkanes, and the fermentation process. About 96% of the world's acetone production is as a by-product of phenol by the cumene process, which uses benzene and propylene, two petrochemical products, as raw materials [206]. Integrated metabolic and evolutionary engineering was applied to *C. cellulovorans* and a 138-fold increase in butanol production was achieved [208]. Due to the current concern about greenhouse gas emissions and the replacement of petrochemical processing by renewable processes and feedstocks, fermentation technologies are of interest to academic and industrial communities [207].

The fermentation of sugar using *Clostridium* strains is a well-known industrial process, for which main products are acetone, butanol, and ethanol (ABE), obtained in a molar ratio of 3:6:1 [207]. Strictly anaerobic Gram-positive microorganisms possess a fermentation metabolism and utilize a variety of sugars, oligosaccharides, and polysaccharides by two distinct phases, acidogenic and solventogenic [209]. Usually, desired products of ABE fermentation are butanol and ethanol; acetone is the major by-product and is produced in the solventogenic phase, along with ethanol and butanol [207]. Glucose is the preferred substrate for ABE fermentation; however, a range of substrates can be used, such as simple sugars (galactose and xylose), disaccharides (maltose, sucrose, and lactose) and complex polysaccharides (cellulose and hemicellulose). Furthermore, direct conversion of starch without any hydrolysis step is possible with *Clostridium* strains [207].

The most widely used species for the production of acetone is *C. acetobutylicum*, although *C. beijerinkii* has also been used with good results. Li et al. [210] reported a fermentation process in a 7 L bioreactor with *C. acetobutylicum* ATCC824, using a cassava-based medium. Six grams per liter of acetone were produced after 50 h of fermentation. The addition of acetate in the culture medium is a new approach to increase acetone and butanol production in ABE fermentation [211,212]. Acetone production doubled with the addition of 4 g/L of acetate in the culture medium containing 50 g/L of glucose with *C. saccharoperbutylacetonicum* N1-4 [211]. For pretreated corncob fermentation by *C. beijerinkii* TISTR1461, an inferior increase (8%) was detected with acetate addition [212].

Batch is the most common operational process for ABE studies, despite its drawbacks, such as low cell density, low reactor productivity, nutritional limitations, and severe product inhibition. An important approach that has attracted interest to achieve greater conversion rates in ABE fermentation is cell immobilization technology. The immobilization can be done by adsorption or entrapment, but due to diffusion limitation problems, passive adhesion to surfaces is preferred [213]. Six grams per liter of acetone was obtained as by-product with *C. acetobutylicum* DSM792 using glucose as a substrate in a continuous column bioreactor with wood pulp as the immobilization material [213]. Kong et al. [39] reported a batch fermentation using free and immobilized *C. acetobutylicum* XY16 with modified sugarcane bagasse as a support for immobilization. A 1.4 times increase in acetone production was achieved with the cell immobilized batch process.

4.4. 1,3-Propanediol

1,3-Propanediol (1,3-PDO), also called trimethylene glycol, 1,3-dihydroxypropane, or propylene glycol, is a three-carbon molecule with the molecular formula $C_3H_8O_2$ [214]. Due to its enormous applications in the synthesis of polyesters, polyethers, polyurethanes, and heterocyclic compounds, such as indole and quinolines, 1,3-PDO is a promising bulk chemical. The chemical industry also uses 1,3-PDO to obtain resins, engine coolants, mortars, and inks [215]. The most important application involves the synthesis of a superior polymer known as polytrimethylene terephthalate (PTT). PTT is a biodegradable polyester with great potential application in textile, carpet, and upholstery manufacturing. Compared to other similar polymers, such as polyethylene terephthalate (PET), polybutylene terephthalate (PBT), or nylon, PTT has better elasticity and better modulus. It shows higher ultraviolet (UV) resistance and tends less to electrostatic charging [214,216].

A number of different chemical and fermentative routes can be used to synthesize 1,3-PDO. The chemical routes to produce 1,3-PDO are hydration of acrolein, followed by hydroformylation and hydrogenation of ethylene oxide. These processes require expensive catalysts, high temperature, and pressure, with several disadvantages, such as toxic intermediates and non-renewable raw materials [214,216]. Alternative fermentative routes are more environmentally friendly compared to chemical processes. Glycerol is the unique substrate that can be directly metabolized to 1,3-PDO. Native microorganisms that convert glycerol into 1,3-PDO belong mainly to the genera *Clostridium*, *Klebsiella*, *Citrobacter*, *Enterobacter* and *Lactobacillus*. All are able to produce 1,3-PDO using glycerol as the sole carbon source except *Lactobacillus*, which produces 1,3-PDO by co-fermentation using both sugar and glycerol as substrates simultaneously [214]. Among these microorganisms, *K. pneumoniae* and *C. butyricum* have been most intensively studied due to their substrate tolerance, high yield, and productivity. Although *C. butyricum* is strictly anaerobic and *K. pneumonia* is facultative anaerobic, *C. butyricum* is more interesting for industrial application because *K. pneumonia* is classified as an opportunistic pathogen [217]. Another possibility of producing 1,3-PDO involves genetically engineered strains that can utilize sugar to glycerol and glycerol to 1,3-PDO consecutively [214,216,217].

Many strategies have been investigated to increase yield and productivity of 1,3-PDO using *Clostridium* strains, including by-product reduction; substrate and product inhibition reduction; different bioreactor operation modes; mutagenesis and genetic modification of strains; cell immobilization; and co-culture fermentation [218–220]. Production of 1,3-PDO from glycerol using different *Clostridium* species and strategies is compared in Table 5.

Table 5. Production of 1,3-PDO from glycerol using different *Clostridium butyricum* strains and strategies.

Strain	FS	Strategy	Scale	$C_{1,3\text{-PDO}}$ [1]	$Y_{1,3\text{-PDO}}$ [2]	$Q_{1,3\text{-PDO}}$ [3]	Ref. [4]
C. butyricum CNCM1211	PG [5]	B [7]	1-L	37.1	0.53	1.32	[221]
C. butyricum CNCM1211	CG [6]	B	1-L	63.4	0.57	0.63	[221]
C. butyricum F2b	CG	B	2-L	47.1	0.52	1.34	[222]
C. butyricum F2b	CG	C [8]	2-L	44.0	0.51	1.76	[222]
C. butyricum VPI 3266	PG	C	2-L	29.7	0.62	2.98	[220]
C. butyricum VPI 3266	CG	C	2-L	31.5	0.50	3.15	[223]
C. butyricum AKR102a	PG	FB [9]	1-L	93.7	0.52	3.30	[40]
C. butyricum AKR102a	CG	FB	1-L	76.2	0.51	2.30	[40]
C. butyricum AKR102a	CG	FB	200-L	61.5	0.53	2.10	[40]
C. butyricum DSP1	CG	B	6.6-L	37.6	0.53	1.12	[224]
C. butyricum DSP1	CG	FB	6.6-L	71.0	0.54	0.68	[224]
C. butyricum DSP1	CG	B	42-L	36.4	0.52	1.13	[224]
C. butyricum DSP1	CG	B	150-L	37.2	0.53	1.33	[224]
C. butyricum VPI1718	CG	RB [10]	1-L	65.5	0.52	1.15	[225]
C. butyricum NCIMB 8082	CG	B	1-L	32.2	0.52	2.38	[226]
C. butyricum NCIMB 8082	CG	FB	1-L	29.8	0.48	2.55	[226]

[1] $C_{1,3\text{-PDO}}$, final 1,3-PDO concentration (g/L); [2] $Y_{1,3\text{-PDO}}$, 1,3-PDO produced per glycerol consumed (g 1,3-PDO/g glycerol); [3] $Q_{1,3\text{-PDO}}$, 1,3-PDO overall productivity (g/L.h); [4] Ref.: reference; [5] PG, pure glycerol; [6] CG, crude glycerol; [7] B, batch fermentation; [8] C, continuous fermentation; [9] FB, fed-batch fermentation; [10] RB, repeated batch cultivation.

4.5. Other Biomolecules

Clostridium spp. may also produce other interesting molecules besides those already mentioned. Butyric acid, isopropyl alcohol, *n*-caproic acid, and hexanol are some of the chemicals that can also be produced by the genus *Clostridium* [198,227–230].

Butyric acid is a four-carbon volatile acid with a global market of more than 80,000 tons per year [227,231]. This carboxylic acid is an important chemical, as its derivatives present several commercial applications. For example, butyrate esters can be applied as fragrance- and flavor-enhancing agents in beverages, food, and cosmetics [227]. Due to its many health benefits, including antineoplastic effects on the large intestine and colon, butyrate and its derivatives have been widely used in pharmaceuticals as drugs for treating hemoglobinopathies, colon cancer, and gastrointestinal diseases; and as prebiotic supplements to animal feeds replacing antibiotics [232]. Another relevant application of butyric acid is butanol production by either biotransformation using microorganisms or catalytic chemical process [230,232].

At present, butanol is mainly produced by the petrochemical industry via oxidation of butyraldehyde obtained from oxosynthesis of propylene [227,232]. However, concerns regarding environmental impacts and the rising desire to use renewable resources have driven industrial attention toward fermentative production of butyric acid, especially for applications in the food and pharmaceutical industries [227,232].

Although butyric acid can be synthesized by various strains belonging to the genera of *Clostridium, Butyrivibrio, Butyribacterium, Sarcina, Eubacterium, Fusobacterium, Megasphaera, Roseburia,*and *Coprococcus* [227,231,232], the preferred strains for potential commercial uses are from the genus *Clostridium* because of their higher and stable productivity and production titers [230,232]. Most butyric acid-producing clostridia ferment glucose, xylose, lactose, starch, and glycerol for cell growth and butyric acid production. Cellulose and CO_2 can also be converted into butyrate by some species [227,232]. *C. carboxidivorans* can utilize CO, CO_2, and H_2 to produce butyric acid. *C. cellulovorans, C. polysaccharolyticum,* and *C. populeti* can use cellulose for butyrate synthesis [232]. Some *Clostridium* species can be manipulated to produce either solvents or acids as the main products, depending on culture conditions. Several species, including *C. butyricum, C. tyrobutyricum,* and *C. thermobutyricum* produce butyrate as the main product with a relatively high productivity and yield; thus, they are the most studied species due to their high commercial potential for butyric acid production. *C. kluyveri* can produce butyric acid as a major product from ethanol and acetate as substrates; however, *n*-caproic acid is produced instead of butyrate when ethanol is present in excess of acetate [227,232].

n-Caproic acid, a six-carbon chain carboxilyc acid, is a versatile platform chemical for producing flavor additives for the food industry and biofuels for aviation. This molecule is also a potential antimicrobial agent that can be used as a green antibiotic [228,233]. *n*-Caproic acid is produced by chain elongation of a carboxylic acid, which is a reversed β-oxidation pathway using ethanol or lactic acid as an electron donor. Short-chain carboxylic acids are elongated by adding two carbons (acetyl-CoA derived from ethanol) each time, converting them into chemicals with six or more carbons [228,234].

C. kluyveri and *Clostridium* sp. BS-1 were identified as *n*-caproic acid producers [233]. The use of *C. kluyveri* pure culture was reported in the caprogenic processes, but open cultures are known to increase metabolite production. Studies have used different substrates, such as acetic acid with ethanol, syngas fermentation effluent with ethanol, yeast-fermentation beer, and lactic acid [233,235].

Another valuable organic molecule that can be obtained by *Clostridium* anaerobic fermentation is isopropanol, also called isopropyl alcohol. Isopropanol is an important organic solvent used in paints and varnishes, removers, antiseptic solutions, printing, perfumery, and cosmetics [236]. This molecule is also an additive for gasoline and diesel, and an important precursor for the production of green propylene [229].

Although ABE fermentations are usually conducted using *C. acetobutylicum*, several *C. beijerinckii* sp., having an additional primary/secondary alcohol dehydrogenase, can convert acetone to isopropanol. Studies using *C. beijerinckii* have reported mainly butanol and isopropanol production with very little ethanol/acetone, unlike other ABE organisms, where the ratio is 6:3:1 (butanol:acetone:ethanol weight ratio) [227,229,237].

Hexanol is also an organic solvent that can be produced by *Clostridium* autotrophic growth on CO and H_2. *C. carboxidivorans* derives energy and carbon for growth from CO and H_2, and produces ethanol, butanol, and hexanol by reduction of organic acids formed in the mechanism of energy conservation and synthesis of cell material [86,197].

5. Concluding Remarks

First-generation biorefinery processes are based on the use of sugar-containing food crops as feedstock and lead to food–fuel competition. The use of lignocellulosic biomass, waste, or waste gases increases the sustainability of biorefineries, thereby overcoming the food–fuel dilemma. These two concepts can be integrated in a biorefinery, reducing the use of food crops, with the help of *Clostridium* species, as proposed in the hypothetical scenario presented in Figure 1. In the example, because sugarcane bagasse can be used to generate fuels and chemicals, reduced amounts of sugarcane juice are needed for this purpose. In this concept, the lignocellulosic biomass (sugarcane bagasse) can be submitted to pre-treatment for enzymatic hydrolysis for microbial transformation (biochemical route), and also be thermally converted to syngas for microbial fermentation (hybrid route). In both routes, significant progress has been made. Economic and market data can be used to choose between these processes, minimizing the disadvantages related to each one. *Clostridium* species are essential in this concept, since they can use syngas for solvent and fuel production, and can also use cellulosic material directly without the need of enzymatic hydrolysis. Further research and development are essential to improve yield and productivity, and to reduce the production costs of syngas fermentation. To achieve this, strategies of developing recombinant clostridia to increase product tolerance and the use of metabolic engineering to direct carbon flow to the production of target molecules must be adopted. In addition, the design of new bioreactor configurations to circumvent inherent problems of gas–liquid mass transfer, along with process optimization and downstream integration, will result in an efficient process with greater scalability potential.

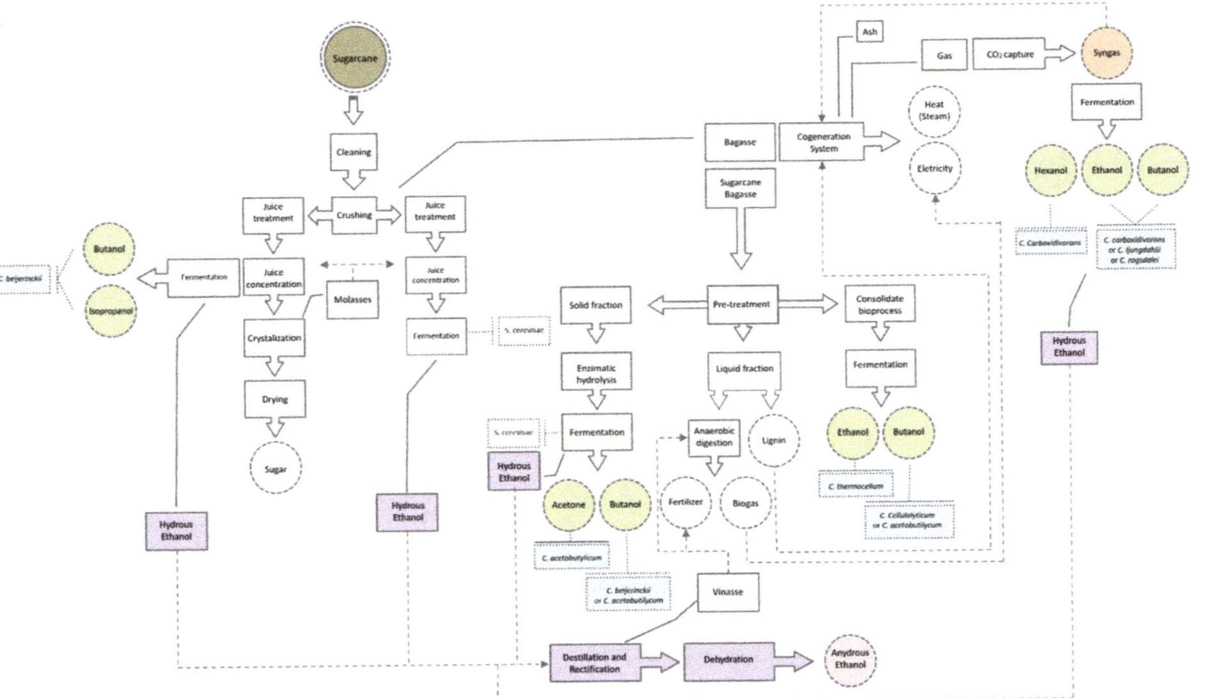

Figure 1. Schematic diagram of a sugarcane biorefinery producing ethanol, sugar, power, butanol, syngas, fertilizer, hexanol, and acetone, based on [238–241] and on information of the present review. Dotted circle—intermediate products and end products; dotted arrows—product destination from one process for use as raw material in another process; blue—*Clostridium* species; green—direct *Clostridium* metabolism products; purple—intermediate product after bioprocess using *Clostridium* species and its process for product refining; red—by-product after bioprocess using *Clostridium* species; and orange—residue used as raw material for fermentation using *Clostridium* species.

Author Contributions: In the present review, all authors contributed substantially to the document in the following parts: conceptualization: V.L., C.B., F.C., A.B., P.A., N.P.J., and T.F.; writing—original draft preparation: V.L., C.B., F.C., A.B., P.A., N.P.J., and T.F.; writing—review and editing: P.A., N.P.J., and T.F.; project administration: P.A., N.P.J., and T.F.; funding acquisition: P.A., N.P.J., and T.F.

Funding: This research was funded Fundação Carlos Chagas Filho de Amparo à Pesquisa do Estado do Rio de Janeiro-Faperj, grant no. E-26/010.002984/2014 (Pensa Rio 2014).Vanessa Liberato received a scholarship from Coordenação de Aperfeiçoamento de Pessoal de Nível Superior (CAPES).

Conflicts of Interest: The authors declare no conflict of interest.

References

1. Earth Overshoot Day Earth Overshoot Day 201 is August 1. Available online: https://www.overshootday.org/newsroom/press-release-july-2018-english/ (accessed on 7 January 2019).

2. Intergovernmental Panel on Climate Change. *Climate Change 2014: Synthesis Report*; Chapter Observed Changes and Their Causes; Intergovernmental Panel on Climate Change: Geneva, Switzerland, 2014; ISBN 9789291691432.

3. Cherubini, F. The biorefinery concept: Using biomass instead of oil for producing energy and chemicals. *Energy Convers. Manag.* **2010**, *51*, 1412–1421. [CrossRef]

4. Sustainable Development Goals The Sustainable Development Agenda. Available online: https://www.un.org/sustainabledevelopment/development-agenda/ (accessed on 15 January 2019).

5. Drosg, B.; Braun, R.; Bochmann, G.; Al Saedi, T. Analysis and characterisation of biogas feedstocks. In *The Biogas Handbook: Science, Production and Applications*; Murphy, J.D., Baxter, D., Eds.; Woodhead Publishing: Sawston/Cambridge, UK, 2013; pp. 52–84, ISBN 9780857097415.

6. Al Seadi, T.; Rutz, D.; Janssen, R.; Drosg, B. Biomass resources for biogas production. In *The Biogas Handbook: Science, Production and Applications*; Murphy, J.D., Baster, D., Eds.; Woodhead Publishing: Sawston/Cambridge, UK, 2013; pp. 19–51, ISBN 9780857097415.

7. Shen, Y.; Jarboe, L.; Brown, R.; Wen, Z. A thermochemical–biochemical hybrid processing of lignocellulosic biomass for producing fuels and chemicals. *Biotechnol. Adv.* **2015**, *33*, 1799–1813. [CrossRef] [PubMed]

8. Latif, H.; Zeidan, A.A.; Nielsen, A.T.; Zengler, K. Trash to treasure: Production of biofuels and commodity chemicals via syngas fermenting microorganisms. *Curr. Opin. Biotechnol.* **2014**, *27*, 79–87. [CrossRef] [PubMed]

9. Murphy, J.D.; Korres, N.E.; Singh, A.; Smyth, B.; Nizami, A.-S.; Thamsiriroj, T. *The Potential for Grass Biomethane as a Biofuel: Compressed Biomethane Generated from Grass, Utilised as a Transport Biofuel*; Environmental Protection Authority, Ireland: Wexford, Ireland, 2011; ISBN 9781840954272.

10. Hassan, S.S.; Willians, G.A.; Jaiswal, A.K. Moving towards the second generation of lignocellulosic biorefineries in the EU: Drivers, challenges, and opportunities. *Renew. Sustain. Energy Rev.* **2018**, *101*, 590–599. [CrossRef]

11. International Energy Agency (IEA) Bioenergy. *IEA Bioenergy Task 42 Biorefining*; IEA: Paris, France, 2014.

12. Parada, M.P.; Asveld, L.; Osseweijer, P.; Duque, J.A.P. An approach for incorporating sustainability in early stages of biorefinery design. In Proceedings of the European Biomass Conference and Exhibition, Stockholm, Sweden, 12–15 June 2017; pp. 105–123.

13. Parajuli, R.; Dalgaard, T.; Jørgensen, U.; Adamsen, A.P.S.; Knudsen, M.T.; Birkved, M.; Gylling, M.; Schjørring, J.K. Biorefining in the prevailing energy and materials crisis: A review of sustainable pathways for biorefinery value chains and sustainability assessment methodologies. *Renew. Sustain. Energy Rev.* **2015**, *43*, 244–263. [CrossRef]

14. Liou, J.S.C.; Balkwill, D.L.; Drake, G.R.; Tanner, R.S. *Clostridium carboxidivorans* sp. nov., a solvent-producing clostridium isolated from an agricultural settling lagoon, and reclassification of the acetogen *Clostridium scatologenes* strain SL1 as *Clostridium drakei* sp. nov. *Int. J. Syst. Evol. Microbiol.* **2005**, *55*, 2085–2091. [CrossRef]

15. Daniell, J.; Köpke, M.; Simpson, S.D. Commercial biomass syngas fermentation. *Energies* **2012**, *5*, 5372–5417. [CrossRef]

16. Schiel-bengelsdorf, B.; Dürre, P. Pathway engineering and synthetic biology using acetogens. *FEBS Lett.* **2012**, *586*, 2191–2198. [CrossRef]

17. Kopke, M.; Held, C.; Hujer, S.; Liesegang, H.; Wiezer, A.; Wollherr, A.; Ehrenreich, A.; Liebl, W.; Gottschalk, G.; Durre, P. *Clostridium ljungdahlii* represents a microbial production platform based on syngas. *Proc. Natl. Acad. Sci. USA* **2010**, *107*, 13087–13092. [CrossRef]

18. Lawson, P.A.; Rainey, F.A. Proposal to restrict the genus *Clostridium* prazmowski to *Clostridium butyricum* and related species. *Int. J. Syst. Evol. Microbiol.* **2016**, *66*, 1009–1016. [CrossRef]

19. Vos, P.; Garrity, G.; Jones, D.; Krieg, N.R.; Ludwig, W.; Rainey, F.A.; Schleifer, K.; Whitman, W.B. (Eds.) *Bergey's Manual of Systematic Bacteriology: Volume 3: The Firmicutes*; Springer Science & Business Media: Heidelberg, Germany; London, UK; New York, NY, USA, 2011; pp. 738–827.

20. Gomes, M.J.P. *Gênero Clostridium spp.*; Favet-Ufrgs: Porto Alegre, Brazil, 2013; pp. 1–67.

21. Dalgleish, T.; Williams, J.M.G.; Golden, A.-M.J.; Perkins, N.; Barrett, L.F.; Barnard, P.J.; Yeung, C.A.; Murphy, V.; Elward, R.; Tchanturia, K.; et al. *Chapter 2 Prokaryotes and Their Habitats*; Springer: Berlin/Heidelberg, Germany, 2007; Volume 136, ISBN 9783662131893.

22. Fernández-Naveira, Á.; Abubackar, H.N.; Veiga, M.C.; Kennes, C. Efficient butanol-ethanol (BE) production from carbon monoxide fermentation by *Clostridium carboxidivorans*. *Appl. Microbiol. Biotechnol.* **2016**, *100*, 3361–3370. [CrossRef] [PubMed]

23. Shen, Y.; Brown, R.; Wen, Z. Syngas fermentation of *Clostridium carboxidivoran* P7 in a hollow fiber membrane biofilm reactor: Evaluating the mass transfer coefficient and ethanol production performance. *Biochem. Eng. J.* **2014**, *85*, 21–29. [CrossRef]

24. Sun, X.; Atiyeh, H.K.; Zhang, H.; Tanner, R.S.; Huhnke, R.L. Enhanced ethanol production from syngas by *Clostridium ragsdalei* in continuous stirred tank reactor using medium with poultry litter biochar. *Appl. Energy* **2019**, *236*, 1269–1279. [CrossRef]

25. Atiyeh, H.K.; Lewis, R.S.; Phillips, J.R.; Huhnke, R.L. Method Improving Producer Gas Fermentation. U.S. Patent 10053711B2, 21 August 2018.

26. Richter, H.; Martin, M.E.; Angenent, L.T. A two-stage continuous fermentation system for conversion of syngas into ethanol. *Energies* **2013**, *6*, 3987–4000. [CrossRef]

27. Phillips, J.R.; Klasson, K.T.; Clausen, E.C.; Gaddy, J.L. Biological production of ethanol from coal synthesis gas—Medium development studies. *Appl. Biochem. Biotechnol.* **1993**, *39–40*, 559–571. [CrossRef]

28. Yin, Y.; Zhang, Y.; Karakashev, D.B.; Wang, J.; Angelidaki, I. Biological caproate production by *Clostridium kluyveri* from ethanol and acetate as carbon sources. *Bioresour. Technol.* **2017**, *241*, 638–644. [CrossRef]

29. Qureshi, N.; Cotta, M.A.; Saha, B.C. Bioconversion of barley straw and corn stover to butanol (a biofuel) in integrated fermentation and simultaneous product recovery bioreactors. *Food Bioprod. Process.* **2014**, *92*, 298–308. [CrossRef]

30. Zhang, J.; Jia, B. Enhanced butanol production using *Clostridium beijerinckii* SE-2 from the waste of corn processing. *Biomass Bioenergy* **2018**, *115*, 260–266. [CrossRef]

31. Khunchantuek, C.; Fiala, K. Optimization of key factors affecting butanol production from sugarcane juice by *Clostridium beijerinckii* TISTR 1461. *Energy Procedia* **2017**, *138*, 157–162. [CrossRef]

32. Zhang, C.; Li, T.; He, J. Characterization and genome analysis of a butanol–isopropanol-producing *Clostridium beijerinckii* strain BGS1. *Biotechnol. Biofuels* **2018**, *11*, 280. [CrossRef]

33. Dong, J.-J.; Ding, J.-C.; Zhang, Y.; Ma, L.; Xu, G.-C.; Han, R.-Z.; Ni, Y. Simultaneous saccharification and fermentation of dilute alkaline-pretreated corn stover for enhanced butanol production by *Clostridium saccharobutylicum* DSM 13864. *FEMS Microbiol. Lett.* **2016**, *363*, fnw003. [CrossRef] [PubMed]

34. Darmayanti, R.F.; Tashiro, Y.; Noguchi, T.; Gao, M.; Sakai, K.; Sonomoto, K. Novel biobutanol fermentation at a large extractant volume ratio using immobilized *Clostridium saccharoperbutylacetonicum* N1-4. *J. Biosci. Bioeng.* **2018**, *126*, 750–757. [CrossRef] [PubMed]

35. Balusu, R.; Paduru, R.R.; Kuravi, S.K.; Seenayya, G.; Reddy, G. Optimization of critical medium components using response surface methodology for ethanol production from cellulosic biomass by *Clostridium thermocellum* SS19. *Process Biochem.* **2005**, *40*, 3025–3030. [CrossRef]

36. Jin, M.; Gunawan, C.; Balan, V.; Dale, B.E. Consolidated bioprocessing (CBP) of AFEX™-pretreated corn stover for ethanol production using *Clostridium phytofermentans* at a high solids loading. *Biotechnol. Bioeng.* **2012**, *109*, 1929–1936. [CrossRef]

37. Lee, S.H.; Kim, S.; Kim, J.Y.; Cheong, N.Y.; Kim, K.H. Enhanced butanol fermentation using metabolically engineered *Clostridium acetobutylicum* with ex situ recovery of butanol. *Bioresour. Technol.* **2016**, *218*, 909–917. [CrossRef]

38. Lehmann, D.; Lütke-Eversloh, T. Switching *Clostridium acetobutylicum* to an ethanol producer by disruption of the butyrate/butanol fermentative pathway. *Metab. Eng.* **2011**, *13*, 464–473. [CrossRef]

39. Kong, X.; He, A.; Zhao, J.; Wu, H.; Jiang, M. Efficient acetone-butanol-ethanol production (ABE) by *Clostridium acetobutylicum* XY16 immobilized on chemically modified sugarcane bagasse. *Bioprocess Biosyst. Eng.* **2015**, *38*, 1365–1372. [CrossRef]

40. Wilkens, E.; Ringel, A.K.; Hortig, D.; Willke, T.; Vorlop, K.D. High-level production of 1,3-propanediol from crude glycerol by *Clostridium butyricum* AKR102a. *Appl. Microbiol. Biotechnol.* **2012**, *93*, 1057–1063. [CrossRef]

41. Li, Y.; Tschaplinski, T.J.; Engle, N.L.; Hamilton, C.Y.; Modriguez, R., Jr.; Liao, J.C.; Schadt, C.W.; Guss, A.M.; Yang, Y.; Graham, D.E. Combined inactivation of the *Clostridium cellulolyticum* lactate and malate dehydrogenase genes substantially increases ethanol yield from cellulose and switchgrass fermentations. *Biotechnol. Biofuels* **2012**, *5*, 2. [CrossRef]

42. Ou, J.; Xu, N.; Ernst, P.; Ma, C.; Bush, M.; Goh, K.; Zhao, J.; Zhou, L.; Yang, S.T.; Liu, X.M. Process engineering of cellulosic n-butanol production from corn-based biomass using *Clostridium cellulovorans*. *Process Biochem.* **2017**, *62*, 144–150. [CrossRef]

43. Malaviya, A.; Jang, Y.S.; Lee, S.Y. Continuous butanol production with reduced byproducts formation from glycerol by a hyper producing mutant of *Clostridium pasteurianum*. *Appl. Microbiol. Biotechnol.* **2012**, *93*, 1485–1494. [CrossRef] [PubMed]

44. Guo, Y.; Dai, L.; Xin, B.; Tao, F.; Tang, H.; Shen, Y.; Xu, P. 1,3-Propanediol production by a newly isolated strain, *Clostridium perfringens* GYL. *Bioresour. Technol.* **2017**, *233*, 406–412. [CrossRef] [PubMed]

45. Bengelsdorf, F.R.; Straub, M.; Dürre, P. Bacterial synthesis gas (syngas) fermentation. *Environ. Technol.* **2013**, *34*, 1639–1651. [CrossRef] [PubMed]

46. Coelho, F.; Nele, M.; Ribeiro, R.; Ferreira, T.; Amaral, P. *Clostridium carboxidivorans*' surface characterization using contact angle measurement (CAM). *Chem. Eng. Trans.* **2016**, *50*, 277–282.

47. Bruant, G.; Lévesque, M.J.; Peter, C.; Guiot, S.R.; Masson, L. Genomic analysis of carbon monoxide utilization and butanol production by *Clostridium carboxidivorans* strain P7T. *PLoS ONE* **2010**, *5*, e13033. [CrossRef] [PubMed]

48. Huhnke, R.; Lewis, R.; Tanner, R.S. Isolation and Characterization of Novel Clostridial Species. U.S. Patent 2008/0057554A1, 6 March 2008.

49. Saxena, J.; Tanner, R.S. Effect of trace metals on ethanol production from synthesis gas by the ethanologenic acetogen, *Clostridium ragsdalei*. *J. Ind. Microbiol. Biotechnol.* **2011**, *38*, 513–521. [CrossRef]

50. Ragsdale, S.W. Life with carbon monoxide. *Crit. Rev. Biochem. Mol. Biol.* **2004**, *39*, 165–195. [CrossRef]

51. Saxena, J.; Tanner, R.S. Optimization of a corn steep medium for production of ethanol from synthesis gas fermentation by *Clostridium ragsdalei*. *World J. Microbiol. Biotechnol.* **2012**, *28*, 1553–1561. [CrossRef]

52. Devarapalli, M.; Atiyeh, H.K.; Phillips, J.R.; Lewis, R.S.; Huhnke, R.L. Ethanol production during semi-continuous syngas fermentation in a trickle bed reactor using *Clostridium ragsdalei*. *Bioresour. Technol.* **2016**, *209*, 56–65. [CrossRef]

53. Kundiyana, D.K.; Wilkins, M.R.; Maddipati, P.; Huhnke, R.L. Effect of temperature, pH and buffer presence on ethanol production from synthesis gas by "*Clostridium ragsdalei*". *Bioresour. Technol.* **2011**, *102*, 5794–5799. [CrossRef]

54. Tanner, R.S.; Miller, L.M.; Yang, D. *Clostridium ljungdahlii* sp. nov., an acetogenic species in clostridial rRNA homology group I. *Int. J. Syst. Bacteriol.* **1993**, *43*, 232–236. [CrossRef] [PubMed]

55. Barker, H.A.; Taha, S.M. *Clostridium kluyverii*, an organism concerned in the formation of caproic acid from ethyl alcohol. *J. Bacteriol.* **1941**, *43*, 347–363.

56. Keis, S.; Shaheen, R.; Jones, D.T. Emended descriptions of *Clostridium acetobutylicum* and Clostridium beijerinckii, and descriptions of *Clostridium saccharoperbutylacetonicum* sp. nov. and *Clostridium saccharobutylicum* sp. nov. *Int. J. Syst. Evol. Microbiol.* **2001**, *51*, 2095–2103. [CrossRef] [PubMed]

57. Trchounian, K.; Müller, N.; Schink, B.; Trchounian, A. Glycerol and mixture of carbon sources conversion to hydrogen by *Clostridium beijerinckii* DSM791 and effects of various heavy metals on hydrogenase activity. *Int. J. Hydrog. Energy* **2017**, *42*, 7875–7882. [CrossRef]

58. Diallo, M.; Simons, A.D.; van der Wal, H.; Collas, F.; Houweling-Tan, B.; Kengen, S.W.M.; López-Contreras, A.M. L-Rhamnose metabolism in *Clostridium beijerinckii* strain DSM 6423. *Appl. Environ. Microbiol.* **2018**, *85*, e02656-18. [CrossRef]

59. Ni, Y.; Xia, Z.; Wang, Y.; Sun, Z. Continuous butanol fermentation from inexpensive sugar-based feedstocks by *Clostridium saccharobutylicum* DSM 13864. *Bioresour. Technol.* **2013**, *129*, 680–685. [CrossRef]

60. Gu, Y.; Feng, J.; Zhang, Z.T.; Wang, S.; Guo, L.; Wang, Y.; Wang, Y. Curing the endogenous megaplasmid in *Clostridium saccharoperbutylacetonicum* N1-4 (HMT) using CRISPR-Cas9 and preliminary investigation of the role of the plasmid for the strain metabolism. *Fuel* **2019**, *236*, 1559–1566. [CrossRef]

61. Wang, S.; Dong, S.; Wang, P.; Tao, Y.; Wang, Y. Genome editing in *Clostridium saccharoperbutylacetonicum* N1-4 with the CRISPR-Cas9 system. *Appl. Environ. Microbiol.* **2017**, *83*, e00233-17. [CrossRef]

62. Viljoen, J.A.; Fred, E.B.; Peterson, W.H. The fermentation of cellulose by thermophilic bacteria. *J. Agric. Sci.* **1926**, *16*, 1–17. [CrossRef]

63. Lo, J.; Olson, D.G.; Murphy, S.J.; Tian, L.; Hon, S.; Lanahan, A.; Guss, A.M.; Lynd, L.R. Engineering electron metabolism to increase ethanol production in *Clostridium thermocellum*. *Metab. Eng.* **2017**, *39*, 71–79. [CrossRef]

64. Dash, S.; Olson, D.G.; Chan, S.H.J.; Amador-Noguez, D.; Lynd, L.R.; Maranas, C.D. Thermodynamic analysis of the pathway for ethanol production from cellobiose in *Clostridium thermocellum*. *Metab. Eng.* **2019**, *55*, 161–169. [CrossRef] [PubMed]

65. Methé, B.A.; Leschine, S.B.; Warnick, T.A. *Clostridium phytofermentans* sp. nov., a cellulolytic mesophile from forest soil. *Int. J. Syst. Evol. Microbiol.* **2002**, *52*, 1155–1160.

66. Fathima, A.A.; Sanitha, M.; Kumar, T.; Iyappan, S.; Ramya, M. Direct utilization of waste water algal biomass for ethanol production by cellulolytic *Clostridium phytofermentans* DSM1183. *Bioresour. Technol.* **2016**, *202*, 253–256. [CrossRef] [PubMed]

67. Algayyim, S.J.M.; Wandel, A.P.; Yusaf, T.; Hamawand, I. Production and application of ABE as a biofuel. *Renew. Sustain. Energy Rev.* **2018**, *82*, 1195–1214. [CrossRef]

68. Tsvetanova, F.; Petrova, P.; Petrov, K. Microbial production of 1-butanol-recent advances and future prospects (review). *J. Chem. Technol. Metall.* **2018**, *53*, 683–696.

69. Zhou, W.; Liu, J.; Fan, S.; Xiao, Z.; Qiu, B.; Wang, Y.; Li, J.; Liu, Y. Biofilm immobilization of *Clostridium acetobutylicum* on particulate carriers for acetone-butanol-ethanol (ABE) production. *Bioresour. Technol. Rep.* **2018**, *3*, 211–217. [CrossRef]

70. Yang, Y.; Nie, X.; Jiang, Y.; Yang, C.; Gu, Y.; Jiang, W. Metabolic regulation in solventogenic clostridia: Regulators, mechanisms and engineering. *Biotechnol. Adv.* **2018**, *36*, 905–914. [CrossRef]

71. Sun, C.; Zhang, S.; Xin, F.; Shanmugam, S.; Wu, Y.R. Genomic comparison of *Clostridium* species with the potential of utilizing red algal biomass for biobutanol production. *Biotechnol. Biofuels* **2018**, *11*, 42. [CrossRef]

72. Li, C.; Wang, Y.; Xie, G.; Peng, B.; Zhang, B.; Chen, W.; Huang, X.; Wu, H.; Zhang, B. Complete genome sequence of *Clostridium butyricum* JKY6D1 isolated from the pit mud of a Chinese flavor liquor-making factory. *J. Biotechnol.* **2016**, *220*, 23–24. [CrossRef]

73. Cardoso, C.J.T.; de Junior, J.S.O.; Kischel, H.; da Silva, W.A.L.; da Arruda, E.D.S.; Souza-Cáceres, M.B.; de Oliveira, F.A.M.; Nogueira, É.; de Nogueira, G.P.; de Melo-Sterza, F.A. Anti-Müllerian hormone (AMH) as a predictor of antral follicle population in heifers. *Anim. Reprod.* **2017**, *15*, 12–16. [CrossRef]

74. Zhang, A.H.; Liu, H.L.; Huang, S.Y.; Fu, Y.S.; Fang, B.S. Metabolic profiles analysis of 1,3-propanediol production process by *Clostridium butyricum* through repeated batch fermentation coupled with activated carbon adsorption. *Biotechnol. Bioeng.* **2017**, *115*, 684–693. [CrossRef]

75. Gaida, S.M.; Liedtke, A.; Jentges, A.H.W.; Engels, B.; Jennewein, S. Metabolic engineering of *Clostridium cellulolyticum* for the production of n-butanol from crystalline cellulose. *Microb. Cell Fact.* **2016**, *15*, 6. [CrossRef]

76. Desvaux, M. *Clostridium cellulolyticum*: Model organism of mesophilic cellulolytic clostridia. *FEMS Microbiol. Rev.* **2005**, *29*, 741–764. [CrossRef] [PubMed]

77. Tomita, H.; Okazaki, F.; Tamaru, Y. Direct IBE fermentation from mandarin orange wastes by combination of *Clostridium cellulovorans* and *Clostridium beijerinckii*. *AMB Express* **2019**, *9*, 1. [CrossRef] [PubMed]

78. Xin, F.; Dong, W.; Zhang, W.; Ma, J.; Jiang, M. Biobutanol production from crystalline cellulose through consolidated bioprocessing. *Trends Biotechnol.* **2019**, *37*, 167–180. [CrossRef] [PubMed]

79. Groeger, C.; Wang, W.; Sabra, W.; Utesch, T.; Zeng, A.P. Metabolic and proteomic analyses of product selectivity and redox regulation in *Clostridium pasteurianum* grown on glycerol under varied iron availability. *Microb. Cell Fact.* **2017**, *16*, 64. [CrossRef]

80. Ahn, J.-H.; Sang, B.-I.; Um, Y. Butanol production from thin stillage using *Clostridium pasteurianum*. *Bioresour. Technol.* **2011**, *102*, 4934–4937. [CrossRef]

81. López-Enríquez, L.; Rodríguez-Lázaro, D.; Hernández, M. Quantitative detection of *Clostridium tyrobutyricum* in milk by real-time PCR. *Appl. Environ. Microbiol.* **2007**, *73*, 3747–3751. [CrossRef]

82. Li, L.; Ai, H.; Zhang, S.; Li, S.; Liang, Z.; Wu, Z.-Q.; Yang, S.-T.; Wang, J.-F. Enhanced butanol production by coculture of *Clostridium beijerinckii* and *Clostridium tyrobutyricum*. *Bioresour. Technol.* **2013**, *143*, 397–404. [CrossRef]

83. Abrini, J.; Naveau, H.; Nyns, E.-J. *Clostridium autoethanogenumt*, sp. nov., an anaerobic bacterium that produces ethanol from carbon monoxide. *Arch. Microbiol.* **1994**, *161*, 345–351. [CrossRef]

84. Valgepea, K.; de Souza Pinto Lemgruber, R.; Abdalla, T.; Binos, S.; Takemori, N.; Takemori, A.; Tanaka, Y.; Tappel, R.; Köpke, M.; Simpson, S.D.; et al. H_2 drives metabolic rearrangements in gas-fermenting *Clostridium autoethanogenum*. *Biotechnol. Biofuels* **2018**, *11*, 55. [CrossRef] [PubMed]

85. Fernández-Naveira, Á.; Abubackar, H.N.; Veiga, M.C.; Kennes, C. Production of chemicals from C_1 gases (CO, CO_2) by *Clostridium carboxidivorans*. *World J. Microbiol. Biotechnol.* **2017**, *33*, 43. [CrossRef] [PubMed]

86. Fernández-Naveira, Á.; Veiga, M.C.; Kennes, C. H-B-E (hexanol-butanol-ethanol) fermentation for the production of higher alcohols from syngas/waste gas. *J. Chem. Technol. Biotechnol.* **2017**, *92*, 712–731. [CrossRef]

87. Jones, D.T.; Woods, D.R. Acetone-butanol fermentation revisited. *Microbiol. Rev.* **1986**, *50*, 484–524.

88. Moon, H.G.; Jang, Y.-S.; Cho, C.; Lee, J.; Binkley, R.; Lee, S.Y. One hundred years of clostridial butanol fermentation. *FEMS Microbiol. Lett.* **2016**, *363*, fnw001. [CrossRef]

89. Noguchi, T.; Tashiro, Y.; Yoshida, T.; Zheng, J.; Sakai, K.; Sonomoto, K. Efficient butanol production without carbon catabolite repression from mixed sugars with *Clostridium saccharoperbutylacetonicum* N1-4. *J. Biosci. Bioeng.* **2013**, *116*, 716–721. [CrossRef]

90. Fast, A.G.; Schmidt, E.D.; Jones, S.W.; Tracy, B.P. Acetogenic mixotrophy: NOVEL options for yield improvement in biofuels and biochemicals production. *Curr. Opin. Biotechnol.* **2015**, *33*, 60–72. [CrossRef]

91. Sun, X.; Atiyeh, H.K.; Adesanya, Y.A.; Zhang, H.; Okonkwo, C.; Ezeji, T. Enhanced acetone-butanol-ethanol production by *Clostridium beijerinckii* using biochar. In Proceedings of the 2019 ASABE Annual International Meeting, Boston, MA, USA, 7–10 July 2019; p. 1.

92. dos Santos Vieira, C.F.; Filho, F.M.; Filho, R.M.; Mariano, A.P. Acetone-free biobutanol production: Past and recent advances in the Isopropanol-Butanol-Ethanol (IBE) fermentation. *Bioresour. Technol.* **2019**, *287*, 121425. [CrossRef]

93. Li, G.; Lee, T.H.; Liu, Z.; Lee, C.F.; Zhang, C. Effects of injection strategies on combustion and emission characteristics of a common-rail diesel engine fueled with isopropanol-butanol-ethanol and diesel blends. *Renew. Energy* **2019**, *130*, 677–686. [CrossRef]

94. Bengelsdorf, F.R.; Dürre, P. Gas fermentation for commodity chemicals and fuels. *Microb. Biotechnol.* **2017**, *10*, 1167–1170. [CrossRef]

95. Raganati, F.; Procentese, A.; Olivieri, G.; Götz, P.; Salantino, P.; Marzocchella, A. Kinetic study of butanol production from various sugars by *Clostridium acetobutylicum* using a dynamic model. *Biochem. Eng.* **2015**, *99*, 156–166. [CrossRef]

96. Shinto, H.; Tashiro, Y.; Yamashita, M.; Kobayashi, G.; Sekiguchi, T.; Hanai, T. Kinetic study of substrate dependency for higher butanol production in acetone–butanol–ethanol fermentation. *Process Biochem.* **2008**, *43*, 1452–1461. [CrossRef]

97. Yu, Y.; Tangney, M.; Aass, H.C.; Mitchell, W.J. Analysis of the mechanism and regulation of lactose transport and metabolism in *Clostridium acetobutylicum* ATCC 824. *Appl. Environ. Microbiol.* **2007**, *73*, 1842–1850. [CrossRef] [PubMed]

98. Abubackar, H.N.; Veiga, M.C.; Kennes, C. Carbon monoxide fermentation to ethanol by *Clostridium autoethanogenum* in a bioreactor with no accumulation of acetic acid. *Bioresour. Technol.* **2015**, *186*, 122–127. [CrossRef]

99. Ciriminna, R.; Pina, C.D.; Rossi, M.; Pagliaro, M. Understanding the glycerol market. *Eur. J. Lipid Sci. Technol.* **2014**, *116*, 1432–1439. [CrossRef]

100. Pagliaro, M.; Rossi, M. Chapter 1. Glycerol: Properties and production. In *The Future of Glycerol*, 2nd ed.; Royal Society of Chemistry (RSC) Publishing: Cambridge, UK, 2010; pp. 1–28.

101. Ayoub, M.; Abdullah, A.Z. Critical review on the current scenario and significance of crude glycerol resulting from biodiesel industry towards more sustainable renewable energy industry. *Renew. Sustain. Energy Rev.* **2012**, *16*, 2671–2686. [CrossRef]

102. Amaral, P.F.F.; Ferreira, T.F.; Fontes, G.C.; Coelho, M.A.Z. Glycerol valorization: New biotechnological routes. *Food Bioprod. Process.* **2009**, *87*, 179–186. [CrossRef]

103. Pradima, J.; Kulkarni, M.R. Review on enzymatic synthesis of value added products of glycerol, a by-product derived from biodiesel production. *Resour. Technol.* **2016**, *3*, 394–405. [CrossRef]

104. Sanguanchaipaiwong, V.; Leksawasdi, N. Using glycerol as a sole carbon source for *Clostridium beijerinckii* fermentation. *Energy Procedia* **2017**, *138*, 1105–1109. [CrossRef]

105. da Silva, G.P.; Mack, M.; Contiero, J. Glycerol: A promising and abundant carbon source for industrial microbiology. *Biotechnol. Adv.* **2008**, *27*, 30–39. [CrossRef]

106. Forsberg, C.W. Production of 1,3-propanediol from glycerol by *Clostridium acetobutylicum* and other *Clostridium* species. *Appl. Environ. Microbiol.* **1987**, *53*, 639–643. [PubMed]

107. Tee, Z.K.; Jahim, J.M.; Tan, J.P.; Kim, B.H. Preeminent productivity of 1,3-propanediol by *Clostridium butyricum* JKT37 and the role of using calcium carbonate as pH neutraliser in glycerol fermentation. *Bioresour. Technol.* **2017**, *233*, 296–304. [CrossRef] [PubMed]

108. García, C.A.; García-Treviño, E.S.; Aguilar-Rivera, N.; Armendáriz, C. Carbon footprint of sugar production in Mexico. *J. Clean. Prod.* **2016**, *112*, 2632–2641. [CrossRef]

109. de Oliveira Lino, F.S.; Basso, T.O.; Sommer, M.O.A. A synthetic medium to simulate sugarcane molasses. *Biotechnol. Biofuels* **2018**, *11*, 221. [CrossRef] [PubMed]

110. Li, H.G.; Luo, W.; Gu, Q.Y.; Wang, Q.; Hu, W.J.; Yu, X.B. Acetone, butanol, and ethanol production from cane molasses using *Clostridium beijerinckii* mutant obtained by combined low-energy ion beam implantation and *N*-methyl-*N*-nitro-*N*-nitrosoguanidine induction. *Bioresour. Technol.* **2013**, *137*, 254–260. [CrossRef] [PubMed]

111. Farooq, U.; Anjum, F.; Zahoor, T.; Rahman, S.; Randhawa, M.; Ahmed, A.; Akram, K. Optimization of lactic acid production from cheap raw material: Sugarcane molasses. *Pak. J. Bot.* **2012**, *44*, 333.

112. Syed, Q.; Nadeem, M.; Nelofer, R. Enhanced butanol production by mutant strains of *Clostridium acetobutylicum* in molasses medium. *Turk. J. Biochem.* **2008**, *33*, 25–30.

113. van der Merwe, A.B.; Cheng, H.; Görgens, J.F.; Knoetze, J.H. Comparison of energy efficiency and economics of process designs for biobutanol production from sugarcane molasses. *Fuel* **2013**, *105*, 451–458. [CrossRef]

114. Jiang, L.; Wang, J.; Liang, S.; Wang, X.; Cen, P.; Xu, Z. Butyric acid fermentation in a fibrous bed bioreactor with immobilized *Clostridium tyrobutyricum* from cane molasses. *Bioresour. Technol.* **2009**, *100*, 3403–3409. [CrossRef]

115. Valli, V.; Gómez-Caravaca, A.M.; di Nunzio, M.; Danesi, F.; Caboni, M.F.; Bordoni, A. Sugar cane and sugar beet molasses, antioxidant-rich alternatives to refined sugar. *J. Agric. Food Chem.* **2012**, *60*, 12508–12515. [CrossRef]

116. Guan, Y.; Tang, Q.; Fu, X.; Yu, S.; Wu, S.; Chen, M. Preparation of antioxidants from sugarcane molasses. *Food Chem.* **2014**, *152*, 552–557. [CrossRef] [PubMed]

117. de Azeredo, L.A.I.; de Lima, M.B.; Coelho, R.R.R.; Freire, D.M.G. A low-cost fermentation medium for thermophilic protease production by *Streptomyces* sp. 594 using feather meal and corn steep liquor. *Curr. Microbiol.* **2006**, *53*, 335. [CrossRef] [PubMed]

118. Maddipati, P.; Atiyeh, H.K.; Bellmer, D.D.; Huhnke, R.L. Ethanol production from syngas by *Clostridium* strain P11 using corn steep liquor as a nutrient replacement to yeast extract. *Bioresour. Technol.* **2011**, *102*, 6494–6501. [CrossRef] [PubMed]

119. Ramachandriya, K.D.; Kundiyana, D.K.; Wilkins, M.R.; Terrill, J.B.; Atiyeh, H.K.; Huhnke, R.L. Carbon dioxide conversion to fuels and chemicals using a hybrid green process. *Appl. Energy* **2013**, *112*, 289–299. [CrossRef]

120. Saha, B.C. Hemicellulose bioconversion. *J. Ind. Microbiol. Biotechnol.* **2003**, *30*, 279–291. [CrossRef]

121. Isikgor, F.H.; Becer, C.R. Lignocellulosic biomass: A sustainable platform for the production of bio-based chemicals and polymers. *Polym. Chem.* **2015**, *6*, 4497–4559. [CrossRef]

122. Tilman, D.; Hill, J.; Lehman, C. Supportin online material for: Carbon-negative biofuels from low-input high-diversity grassland biomass. *Science* **2006**, *314*, 1598–1600. [CrossRef]

123. Ahn, Y.; Lee, S.H.; Kim, H.J.; Yang, Y.H.; Hong, J.H.; Kim, Y.H.; Kim, H. Electrospinning of lignocellulosic biomass using ionic liquid. *Carbohydr. Polym.* **2012**, *88*, 395–398. [CrossRef]

124. Menon, V.; Rao, M. Trends in bioconversion of lignocellulose: Biofuels, platform chemicals & biorefinery concept. *Prog. Energy Combust. Sci.* **2012**, *38*, 522–550.

125. Nielsen, J.; Villadsen, J.; Lidén, G. *Principles, Bioreaction Engineering*; Springer: Berlin, Germany, 2014; ISBN 9781441996879.

126. Fengel, D.; Wegener, G. *Wood—Chemistry, Ultrastructure, Reactions*; Walter de Gruyter: Berlin, Germany, 1989; ISBN 9783935638395.

127. Hendriks, A.T.W.M.; Zeeman, G. Pretreatments to enhance the digestibility of lignocellulosic biomass. *Bioresour. Technol.* **2009**, *100*, 10–18. [CrossRef]

128. Li, F.; Cheng, S.; Yu, H.; Yang, D. Waste from livestock and poultry breeding and its potential assessment of biogas energy in rural China. *J. Clean. Prod.* **2016**, *126*, 451–460. [CrossRef]

129. Ruiz, H.A.; Rodríguez-Jasso, R.M.; Fernandes, B.D.; Vicente, A.A.; Teixeira, J.A. Hydrothermal processing, as an alternative for upgrading agriculture residues and marine biomass according to the biorefinery concept: A review. *Renew. Sustain. Energy Rev.* **2013**, *21*, 35–51. [CrossRef]

130. Amiri, H.; Karimi, K. Pretreatment and hydrolysis of lignocellulosic wastes for butanol production: Challenges and perspectives. *Bioresour. Technol.* **2018**, *270*, 702–721. [CrossRef] [PubMed]

131. Qureshi, N.; Ezeji, T.C.; Ebener, J.; Dien, B.S.; Cotta, M.A.; Blaschek, H.P. Butanol production by *Clostridium beijerinckii*. Part I: Use of acid and enzyme hydrolyzed corn fiber. *Bioresour. Technol.* **2008**, *99*, 5915–5922. [CrossRef]

132. Bellido, C.; Pinto, M.L.; Coca, M.; González-Benito, G.; García-Cubero, M.T. Acetone–butanol–ethanol (ABE) production by *Clostridium beijerinckii* from wheat straw hydrolysates: Efficient use of penta and hexa carbohydrates. *Bioresour. Technol.* **2014**, *167*, 198–205. [CrossRef] [PubMed]

133. Sun, Z.; Liu, S. Production of n-butanol from concentrated sugar maple hemicellulosic hydrolysate by *Clostridia acetobutylicum* ATCC824. *Biomass Bioenergy* **2012**, *39*, 39–47. [CrossRef]

134. Ruiz-Dueñas, F.J.; Martínez, Á.T. Microbial degradation of lignin: How a bulky recalcitrant polymer is efficiently recycled in nature and how we can take advantage of this. *Microb. Biotechnol.* **2009**, *2*, 164–177. [CrossRef]

135. Orgill, J.J.; Atiyeh, H.K.; Devarapalli, M.; Phillips, J.R.; Lewis, R.S.; Huhnke, R.L. A comparison of mass transfer coefficients between trickle-bed, Hollow fiber membrane and stirred tank reactors. *Bioresour. Technol.* **2013**, *133*, 340–346. [CrossRef]

136. Mohammadi, M.; Najafpour, G.D.; Younesi, H.; Lahijani, P.; Uzir, M.H.; Mohamed, A.R. Bioconversion of synthesis gas to second generation biofuels: A review. *Renew. Sustain. Energy Rev.* **2011**, *15*, 4255–4273. [CrossRef]

137. Henstra, A.M.; Sipma, J.; Rinzema, A.; Stams, A.J. Microbiology of synthesis gas fermentation for biofuel production. *Curr. Opin. Biotechnol.* **2007**, *18*, 200–206. [CrossRef]

138. Huber, G.W.; Iborra, S.; Corma, A. Synthesis of transportation fuels from biomass: Chemistry, catalysts, and engineering. *Chem. Rev.* **2006**, *106*, 4044–4098. [CrossRef] [PubMed]

139. Woolcock, P.J.; Brown, R.C. A review of cleaning technologies for biomass-derived syngas. *Biomass Bioenergy* **2013**, *52*, 54–84. [CrossRef]

140. Dry, M.E. The fischer-tropsch process: 1950–2000. *Catal. Today* **2002**, *71*, 227–241. [CrossRef]

141. Phillips, J.R.; Clausen, E.C.; Gaddy, J.L. Synthesis gas as substrate for the biological production of fuels and chemicals. *Appl. Biochem. Biotechnol.* **1994**, *45–46*, 145–157. [CrossRef]

142. Vega, J.L.; Clausen, E.C.; Gaddy, J.L. Design of bioreactors for coal synthesis gas fermentations. *Resour. Conserv. Recycl.* **1990**, *3*, 149–160. [CrossRef]

143. Worden, R.M.; Grethlein, A.J.; Jain, M.K.; Datta, R. Production of butanol and ethanol from synthesis gas via fermentation. *Fuel* **1991**, *70*, 615–619. [CrossRef]

144. Brown, R.C. Hybrid thermochemical/biological processing: Putting the cart before the horse? *Appl. Biochem. Biotechnol.* **2007**, *137–140*, 947–956. [CrossRef]

145. Yasin, M.; Jeong, Y.; Park, S.; Jeong, J.; Lee, E.Y.; Lovitt, R.W.; Kim, B.H.; Lee, J.; Chang, I.S. Microbial synthesis gas utilization and ways to resolve kinetic and mass-transfer limitations. *Bioresour. Technol.* **2015**, *177*, 361–374. [CrossRef]

146. Munasinghe, P.C.; Khanal, S.K. Syngas fermentation to biofuel: Evaluation of carbon monoxide mass transfer and analytical modeling using a composite hollow fiber (CHF) membrane bioreactor. *Bioresour. Technol.* **2012**, *122*, 130–136. [CrossRef]

147. Bredwell, M.D.; Worden, R.M. Mass-transfer properties of microbubbles. 1. Experimental studies. *Biotechnol. Prog.* **1998**, *14*, 31–38. [CrossRef]

148. Bredwell, M.D.; Srivastava, P.; Worden, R.M. Reactor design issues for synthesis-gas fermentations. *Biotechnol. Prog.* **1999**, *15*, 834–844. [CrossRef] [PubMed]

149. Worden, R.M.; Bredwell, M.D. Mass-transfer properties of microbubbles. 2. Analysis using a dynamic model. *Biotechnol. Prog.* **1998**, *14*, 39–46. [CrossRef]

150. Klasson, K.T.; Ackerson, M.D.; Clausen, E.C.; Gaddy, J.L. Bioreactor design for synthesis gas fermentations. *Fuel* **1991**, *70*, 605–614. [CrossRef]

151. Christi, M.Y. *Airlift Bioreactors*; Elsevier Science Publishing Co.: New York, NY, USA, 1989; ISBN 1851663207.

152. Kadic, E.; Heindel, T.J. *An Introduction to Bioreactor Hydrodynamics and Gas-Liquid Mass Transfer*; Wiley: Hoboken, NJ, USA, 2017; ISBN 9781118104019.

153. Kumaresan, T.; Nere, N.K.; Joshi, J.B. Effect of internals on the flow pattern and mixing in stirred tanks. *Ind. Eng. Chem. Res.* **2005**, *44*, 9951–9961. [CrossRef]

154. Munasinghe, P.C.; Khanal, S.K. Evaluation of hydrogen and carbon monoxide mass transfer and a correlation between the myoglobin-protein bioassay and gas chromatography method for carbon monoxide determination. *RSC Adv.* **2014**, *4*, 37575–37581. [CrossRef]

155. Ungerman, A.J.; Heindel, T.J. Carbon monoxide mass transfer for syngas fermentation in a stirred tank reactor with dual impeller configurations. *Biotechnol. Prog.* **2007**, *23*, 613–620. [CrossRef]

156. Yasin, M.; Park, S.; Jeong, Y.; Lee, E.Y.; Lee, J.; Chang, I.S. Effect of internal pressure and gas/liquid interface area on the CO mass transfer coefficient using hollow fibre membranes as a high mass transfer gas diffusing system for microbial syngas fermentation. *Bioresour. Technol.* **2014**, *169*, 637–643. [CrossRef]

157. Coelho, F.M.B.; Botelho, A.M.; Ivo, O.F.; Amaral, P.F.F.; Ferreira, T.F. Volumetric mass transfer coefficient for carbon monoxide in a dual impeller stirred tank reactor considering a perfluorocarbon–water mixture as liquid phase. *Chem. Eng. Res. Des.* **2019**, *143*, 160–169. [CrossRef]

158. Orgill, J.J.; Abboud, M.C.; Atiyeh, H.K.; Devarapalli, M.; Sun, X.; Lewis, R.S. Measurement and prediction of mass transfer coefficients for syngas constituents in a hollow fiber reactor. *Bioresour. Technol.* **2019**, *276*, 1–7. [CrossRef]

159. Molitor, B.; Richter, H.; Martin, M.E.; Jensen, R.O.; Juminaga, A.; Mihalcea, C.; Angenent, L.T. Carbon recovery by fermentation of CO-rich off gases—Turning steel mills into biorefineries. *Bioresour. Technol.* **2016**, *215*, 386–396. [CrossRef]

160. Chang, I.S.; Kim, B.H.; Lovitt, R.W.; Bang, J.S. Effect of CO partial pressure on cell-recycled continuous CO fermentation by Eubacterium limosum KIST612. *Process Biochem.* **2001**, *37*, 411–421. [CrossRef]

161. Klasson, K.T.; Gupta, A.; Clausen, E.C.; Gaddy, J.L. Evaluation of mass-transfer and kinetic parameters for *Rhodospirillum rubrum* in a continuous stirred tank reactor. *Appl. Biochem. Biotechnol.* **1993**, *39–40*, 549–557. [CrossRef]

162. Cowger, J.P.; Klasson, K.T.; Ackerson, M.D.; Clausen, E.; Caddy, J.L. Mass-transfer and kinetic aspects in continuous bioreactors using *Rhodospirillum rubrum*. *Appl. Biochem. Biotechnol.* **1992**, *34–35*, 613–624. [CrossRef]

163. Lee, P.H.; Ni, S.Q.; Chang, S.Y.; Sung, S.; Kim, S.H. Enhancement of carbon monoxide mass transfer using an innovative external hollow fiber membrane (HFM) diffuser for syngas fermentation: Experimental studies and model development. *Chem. Eng. J.* **2012**, *184*, 268–277. [CrossRef]

164. Riggs, S.S.; Heindel, T.J. Measuring carbon monoxide gas-liquid mass transfer in a stirred tank reactor for syngas fermentation. *Biotechnol. Prog.* **2006**, *22*, 903–906. [CrossRef] [PubMed]

165. Kapic, A.; Jones, S.T.; Heindel, T.J. Carbon monoxide mass transfer in a syngas mixture. *Ind. Eng. Chem. Res.* **2006**, *45*, 9150–9155. [CrossRef]

166. International Energy Agency (IEA). *World Energy Outlook*; IEA: Paris, France, 2018.

167. International Energy Agency (IEA). *Biofuels for Transport: Tracking Clean Energy Progress*; IEA: Paris, France, 2018.

168. Sun, Y.; Cheng, J. Hydrolysis of lignocellulosic materials for ethanol production: A review. *Bioresour. Technol.* **2002**, *83*, 1–11. [CrossRef]

169. Cheng, J.J.; Timilsina, G.R. Status and barriers of advanced biofuel technologies: A review. *Renew. Energy* **2011**, *36*, 3541–3549. [CrossRef]

170. Gallo, J.M.R.; Bueno, J.M.C.; Schuchardt, U. Catalytic transformations of ethanol for biorefineries. *J. Braz. Chem. Soc.* **2014**, *25*, 2229–2243. [CrossRef]

171. Roozbehani, B.; Mirdrikvand, M.; Imani, S.; Roshan, A.C. Synthetic ethanol production in the Middle East: A way to make environmentally friendly fuels. *Chem. Technol. Fuels Oils* **2013**, *49*, 115–124. [CrossRef]

172. Kundiyana, D.K.; Huhnke, R.L.; Wilkins, M.R. Syngas fermentation in a 100-L pilot scale fermentor: Design and process considerations. *J. Biosci. Bioeng.* **2010**, *109*, 492–498. [CrossRef] [PubMed]

173. Martin, M.E.; Richter, H.; Saha, S.; Angenent, L.T. Traits of selected *Clostridium* strains for syngas fermentation to ethanol. *Biotechnol. Bioeng.* **2016**, *113*, 531–539. [CrossRef] [PubMed]

174. Shen, S.; Gu, Y.; Chai, C.; Jiang, W.; Zhuang, Y.; Wang, Y. Enhanced alcohol titre and ratio in carbon monoxide-rich off-gas fermentation of *Clostridium carboxidivorans* through combination of trace metals optimization with variable-temperature cultivation. *Bioresour. Technol.* **2017**, *239*, 236–243. [CrossRef] [PubMed]

175. Kim, Y.K.; Lee, H. Use of magnetic nanoparticles to enhance bioethanol production in syngas fermentation. *Bioresour. Technol.* **2016**, *204*, 139–144. [CrossRef]

176. Li, D.; Meng, C.; Wu, G.; Xie, B.; Han, Y.; Guo, Y.; Song, C.; Gao, Z.; Huang, Z. Effects of zinc on the production of alcohol by *Clostridium carboxidivorans* P7 using model syngas. *J. Ind. Microbiol. Biotechnol.* **2018**, *45*, 61–69. [CrossRef]

177. Abubackar, H.N.; Fernández-naveira, Á.; Veiga, M.C.; Kennes, C. Impact of cyclic pH shifts on carbon monoxide fermentation to ethanol by *Clostridium autoethanogenum*. *Fuel* **2016**, *178*, 56–62. [CrossRef]

178. Wang, S.; Dong, S.; Wang, Y. Enhancement of solvent production by overexpressing key genes of the acetone-butanol-ethanol fermentation pathway in *Clostridium saccharoperbutylacetonicum* N1-4. *Bioresour. Technol.* **2017**, *245*, 426–433. [CrossRef]

179. Fernández-Naveira, Á.; Veiga, M.C.; Kennes, C. Glucose bioconversion profile in the syngas-metabolizing species *Clostridium carboxidivorans*. *Bioresour. Technol.* **2017**, *244*, 552–559. [CrossRef]

180. Singh, N.; Mathur, A.S.; Gupta, R.P.; Barrow, C.J.; Tuli, D.; Puri, M. Enhanced cellulosic ethanol production via consolidated bioprocessing by *Clostridium thermocellum* ATCC 31924. *Bioresour. Technol.* **2018**, *250*, 860–867. [CrossRef]

181. Yang, X.; Xu, M.; Yang, S.T. Metabolic and process engineering of *Clostridium cellulovorans* for biofuel production from cellulose. *Metab. Eng.* **2015**, *32*, 39–48. [CrossRef]

182. Khanna, S.; Goyal, A.; Moholkar, V.S. Effect of fermentation parameters on bio-alcohols production from glycerol using immobilized *Clostridium pasteurianum*: An optimization study. *Prep. Biochem. Biotechnol.* **2013**, *43*, 828–847. [CrossRef] [PubMed]

183. Vohra, M.; Manwar, J.; Manmode, R.; Padgilwar, S.; Patil, S. Bioethanol production: Feedstock and current technologies. *J. Environ. Chem. Eng.* **2014**, *2*, 573–584. [CrossRef]

184. Devarapalli, M.; Atiyeh, H.K. A review of conversion processes for bioethanol production with a focus on syngas fermentation. *Biofuel Res. J.* **2015**, *2*, 268–280. [CrossRef]

185. Lynd, L.R.; Weimer, P.J.; van Zyl, W.H.; Pretorius, I.S. Microbial cellulose utilization: Fundamentals and biotechnology. *Microbiol. Mol. Biol. Rev.* **2003**, *66*, 506–577. [CrossRef] [PubMed]

186. Margeot, A.; Hahn-Hagerdal, B.; Edlund, M.; Slade, R.; Monot, F. New improvements for lignocellulosic ethanol. *Curr. Opin. Biotechnol.* **2009**, *20*, 372–380. [CrossRef]

187. Lynd, L.R.; van Zyl, W.H.; McBride, J.E.; Laser, M. Consolidated bioprocessing of cellulosic biomass: An update. *Curr. Opin. Biotechnol.* **2005**, *16*, 577–583. [CrossRef]

188. Sasaki, C.; Kushiki, Y.; Asada, C.; Nakamura, Y. Acetone-butanol-ethanol production by separate hydrolysis and fermentation (SHF) and simultaneous saccharification and fermentation (SSF) methods using acorns and wood chips of *Quercus acutissima* as a carbon source. *Ind. Crop. Prod.* **2014**, *62*, 286–292. [CrossRef]

189. Shen, Y.; Brown, R.C.; Wen, Z. Syngas fermentation by *Clostridium carboxidivorans* P7 in a horizontal rotating packed bed biofilm reactor with enhanced ethanol production. *Appl. Energy* **2017**, *187*, 585–594. [CrossRef]

190. Qureshi, N.; Blaschek, H.P. ABE production from corn: A recent economic evaluation. *J. Ind. Microbiol. Biotechnol.* **2001**, *27*, 292–297. [CrossRef]

191. Lee, S.Y.; Park, J.H.; Jang, S.H.; Nielsen, L.K.; Kim, J.; Jung, K.S. Fermentative butanol production by clostridia. *Biotechnol. Bioeng.* **2008**, *101*, 209–228. [CrossRef]

192. Jin, C.; Yao, M.; Liu, H.; Lee, C.F.F.; Ji, J. Progress in the production and application of n-butanol as a biofuel. *Renew. Sustain. Energy Rev.* **2011**, *15*, 4080–4106. [CrossRef]

193. Zheng, J.; Tashiro, Y.; Wang, Q.; Sonomoto, K. Recent advances to improve fermentative butanol production: Genetic engineering and fermentation technology. *J. Biosci. Bioeng.* **2015**, *119*, 1–9. [CrossRef] [PubMed]

194. Nicolaou, S.A.; Gaida, S.M.; Papoutsakis, E.T. A comparative view of metabolite and substrate stress and tolerance in microbial bioprocessing: From biofuels and chemicals, to biocatalysis and bioremediation. *Metab. Eng.* **2010**, *12*, 307–331. [CrossRef] [PubMed]

195. Gallardo, R.; Alves, M.; Rodrigues, L.R. Modulation of crude glycerol fermentation by *Clostridium pasteurianum* DSM 525 towards the production of butanol. *Biomass Bioenergy* **2014**, *71*, 134–143. [CrossRef]

196. Lu, C.; Zhao, J.; Yang, S.T.; Wei, D. Fed-batch fermentation for n-butanol production from cassava bagasse hydrolysate in a fibrous bed bioreactor with continuous gas stripping. *Bioresour. Technol.* **2012**, *104*, 380–387. [CrossRef]

197. Phillips, J.R.; Atiyeh, H.K.; Tanner, R.S.; Torres, J.R.; Saxena, J.; Wilkins, M.R.; Huhnke, R.L. Butanol and hexanol production in *Clostridium carboxidivorans* syngas fermentation: Medium development and culture techniques. *Bioresour. Technol.* **2015**, *190*, 114–121. [CrossRef]

198. Zhang, J.; Yu, L.; Lin, M.; Yan, Q.; Yang, S.-T. n-Butanol production from sucrose and sugarcane juice by engineered *Clostridium tyrobutyricum* overexpressing sucrose catabolism genes and adhE2. *Bioresour. Technol.* **2017**, *233*, 51–57. [CrossRef]

199. Maiti, S.; Gallastegui, G.; Sarma, S.J.; Brar, S.K.; le Bihan, Y.; Drogui, P.; Buelna, G.; Verma, M. A re-look at the biochemical strategies to enhance butanol production. *Biomass Bioenergy* **2016**, *94*, 187–200. [CrossRef]

200. Ahlawat, S.; Kaushal, M.; Palabhanvi, B.; Muthuraj, M.; Goswami, G.; Das, D. Nutrient modulation based process engineering strategy for improved butanol production from *Clostridium acetobutylicum*. *Biotechnol. Prog.* **2019**, *35*, e2771. [CrossRef]

201. Dong, Z.; Liu, G.; Liu, S.; Liu, Z.; Jin, W. High performance ceramic hollow fiber supported PDMS composite pervaporation membrane for bio-butanol recovery. *J. Membr. Sci.* **2014**, *450*, 38–47. [CrossRef]

202. Raganati, F.; Procentese, A.; Olivieri, G.; Russo, M.E.; Salatino, P.; Marzocchella, A. Bio-butanol separation by adsorption on various materials: Assessment of isotherms and effects of other ABE-fermentation compounds. *Sep. Purif. Technol.* **2018**, *191*, 328–339. [CrossRef]

203. Research&Market. Acetone: Worldwide Market Outlook to 2023—Expected to Reach ~7666 Thousand Tons by the End of 2023. 2018. Available online: https://www.prnewswire.com/news-releases/acetone-worldwide-market-outlook-to-2023---expected-to-reach-7-666-thousand-tons-by-the-end-of-2023--300757524.html (accessed on 6 June 2019).

204. Sauer, M. Industrial production of acetone and butanol by fermentation—100 years later. *FEMS Microbiol. Lett.* **2016**, *363*, fnw134. [CrossRef]

205. Sreekumar, S.; Baer, Z.C.; Pazhamalai, A.; Gunbas, G.; Grippo, A.; Blanch, H.W.; Clark, D.S.; Toste, F.D. Production of an acetone-butanol-ethanol mixture from *Clostridium acetobutylicum* and its conversion to high-value biofuels. *Nat. Protoc.* **2015**, *10*, 528–537. [CrossRef]

206. HIS Markit. Acetone. Chemical Economics Handbook. Available online: https://ihsmarkit.com/products/acetone-chemical-economics-handbook.html (accessed on 9 June 2019).

207. Luo, H.; Ge, L.; Zhang, J.; Ding, J.; Chen, R.; Shi, Z. Enhancing acetone biosynthesis and acetone-butanol-ethanol fermentation performance by co-culturing *Clostridium acetobutylicum/Saccharomyces cerevisiae* integrated with exogenous acetate addition. *Bioresour. Technol.* **2016**, *200*, 111–120. [CrossRef]

208. Wen, Z.; Ledesma-Amaro, R.; Lin, J.; Jiang, Y.; Yang, S. Improved n-butanol production from *Clostridium cellulovorans* by integrated metabolic and evolutionary engineering. *Appl. Environ. Microbiol.* **2019**, *85*, e02560-18. [CrossRef]

209. Millat, T.; Winzer, K. Mathematical modelling of clostridial acetone-butanol-ethanol fermentation. *Appl. Microbiol. Biotechnol.* **2017**, *101*, 2251–2271. [CrossRef]

210. Li, X.; Shi, Z.; Li, Z. Increasing butanol/acetone ratio and solvent productivity in ABE fermentation by consecutively feeding butyrate to weaken metabolic strength of butyrate loop. *Bioprocess Biosyst. Eng.* **2014**, *37*, 1609–1616. [CrossRef]

211. Gao, M.; Tashiro, Y.; Yoshida, T.; Zheng, J.; Wang, Q.; Sakai, K.; Sonomoto, K. Metabolic analysis of butanol production from acetate in *Clostridium saccharoperbutylacetonicum* N1-4 using 13C tracer experiments. *RSC Adv.* **2015**, *5*, 8486–8495. [CrossRef]

212. Boonsombuti, A.; Komolpis, K.; Luengnaruemitchai, A.; Wongkasemjit, S. Enhancement of ABE fermentation through regulation of ammonium acetate and D-xylose uptake from acid-pretreated corncobs. *Ann. Microbiol.* **2014**, *64*, 431–439. [CrossRef]

213. Survase, S.A.; van Heiningen, A.; Granström, T. Continuous bio-catalytic conversion of sugar mixture to acetone-butanol-ethanol by immobilized *Clostridium acetobutylicum* DSM 792. *Appl. Microbiol. Biotechnol.* **2012**, *93*, 2309–2316. [CrossRef]

214. Tjahjasari, D.; Kaeding, T.; Zeng, A.P. 1,3-Propanediol and polytrimethyleneterephthalate. In *Comprehensive Biotechnology. Volume 3: Ind Biotechnol Commodity Products*; Elsevier: Pergamon, Turkey, 2011; pp. 229–242.

215. Selembo, P.A.; Perez, J.M.; Lloyd, W.A.; Logan, B.E. Enhanced hydrogen and 1,3-propanediol production from glycerol by fermentation using mixed cultures. *Biotechnol. Bioeng.* **2009**, *104*, 1098–1106. [CrossRef] [PubMed]

216. Sun, Y.; Ma, C.; Fu, H.; Mu, Y.; Xiu, Z. *1,3-Propanediol. Bioprocessing of Renewable Resources to Commodity Bioproducts*, 1st ed.; John Wiley & Sons, Inc.: Hoboken, NJ, USA, 2014; pp. 289–326.

217. Zhuge, B.; Zhang, C.; Fang, H.; Zhuge, J.; Permaul, K. Expression of 1,3-propanediol oxidoreductase and its isoenzyme in *Klebsiella pneumoniae* for bioconversion of glycerol into 1,3-propanediol. *Appl. Microbiol. Biotechnol.* **2010**, *87*, 2177–2184. [CrossRef] [PubMed]

218. Zeng, A.-P. Pathway and kinetic analysis of 1,3-propanediol production from glycerol fermentation by *Clostridium butyricum*. *Bioprocess Eng.* **1996**, *14*, 169–175. [CrossRef]

219. Lee, C.S.; Aroua, M.K.; Dauda, W.M.A.W.; Cognet, P.; Pérès-Lucchese, Y.; Fabre, P.-L.; Reynes, O.; Latapie, L. A review: Conversion of bioglycerol into 1,3-propanediol via biological and chemical method. *Renew. Sustain. Energy Rev.* **2015**, *42*, 963–972. [CrossRef]

220. Jensen, T.O.; Kvist, T.; Mikkelsen, M.J.; Westermann, P. Production of 1,3-PDO and butanol by a mutant strain of *Clostridium pasteurianum* with increased tolerance towards crude glycerol. *AMB Express* **2012**, *2*, 44. [CrossRef] [PubMed]

221. Barbirato, F.; Himmi, E.-H.; Conte, T.; Bories, A. 1,3-Propanediol production by fermentation: An interesting way to valorize glycerin from the ester and ethanol industries. *Ind. Crop. Prod.* **1998**, *7*, 281–289. [CrossRef]

222. Papanikolaou, S.; Fick, M.; Aggelis, G. The effect of raw glycerol concentration on the production of 1,3-propanediol by *Clostridium butyricum*. *J. Chem. Technol. Biotechnol.* **2004**, *79*, 1189–1196. [CrossRef]

223. González-Pajuelo, M.; Andrade, J.C.; Vasconcelos, I. Production of 1,3-propanediol by *Clostridium Butyricum* VPI 3266 using a synthetic medium and raw glycerol. *J. Ind. Microbiol. Biotechnol.* **2004**, *31*, 442–446. [CrossRef]

224. Szymanowska-Powałowska, D.; Białas, W. Scale-up of anaerobic 1,3-propanediol production by *Clostridium Butyricum* DSP1 from crude glycerol. *BMC Microbiol.* **2014**, *14*, 45. [CrossRef]

225. Chatzifragkou, A.; Papanikolaou, S.; Kopsahelis, N.; Kachrimanidou, V.; Dorado, M.P.; Koutinas, A.A. Biorefinery development through utilization of biodiesel industry by-products as sole fermentation feedstock for 1,3-propanediol production. *Bioresour. Technol.* **2014**, *159*, 167–175. [CrossRef]

226. Martins, F.F.; Saab, V.S.; Ribeiro, C.M.S.; Coelho, M.A.Z.; Ferreira, T.F. Production of 1,3-propanediol by *Clostridium butyricum* growing on biodiesel derived glycerol. *Chem. Eng. Trans.* **2016**, *50*, 289–294.

227. Wang, J.; Lin, M.; Xu, M.; Yang, S.T. Anaerobic fermentation for production of carboxylic acids as bulk chemicals from renewable biomass. *Adv. Biochem. Eng. Biotechnol.* **2016**, *156*, 323–362. [PubMed]

228. Agler, M.T.; Spirito, C.M.; Usack, J.G.; Werner, J.J.; Angenent, L.T. Development of a highly specific and productive process for n-caproic acid production: Applying lessons from methanogenic microbiomes. *Water Sci. Technol.* **2013**, *69*, 62–68. [CrossRef] [PubMed]

229. Liu, L.; Yang, J.; Yang, Y.; Luo, L.; Wang, R.; Zhang, Y.; Yuan, H. Consolidated bioprocessing performance of bacterial consortium EMSD5 on hemicellulose for isopropanol production. *Bioresour. Technol.* **2019**, *292*, 121965. [CrossRef] [PubMed]

230. Saini, M.; Wang, Z.W.; Chiang, C.J.; Chao, Y.P. Metabolic engineering of *Escherichia coli* for production of butyric acid. *J. Agric. Food Chem.* **2014**, *62*, 4342–4348. [CrossRef] [PubMed]

231. Luo, H.; Yang, R.; Zhao, Y.; Wang, Z.; Liu, Z.; Huang, M.; Zeng, Q. Recent advances and strategies in process and strain engineering for the production of butyric acid by microbial fermentation. *Bioresour. Technol.* **2018**, *253*, 343–354. [CrossRef]

232. Yang, S.T.; Yu, M.; Chang, W.L.; Tang, I.C. Anaerobic fermentations for the production of acetic and butyric acids. In *Bioprocessing Technologies in Biorefinery for Sustainable Production of Fuels, Chemicals, and Polymers*; John Wiley & Sons, Inc.: Hoboken, NJ, USA, 2013.

233. Cavalcante, W.A.; Leitão, R.C.; Gehring, T.A.; Angenent, L.T.; Santaella, S.T. Anaerobic fermentation for n-caproic acid production: A review. *Process Biochem.* **2017**, *54*, 106–119. [CrossRef]

234. Zhu, X.; Zhou, Y.; Wang, Y.; Wu, T.; Li, X.; Li, D.; Tao, Y. Production of high-concentration n-caproic acid from lactate through fermentation using a newly isolated *Ruminococcaceae* bacterium CPB6. *Biotechnol. Biofuels* **2017**, *10*, 102. [CrossRef]

235. Ge, S.; Usack, J.G.; Spirito, C.M.; Angenent, L.T. Long-term n-caproic acid production from yeast-fermentation beer in an anaerobic bioreactor with continuous product extraction. *Environ. Sci. Technol.* **2015**, *49*, 8012–8021. [CrossRef]

236. Yang, Y.; Hoogewind, A.; Moon, Y.H.; Day, D. Production of butanol and isopropanol with an immobilized *Clostridium*. *Bioprocess Biosyst. Eng.* **2016**, *39*, 421–428. [CrossRef] [PubMed]

237. Moon, Y.H.; Han, K.J.; Kim, D.; Day, D.F. Enhanced production of butanol and isopropanol from sugarcane molasses using *Clostridium beijerinckii* optinoii. *Biotechnol. Bioprocess Eng.* **2015**, *20*, 871–877. [CrossRef]

238. Mariano, A.P.; Dias, M.O.; Junqueira, T.L.; Cunha, M.P.; Bonomi, A.; Filho, R.M. Butanol production in a first-generation Brazilian sugarcane biorefinery: Technical aspects and economics of greenfield projects. *Bioresour. Technol.* **2013**, *135*, 316–323. [CrossRef] [PubMed]

239. Moncada, J.; Tamayo, J.A.; Cardona, C.A. Integrating first, second, and third generation biorefineries: Incorporating microalgae into the sugarcane biorefinery. *Chem. Eng. Sci.* **2014**, *118*, 126–140. [CrossRef]

240. Rabelo, S.C.; Carrere, H.; Filho, R.M.; Costa, A.C. Production of bioethanol, methane and heat from sugarcane bagasse in a biorefinery concept. *Bioresour. Technol.* **2011**, *102*, 7887–7895. [CrossRef] [PubMed]

241. Cavalett, O.; Junqueira, T.L.; Dias, M.O.; Jesus, C.D.; Mantelatto, P.E.; Cunha, M.P.; Franco, H.C.J.; Cardoso, T.F.; Filho, R.M.; Bonomi, A.; et al. Environmental and economic assessment of sugarcane first generation biorefineries in Brazil. *Clean Technol. Environ. Policy* **2012**, *14*, 399–410. [CrossRef]

© 2019 by the authors. Licensee MDPI, Basel, Switzerland. This article is an open access article distributed under the terms and conditions of the Creative Commons Attribution (CC BY) license (http://creativecommons.org/licenses/by/4.0/).

MDPI

St. Alban-Anlage 66

4052 Basel

Switzerland

Tel. +41 61 683 77 34

Fax +41 61 302 89 18

www.mdpi.com

Catalysts Editorial Office

E-mail: catalysts@mdpi.com

www.mdpi.com/journal/catalysts

www.ingramcontent.com/pod-product-compliance
Lightning Source LLC
LaVergne TN
LVHW071010170726
843515LV00006B/1129